Electronic Imaging Technology

Edward R. Dougherty
Editor

SPIE OPTICAL ENGINEERING PRESS
A Publication of SPIE—The International Society for Optical Engineering
Bellingham, Washington USA

Library of Congress Cataloging-in-Publication Data

Dougherty, Edward R.
    Electronic imaging technology / Edward R. Dougherty.
        p. cm.
    Includes bibliographical references and index.
    ISBN 0-8194-3037-4 (softcover)
        1. Image processing—Digital techniques.   2. Video compression. 3. Multimedia systems.
    4. Computer architecture. I. Title.
TA1637.D682 1999
621.36'7—dc21   98-53084
                                                                            CIP

Published by

SPIE—The International Society for Optical Engineering
P.O. Box 10
Bellingham, Washington 98227-0010
Phone: 360/676-3290
Fax: 360/647-1445
Email: spie@spie.org
WWW: http://www.spie.org/

Printed in the United States of America.

MC600t1.

Electronic Imaging Technology

CONTENTS

Chapter 5   Color Image Processing / 121
Arthur Robert Weeks

Chapter 6   Enhancement of Digital Documents / 167
Robert P. Loce and Edward R. Dougherty

Chapter 11 Hardware Architecture for Image Processing / 387
Divyendu Sinha and Edward R. Dougherty

PREFACE

This book is composed of chapters covering different areas within electronic imaging, with the focus being on system overview, technology, and practical application. The book commences with an introductory chapter and after that each chapter is essentially self-contained. The authors have practical experience in the individual topics they cover and each chapter provides insight into its particular application domain and how the various system components contribute to problem solutions. Many of the authors are editors of the *SPIE/IS&T Journal of Electronic Imaging*.

The introductory chapter provides a brief history of the subject and an overview of applications. It goes on to provide some basic material regarding digital imaging that is meant to give a backdrop to those reading the book who are not necessarily familiar with the basic structure of digital images.

The second chapter treats an extremely important topic in electronic imaging, video compression. Owing to the need to move vast amounts of image data, compression is an absolutely necessity. The authors introduce a number of coding algorithms and then turn their attention to video compression standards. Four current standards are covered: H.261, MPEG-1, MPEG-2, and H.263. The chapter also discusses the development of MPEG-4, in particular, the functionalities important to MPEG-4.

The third chapter defines multimedia in a wide sense as, "a new technology merging computer, communication, and consumer electronics with content provided by the media industry." The general concept of a multimedia system is introduced, along with applications, overall system requirements, and sections on various system components and their integration.

Chapter four discusses visualization, the creation of images from data. The task is to produce imagery that is compatible with the human visual system so that understanding is facilitated. This can only be accomplished with an understanding of the human visual system, and the chapter begins by covering basic aspects of that system. It then discusses different types of visualization and mappings. It concludes with perceptual principles for visualization design.

Color is becoming increasingly important for electronic image processing systems and there are a number of attributes of image processing that are unique to color. The fifth chapter begins by introducing some basic principles of color science and then discussing color models. It provides examples of color image processing and finishes with some remarks on color displays.

Digital document processing is a field within electronic image processing that has developed special methods to address its own unique problems. The human eye is particularly sensitive to certain effects created by digitally sampling binary and low-bit images. Therefore it is common to enhance digital documents prior to display or printing. Chapter six discusses the construction of binary digital filters for document enhancement. Its major focus is on resolution conversion.

The seventh chapter covers halftoning, which consists of various coding methods to reduce the number of quantization levels per pixel without too greatly impacting the appearance to a human observer. The chapter gives a history of halftoning and then reviews the relevant aspects of visual perception. A number of halftoning methods are discussed, including noise encoding, ordered dither, and error diffusion.

Chapter eight discusses document recognition. Upon scanning a document, characters and other objects need to be transformed from image data to symbolic representation for further processing. A document is segmented into blocks and the blocks are then classified into text, graphics, etc. A number of topics are treated, including runlength smoothing, background covering, document decoding, table recognition, form recognition, and font identification.

Handwriting recognition is quite different than recognition of machine fonts. It is often true in pattern recognition that segmentation and recognition are not fully separable, and this is particularly valid in handwriting recognition. The approach taken in the ninth chapter is to consider recognition at the word level, utilize a lexicon from which each word must be chosen, and rate an observed word relative to words in the lexicon. The chapter explains the overall structure of lexicon-driven, segmentation-based handwritten word recognition. It treats segmentation approaches, character level processing, and word level processing, with its focus being at the word level.

Given a hardcopy image that one wishes to process, that hardcopy needs to be scanned into a digital format. Chapter ten covers image scanning, beginning with the basics of digitization and going on to explain, in detail, optical and mechanical systems, sensors (including color CCD sensors), and electronics. Owing to the importance of image quality, significant attention is paid to factors affecting the quality of scanned images, such as the scanner modulation transfer function and color filtering and calibration. The chapter concludes with brief coverage of some image processing tasks typically applied to a scanned image prior to printing.

Ultimately, practical digital image processing depends on a computer system. Successful implementation often requires an appropriate computational environment. This is especially true for meeting real-time requirements. The last chapter of the book treats hardware architecture for image processing. It introduces the general concept of parallel processing and then discusses pipelining and the use of multiple processors, with application to convolution and morphological filtering. There are two sections giving detailed accounts of commercial image processing systems: AISI's AIS processors and ITI's MVC 150/40 processor.

The book is meant for the general body of people interested in applied, industrial electronic imaging technology. Hopefully, one whose specialty is in one domain can read various chapters to obtain a broader view of the general subject. Those desiring a more in-depth account of a particular topic can proceed to a full text, or to the research literature. It is my hope that the book proves useful to a wide range of practitioners within the electronic imaging industry and to those people who are just becoming involved with the field.

Edward R. Dougherty
December 1998

Electronic Imaging Technology

1

INTRODUCTION

Arthur Robert Weeks
University of Central Florida
Orlando, Florida

This chapter gives a historical background of image storage and retrieval and the development of electronic image processing. The applications and use of modern electronic image processing are also presented.

1.1 HISTORICAL BACKGROUND

Throughout the history of mankind, there has been the desire of the human race to record an instance of time for future generations via the use of pictures (images). Pictures have also been used in various early languages, providing an easy method of communicating information from one human to the next. The earliest documented use of images to communicate an idea is seen in the drawings created by the early cavemen. Using primitive tools, images were recorded into stone describing the details of everyday life. Important events such as the winning or losing of a battle were often recorded with these man-made images. Even though these images were used for communication purposes, today they give a historical record of early human civilization, providing the details of everyday life. The importance of pictures used as a language to represent ideas is also well illustrated by Egyptian hieroglyphics. The prolific preservation of this language engraved in the stone of various Egyptian remains has provided historians with a detailed story of the Egyptian civilization.

Another early form of image generation and storage was the use of various color paints/inks to record scenes observed by the human eye. This required the talent and expertise of humans who could create an image using various colors of paints/inks from a scene that was visualized and perceived by them. For many centuries, this was the only means of recording images. It was not uncommon that an artist would accompany soldiers into battle to record the historic event. During the Middle Ages and the Renaissance, the creation of man-made visual images played an important part in conveying important religious concepts to the average person. Many artists such as Leonardo da Vinci and Michelangelo were commissioned to produce paintings of religious scenes within various churches. Artists were also commissioned to produce portrait paintings of royalty. During this time, this was the only means by which the image of a person could be passed from generation to generation. These man-made images had their limitations in that all were created after the visualization and interpretation by another human. The resemblance of the final image to that of the actual scene depended on the artist producing the image. Very seldom would several artists commissioned to produce a painting of the same scene produce exactly the same painting. Each would visualize and perceive something different in the scene. In fact, the differences between the paintings were attributed to an artist's style and perception of the scene being painted.

Since the beginning of time, it has been desired by human civilization to record images of scenes as accurately as possible, removing any human interpretation. During the second half of the sixteenth century an Italian philosopher, J. B. Porta, made an important discovery by accident. He discovered that light rays penetrating a small hole in a door enclosing a dark room produced an upside down image of the exterior scene on a white screen placed behind the door in the dark room. His discovery was the forerunner to the modern day pinhole camera. Amazed at his discovery, he soon modified his experiment, replacing the small hole with a convex lens. With the additional use of a mirror, Porta was able to produce a right side up image of an exterior scene. Porta realized the importance of his discovery and immediately recommended his device (camera obscura) to artists wishing to record exact images of scenes onto oil paintings. It was an artist, Canaletto, who first used Porta's discovery to produce paintings of Venice.

The discovery of the camera by Porta was the first step toward modern day photography. At about the same time as Porta's camera, it was discovered by a chemist in France that silver chloride changes its characteristics from clear to black when exposed to light. It was not until two centuries later that Jacques Alexandre Charles, the inventor of the hydrogen balloon, produced simple photographs of silhouettes of his students' heads using writing paper immersed in silver chloride. Unfortunately, Jacques Alexandre Charles, like many before him, did not have a way to stop the development process. Once a sheet of silver chloride paper was exposed to a light source to produce an image, it could be viewed only in very low light levels such as candle light. Eventually, the long term exposure to light would turn the whole silver chloride paper to black. The last piece of the puzzle remained to be discovered, that of stopping (fixing) the development action.

It was not until 1835 that Henry Fox Talbot was able to stop the development process. His first images were of leaves and flowers, formed by pressing them against silver nitrate immersed paper and then exposing the paper and the objects to sunlight. Once the paper was exposed to light, producing the image of a leaf or flower, Talbot's final process was to "fix" the paper to prevent any further degradation of the image. Later, in the summer of 1835, Talbot used the camera obscura to produce images of his house. On this day, the age of modern photography was born. It was not until 1839, when

Talbot perfected his approach, that his work was presented to the Royal Society. It should be pointed out that even though Talbot was able to produce a stored image without it first being interpreted by a human, his images were negatives of the original objects. It was not until years later that the two step process was used of producing a negative and repeating the exposure and development processing one more time to produce a positive image. Even though the earliest photographs were on paper, in the middle part of the nineteenth century glass plates were used as a means of recording an image using chemical photography. The stability of these plates are evident in the number that remain over 100 years later. Today they are collector items among historians and photographers.

The invention of modern photography brought with it the desire of the human civilization to record images for observation by future generations and to provide the means of freezing the "images" of history. For example, the importance of early photography is evident in the still photographs that remain and clearly document the U.S. Civil War. This war was one of the first major conflicts in history to be completely documented by modern photography. Historians still refer to these photographs to describe key battles. For most of the nineteenth century, photography was limited to a specialized few who were trained to use the various complex cameras and were familiar with the use of the complex chemical process used in the development of photographs. Photography was used only for special events or by the wealthy as a mean of capturing images to record key events or people. It was the invention of the simple roll-camera ("Kodak" camera) in 1884 by George Eastman that made modern photography available to the average individual. What this camera accomplished was the replacement of large photo-sensitive glass plates with a flexible roll of film that was easy to load into the camera. George Eastman also provided a network of laboratories that would develop and process these film-rolls making the use of modern photography as easy as possible for the general public. The technical knowledge of the chemical development process was no longer required to produce photographs using modern photography.

As the field of black and white photography was maturing in the last part of the nineteenth century, researchers were pursuing the generation of color images. It was the work of James Clerk Maxwell and James D. Forbes that showed that the infinite number of colors available in the visible spectrum could be reproduced using a three primary color system. Using a spinning top, they placed circles of different color papers on top of the top. When the top was spinning at a high rate, the different color circles blurred together, producing a new observable color. Maxwell and Forbes were able to produce the color yellow from a red and a green circle. In fact, they found that they could produce any color from a combination of three colors, red, blue, and green. Maxwell suggested a method of proving his three color theory, by photographing a color image using three (black and white) photographic plates taken of exactly the same scene but with a red, blue, or green color filter placed in front of each plate. The three plates, which represented the three primary color images of red, green and blue, were then used to produce three images that were overlaid on top of each other on a white screen. In the generation of the images, a red light source was used to illuminate the red filtered plate, a green light was used to illuminate the green, and a blue light was used to illuminate the blue. Essentially, overlaying the three images produced a composite image composed of a red, a blue, and a green image of the original color scene. On May 17, 1861, Maxwell gave a lecture to the Royal Institute, where he stunned the audience with the presentation of a color image of a plaid ribbon generated from three red, green, and blue photographic plates. Maxwell's work not only laid the groundwork for modern color photography, but as a result of his research, every color television/computer system today uses the three primary colors of red, green, and blue to generate color images.

It was also during this period that research began on combining still photographic images to produce moving images. The concept of motion pictures is based on the visual phenomenon known as persistence of vision. The images received by the brain from the eye are stored for about 60 milliseconds after viewing by the eye. In this way, objects moving across the field of view of the eye in faster than 60 milliseconds are ignored by the brain. For example, consider a set of cards containing a set of images of a person walking. Each adjacent card contains the image of the person walking as he or she moves just slightly to the left. Placing these cards in front of an observer and fanning them at a rate faster than about 15 cards per second produces the illusion that the person is walking smoothly to the left. This is the fundamental concept of motion pictures.

The goal of Thomas A. Edison and his assistant William Kennedy Laurie Dickson was to produce a set of time elapsed photographs to give the illusion of a moving image. It was the invention of the roll-film by George Eastman (modified for a positive image) that made the invention of motion pictures possible. In 1889, Edison designed a camera (kinescope) that automatically advanced a roll of film past a shutter and a lens. At a periodic rate of 48 images per second, the film was held still and the shutter was opened, imaging a scene onto the roll of film. The film was then advanced and the process was repeated. When finished, the roll of film contained a time sequence of images, presenting the evolution of a scene as a function of time. After successful use of their camera, Edison and Dickson designed a projector that rapidly projected the time sequence of images saved on the roll of film onto a screen. What they observed was a smoothly moving image. Within 60 years of the invention of still photography, the recording of real scenes as a function of time was accomplished. With the advent of combining sound with moving pictures, a complete record of an important historical event and prominent people was now possible. An example is the millions of feet of newsreel film that was generated in the early part of the twentieth century. These films clearly document the important historical events of time for future generations.

Before moving onto the discussion of still images, which is the emphasis of this text, a brief history of television is in order. This invention, and the efforts to transmit images across the Atlantic Ocean began the history of transmitting images via electronic methods. These efforts laid the groundwork for the electronic generation, manipulation, transmission, and storage of still images. One of the earliest accounts of an electronic television system was given in a letter to *Nature* in April, 1880, by John Perry. He proposed an electronic camera that used an array of selenium detectors as a means of converting an image intensity into an array of electrical signals. The interesting thing about this camera was that it did not use any type of scanning mechanism typically found in many early electronic cameras. In fact, the camera proposed by Perry's paper would be referred to as a focal-plane camera and is very similar to the modern charge coupled device, CCD, camera.

There were also two types of receivers described. The first used an array of magnetic needles, one for each selenium detector, to open and close a set of apertures. The size of the aperture was directly proportional to the light incident on its corresponding selenium detector. The second system used the Kerr effect to rotate the polarization of linearly polarized light passing through a small crystal. As an electric field is placed on a Kerr cell, the angle in which the polarization of light is rotated is varied. In this system, an array of small crystals equal to the number of selenium detectors was proposed. In front of each Kerr cell, a polarizer was placed so that the combination of the linearly polarized light intensity emerging from the Kerr cell and the polarizer produced an intensity variation proportional to the electric field on the Kerr cell. The goal of this receiver was to have the electric field induced on the Kerr cell be proportional to the light intensity incident on its respective selenium detector. It is interesting to note that this is the same concept used for flat screen liquid crystal displays, except the angle in which the polarization light is rotated is controlled by the liquid crystal elements. It took approximately 100 years after the original concept paper by Perry before this type of receiver system was implemented. Unlike modern television systems, Perry's proposed system requires no point-by-point scanning of a scene. The major limitation of his system was that only a limited number of array elements could be used because of the complexity of the wiring and the number of parts required.

The importance of Perry's paper is that it proposed to sample spatially a continuously varying spatial image using an array of detectors to produce an array of electric signals that contained all the information necessary to regenerate an image of the original scene. Many of the other early proposed systems were based on electro-mechanical systems that used the concept of scanning an image element by element. One of the first systems, designed by Paul Nipkow, used two rotating discs (one located at the transmitter and one located at the receiver) containing 24 holes oriented in helical fashion on each disc. The discs were then synchronized and rotated at a rate faster than the persistence rate of the eye. Located at the transmitter was a selenium detector that produced an electrical signal that varied as the scene was sequentially scanned. This electrical signal then modulated a light source at the receiver to produce a modulated intensity of light that was synchronized to the sampling of the original scene. The net effect was a perceived image of the original scene at the receiver.

Modern day television had to wait for the electronic vacuum tube invented by Lee De Forest in 1906. On December 29, 1923, Vladimir K. Zworykin of the Westinghouse Electric and Manufacturing Company applied for a patent for a complete electric television system containing no moving parts. Key to a complete electric television was the development of the cathode ray tube (picture tube) used to convert electrical signals into varying light intensities, an all electronic scanning method, and an electronic camera. It was not until 1929 that Zworykin produced a satisfactory picture tube (kinescope) that enabled the development and design of affordable television receivers. By 1933, RCA built and operated a complete television system using an improved electronic camera, which was developed by Zworykin who was now at RCA Incorporated. This system was completely electronic except for the use of a mechanical synchronization generator. By 1940, the mechanical synchronization system was replaced by an electronic system, providing the way for the design of a complete electronic television. During the week of June 25 to 28, 1940, RCA demonstrated the use of its television system by broadcasting the Republican National Convention in Philadelphia to the NBC studios and transmitting this signal from the Empire State Building to approximately 4000 television receivers located throughout New York City.

On January 27, 1941, the National Television Standards Committee (NTSC) sent its recommendations to the Federal Communications Commission (FCC) for the approval of a commercial television system. This committee recommended that each channel be 6 MHz in frequency, with the sound carrier placed 4.5 MHz above the video carrier. The recommendation specified that the video information as well as the synchronization signals be transmitted using amplitude modulation, AM, while the sound be transmitted using the newly developed (by Armstrong) frequency modulation, FM. It also recommended that a 4 (horizontal) by 3 (vertical) aspect ratio as well as a vertical resolution of 441 horizontal lines scanned at a rate of 30 frames per second be used in the transmitting of the picture. Finally, the committee recommended that interlaced scanning should be used, generating two fields (one for the even lines and one for the odd lines) at the rate of 60 times per second. On May 2, 1941, the FCC adopted the recommendations of the NTSC and set the date of July 1, 1941, as the official start date for commercial television in the United States. However, the start of World War II delayed the introduction of commercial television into the American household until the end of the 1940s. By the middle of the 1950s, with the invention of color television, a second set of NTSC standards was generated. These required that the transmission of color images be compatible with existing black and white television standards set forth by the original NTSC committee in 1941, which are the standards still in practice today, some 40 years later.

The desire to transmit an image to a distant location was not limited to motion pictures alone. Several of the initial concepts that led to the invention of the television dealt directly with the transmission of still photographs. On March 17, 1891, Noah Steiner Amstutz applied for a patent for the first device that would transmit photographs. The key to Amstutz's system was the special preparation of photographs to be transmitted. Prior to transmission, photographs underwent a chemical process to provide a surface height variation proportional to the silver density on the photograph. Amstutz's system used a needle attached to a variable resistor that produced varying voltage output as the needle was scanned across the surface of the specially prepared photograph. The receiver discussed in the patent used a special material that when developed produced dark and light areas proportional to the height distribution of the material. Amstutz's receiver was simply a tracing tool that varied the height of this special material proportional to the height of the modified photograph. A cutting tool scanned and cut this special material in synchronization with the scanning of the photograph. The final process was to develop this cut material chemically, producing a copy of the original photograph. In May 1891, Amstutz successfully sent pictures over a 25-mile line using his system.

Prior to Amstutz's system, it took approximately a week for photographs to cross the Atlantic Ocean from Europe to the United States. In the early part of 1920, several leading newspaper companies in London and New York established a system to transmit photographs across the Atlantic Ocean in approximately two to three hours, which became known as the Bartlane system. This system scanned the input image element by element, producing a paper tape record of the gray tones within the original photograph. The first Bartlane system used a standard Baudot 5-bit telegraph with modified typeface keys. Similar to the concepts of dithered printing methods used today, each key had a different impact width, so when impacted against a piece of paper each produced a different sized ink dot. By 1921, this system was abandoned in favor of an off-line photographic method that was much more accurate at reproducing the gray tones of the original photograph after it had been converted to paper tape. The corresponding holes in the paper tape controlled the opening and closing of a set of shutters that in turn controlled the exposure on an unexposed photographic negative. As the density of holes increased on the paper tape, the light illuminating the unexposed negative increased. By scanning the unexposed photograph element by element and modulating the light illuminating the unexposed photograph, accurate reproduction of the gray tones contained within the original photograph was possible.

To convert a photograph into a digital number for storage on paper tape required the use of photoengraving techniques. Several metal plates were coated with a photosensitive insulating material and then exposed to the original photograph using different exposures. For each successive exposure, the light used to illuminate the photographic plate was increased by a factor of two. After chemically developing the exposed photographic plates, the insulating material that remained on the plate exposed using the lowest light level corresponded to the black gray tones of the original photograph. The plate exposed using the next lowest light level had both black and dark gray tones recorded in the insulating area. Hence, each successive plate had more gray tones stored in its insulating area. Essentially, the binary pattern of insulating material and metal on each plate corresponded to the bit pattern representation of the gray tones of the original photograph. The number of plates represented the number of bits used to encode the different gray tones. This system, as incorporated into the Bartlane system, used five plates (5 bits) to encode the different gray tones. By 1927, the technology existed to record electronically gray tone variations within a photograph using a photoelectric cell. The photograph was scanned element by element and the light reflected off the photograph was recorded, converted to a digital number, and then saved on paper tape.

The requirement for good imagery for military reconnaissance during World War II generated further advancements in photographic methods. As an example, the new technical field of high speed photography was developed and used extensively for aerial reconnaissance. During this period, the use of photographic manipulation (the predecessor to modern image processing) became important. The first of these methods simply corrected for poor camera exposures due to poor lighting and weather conditions. Individuals specially trained in the art of object recognition and identification were used to locate enemy targets from aerial photographs. A common practice was to enlarge and contrast enhance aerial photographs to highlight key military targets, making the job of the photographic interpreter easier.

It was not until the invention of the digital computer and the requirement of the NASA space program in the early 1960s that digital image processing came into existence. After several earlier unsuccessful unmanned Ranger spacecraft missions, Ranger 7 successfully sent thousands of television images of the lunar surface back to earth. These images were then converted into digital format and digitally processed to remove geometrical and camera response distortions. This initial image processing was done at NASA's Jet Propulsion Laboratory (JPL) in Pasadena, California. The initial digital image processing results of these images were so good that NASA continued its funding, resulting in the development of new image processing methods.

During the mid-1960s, NASA had the requirement of photographing the lunar surface to determine valid landing sites for the manned Apollo program. The Surveyor series of unmanned spacecraft were designed to survey the lunar surface to locate and verify valid landing sites. Surveyor 7 sent several thousand television images of the lunar surface from its landing site back to earth. Many of these images were converted to digital format and image processed in an attempt to determine the composition and structure of the lunar surface.

The Mariner series of unmanned spacecraft returned digital images of Mars, Venus, and Mercury during the late 1960s and early 1970s. Mariners 4, 6, and 7 returned digital images of Mars as they flew by that planet. Mariner 9 was placed into orbit around the planet Mars and sent back to earth digital images of the Martian surface that were used to map its surface. The Mariner 10 mission was designed so that the spacecraft would fly by Venus and Mercury and around the sun. Thousands of images were obtained and enhanced using digital image processing methods during the life of the Mariner program. The Viking unmanned spacecraft series also used digital image processing techniques to enhance images that were sent back to earth from the spacecrafts during the middle and late 1970s. Two orbiting and two landing Viking spacecrafts sent back high resolution images of the Martian surface, while two other Voyager spacecrafts flew by Jupiter and Saturn, sending detailed images of these planets never seen before.

The use of digital image processing techniques was initially limited to NASA's unmanned space program to evaluate the surface terrain of other planets and the moon, but with the large success of electronic image processing of images obtained from these missions, the tools of electronic image processing were extended to other NASA programs. NASA also launched a series of satellites that orbit the earth, which include LANDSAT, SEASAT, TIROS, GEOS, and NIMBUS, providing multispectral images of the earth's surface. These satellites include spacecrafts. These satellites provide detailed images of the earth's surface and weather information on a daily basis. For example, the formerly difficult task of predicting the formation and path of a hurricane has become routine with the use of weather satellite imagery.

1.2 APPLICATIONS OF IMAGE PROCESSING

The use of image processing techniques to enhance gray tone images has found applications beyond the initial use by NASA. Today, image processing is used in the fields of astronomy, medicine, crime and fingerprint analysis, remote sensing, manufacturing, aerospace and defense, movies and entertainment, and multimedia. Image processing has benefited the field of astronomy by partially removing the effect of the blurring of astronomical images by the earth's atmosphere using restoration and phase estimation techniques. An excellent example of the efficacy of digital image processing is in the restoring of images obtained from the Hubble telescope. After initial deployment, it was found that one of its optical telescope mirrors was incorrectly designed and contained aberrations. Images received from the telescope were badly blurred. Until NASA could design a set of corrective optics, which took several years, the use of the Hubble was limited. As a stopgap measure, image restoration methods were applied to the blurred images, removing a majority of the blurring that was present and making the incoming images usable until the corrective lenses were inserted in the optical train of the Hubble telescope.

The medical field has benefited greatly from the use of digital image processing. Image processing techniques have been applied to ultrasonic imagery, which has improved the evaluation and monitoring of the fetus during prenatal care. Image processing has also made the early diagnosis of breast cancer much easier by enhancing X-ray images of the chest. With the suspect tissue highlighted, a medical technician can concentrate on these areas, improving the accuracy of the diagnosis. Image processing methods have become standard practice in the generation of magnetic resonance images. Digital image processing methods have also improved standard diagnostic procedures that previously required several days before a medical doctor could make a decision. For example, the upper-GI imaging series that is commonly used to evaluate stomach difficulties used to take several days before the doctor received the developed X-rays for evaluation. Now, using digital imaging methods, real time video is available for evaluation during the procedure, providing immediate feedback to the doctor. X-rays are also collected during the procedure as a double check because of their high spatial resolution. Finally, image processing has been used extensively in the various microsurgery techniques. For example, in vitro fertilization techniques use a fiber optic imaging system to provide real time video so that a medical doctor can easily locate the ovaries for egg extraction.

Both crime and fingerprint analysis have benefited from the various image processing techniques that exist. For example, given an old photograph of a young child who has been reported missing, image processing methods can be used to predict the changes in the facial features as the child ages. This allows law enforcement agents to estimate the appearance of a child who has been missing for many years. Image processing methods have also been used to enhance slow rate surveillance systems used in banks and stores. These low resolution systems are key in many instances in providing the evidence necessary to convict the appropriate criminal. During a bank robbery, these surveillance systems provide images of the robbers. Image enhancement is used to highlight their faces, providing the police with an actual image that can be used as evidence. Automatic fingerprint analysis is another area that has received a lot of attention in the field of image processing. Due to the large number of fingerprints that are kept on file, techniques have been developed that help in their automatic classification, reducing the time required to identify an unknown fingerprint.

Both manufacturing and the aerospace and defense industries have benefited from modern image processing techniques. Such methods are routinely used in the automatic recognition and identification of faulty parts during the manufacturing process. The use of "smart" bombs throughout the

1991 Persian Gulf War has shown the decisive advantage of incorporating state-of-the-art image processing methods into weapons design. The pinpoint accuracy that was achieved by these weapons easily surpassed the accuracy of conventional weapons, as seen by daily new releases of the various bombing missions.

In recent years, the motion picture industry has applied image processing methods in the restoration of old films and in the use of creating special effects in new movies. During the initial part of the twentieth century, millions of feet of newsreel film and silent motion pictures were made on a film substrate that is now physically degrading. In the last several years, many of these original works have been completely restored using image processing methods. For example, before the last release of the movie *Snow White* by Walt Disney Productions, this film underwent a complete restoration. When the master photographic plates for the movie were created, they contained dust spots that became observable in the movie. In addition, since the creation of these final print plates, the original colors had yellowed. This was more evident in the white areas of the movie. During the restoration process, the spots were removed and the color was readjusted, correcting for the yellowing process. This was a tedious, difficult process, because over 200,000 individual plates that comprised the movie were restored. Moviemakers are just now learning what they can do with image processing. Recent movies have incorporated special effects that would not be possible without the use of image processing. For example, "morphing" of one object into another is easily accomplished using digital techniques.

With the advancement in the computing power of desktop computers, the interest in incorporating still images as well as video clips in presentations has increased drastically in the last couple of years. Still image digitizer hardware that acquires a single frame of color video from a camera and stores the image within a computer is available today for as little as $200. A few years ago this same hardware cost several thousand dollars. The prices of color desktop scanners have dropped to levels low enough that these devices are becoming standard within an office environment. The ability to display true-color images on a standard desktop computer is now very inexpensive ($150). Just a few years ago, image processing researchers had to buy very expensive dedicated computer hardware and image display hardware. The digitization of real time video has also become very popular. Hardware that was limited to video production studios a few years ago is now affordable and readily available. It is now possible for an average individual to create and edit a complete video clip using digital techniques within his/her computer.

Recently there has been a large number of manufacturers producing digital still frame cameras. These cameras use a CCD detector as the electronic detector and provide the analog to digital conversion of the acquired image within the camera electronics. Several of these cameras contain some form of magnetic media, allowing for the onboard storage of several images. Software and interface hardware are included that allow downloading of the images to a standard desktop computer and the storage of the images in many of the popular image formats. The new PhotoCD© system created by Kodak Inc. has generated a lot of interest in digital image capture and storage. This system provides an easy and inexpensive method of digitizing (at high resolution) photographs from 35 mm film. A photographer uses a standard 35 mm camera to shoot photographs and then sends the exposed negatives to a Kodak laboratory for development. Upon completion of the development process, the photographer receives both the negatives and a CD-ROM containing digitized images of the photographs. The ease of use of this system has made it very popular.

After an image has been acquired and converted into a digital format, it can then be modified using electronic image processing techniques. In many instances, the goal is to use this image in the automatic recognition of objects present within the image. One example could be the automatic recognition and removal of a faulty part during a manufacturing process. Figure 1.1 is a block diagram describing the typical steps used in the processing of an image using digital image processing methods. The final output described in the block diagram is to perform a task based on the recognition of objects present within the acquired image. The first step is to acquire an image using an electronic camera. Next, the analog signal(s) from the camera is converted to a digital format using digitizing electronics. This piece of hardware can be a standard NTSC video acquisition system or it can be a dedicated piece of hardware that interfaces to a specific vendor's camera system. Once the image is digitized and stored within the computer, image preprocessing techniques can be used to improve the quality of the image.

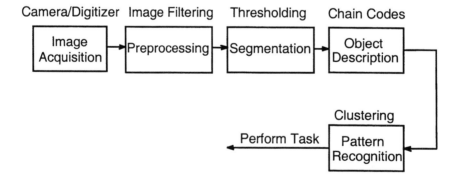

Figure 1.1 Acquisition and processing of an image.

Figure 1.2(b) shows the result of using histogram equalization on the low contrast image shown in Figure 1.2(a). Features not observable in the original

(a)

(b)

Figure 1.2 An example of image enhancement: (a) the original low
contrast image and (b) the enhanced image.

image are easily seen in the output image. In particular, the details of the fruit in the basket are now much more evident. Another example of image preprocessing is the use of image restoration techniques to remove blurring from an image. Figure 1.3(a) is a degraded image that has been blurred in the horizontal direction. In this example, the source and type of the blurring degradation is given and is used in the restoration process. Figure 1.3(b) is the restored, unblurred image after application of Wiener filtering. Notice the increased sharpness of this image over the original image given in Figure 1.3(a).

After preprocessing of an image to enhance or restore key features that were degraded during the acquisition process, the next step in building an autonomous recognition system is to segment objects/features from the image background using segmentation techniques. A very common method is to base the segmentation decision on a gray tone threshold value. For example, gray tone values within the image greater than the threshold value are considered object elements. The next step in the recognition system given in Figure 1.1 uses object description techniques to reduce the amount of data that must be recognized. In many situations, only the contours of the object are needed for classification. Keeping only the contours reduces the amount of data that must be classified by the pattern recognition system. This reduces the complexity of the recognition process. For example, a set of fixed contours representing a set of known objects can be stored by the pattern recognition system and compared against an unknown contour in the input image. The contour that most closely matches the unknown contour is then classified as the object represented by the unknown contour. The output of the pattern recognition stage is a decision that is performed by the overall autonomous system.

1.3 IMAGE FORMATION

An image is a distribution of light energy as a function of spatial position. Figure 1.4 depicts the formation of an image on a video camera. A light source (the sun in Figure 1.4) emits light energy that is incident on an object that will form an image. Light energy can do one of three things when incident on an object: be absorbed by the object, be transmitted through the object, or be reflected from the object. By conservation of energy, the total light energy incident on an object must be conserved. Let R be the percentage of light that is reflected, A be the percentage of light that is absorbed, and T be the percentage of light that is transmitted, then

$$R + T + A = 1 . \tag{1.1}$$

If an object transmits most of the light energy, it is referred to as a clear object because other objects behind this object can still be seen. Glass is an excellent

(a)

(b)

Figure 1.3 An example of image restoration: (a) the original blurred
image and (b) the restored image.

example of a clear object in that most of the light energy incident on it is transmitted. An opaque object transmits no light energy and simply either reflects or absorbs the incident light. If an object absorbs all light energy, then by the conservation of energy, this object re-radiates this light energy in the form of heat.

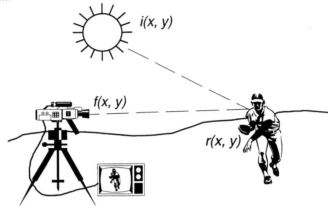

Figure 1.4 An illustration depicting the formation of an image (images, ©New Vision Technologies).

Of the three possible methods in which light energy interacts with an object, light reflected from its surface is the most important in the formation of an image. Figure 1.4 shows that light radiating from the sun is reflected off the surface of the baseball player and then received by the input lens of the camera that forms an image on the camera's electronic detector. The image formed at the camera's lens can be expressed mathematically as

$$f(x, y) = i(x, y) \cdot r(x, y) \ . \tag{1.2}$$

Equation (1.2) gives the image model at the camera as a function of the light illumination $i(x, y)$ and the surface reflection of an object $r(x, y)$, where $r(x, y)$ varies between 0 and 1. The function $f(x, y)$ describes the light energy of the image at the spatial coordinate x, y. Since an image is the spatial distribution of light energy $f(x, y)$, it can take on only positive, real values. Finally, the image formed at the camera in Figure 1.4 is converted to an electrical signal, which produces the image seen on the television.

1.4 SAMPLING AND QUANTIZATION

Before an image can be manipulated using the various image processing techniques, it must be spatially sampled. The process of sampling an image is the process of applying a two-dimensional grid to a spatially continuous image to divide it into a two-dimensional array of elements. Figure 1.5 shows a sampled image containing a total of *NM* sampled elements using a rectangular grid. Another commonly used sampling grid is the hexagonal grid. Any type of sampling grid can be used, but the rectangular grid is by far the most common simply because of its relationship to two-dimensional arrays. The fundamental unit of a sampled image is a picture element and is typically referred to as a *pel* or a *pixel*. The value of each pixel is equal to the average intensity of the continuous spatial image covered by that pixel.

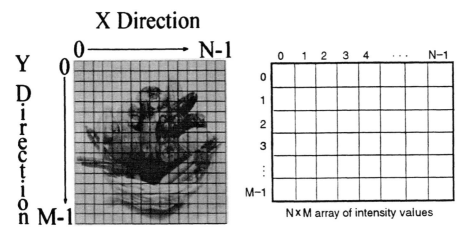

Figure 1.5 A spatially sampled image containing $N \times M$ picture elements.

The result of sampling produces a two-dimensional array of numbers that is directly proportional to the intensity levels of the continuous spatial image. Figure 1.6 shows a two-dimensional array of $N \times M$ elements used to represent a sampled image. The array $G(x, y)$ defines that intensity value of the sampled image at the pixel coordinate x, y. Figure 1.7 shows a simple pseudo program that accesses each pixel within a sampled image $G(x, y)$ and then multiplies the intensity of each pixel by a constant K. The results of this operation produce a new image $R(x, y)$. The program accesses each pixel in the sampled image $G(x, y)$ using two FOR loops: one in the x direction and one in the y direction.

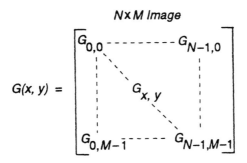

Figure 1.6 A two-dimensional array representing a sampled image.

Today there are a number of image sizes that are standard. Many of these were chosen to be compatible with the spatial size of NTSC video and to meet the storage size requirements of digital memory. In addition, image sizes that are powers of two exist because of the requirements for computing the fast Fourier transform (FFT), to be presented in Chapter 2. Typical spatial sizes are 256×256, 320×240, 512×512, 640×480, 1024×1024, 1024×768, 2048×2048, and 4096×4096. Until recently, because of the high cost of memory, the smaller sizes such as 256×256, 320×240, and 512×512 were standard. For compatibility with black and white NTSC video, an image resolution of 640×480 is used. A black and white NTSC video image of this size is also given the designation of RS170. Figure 1.8 gives an example of the same image presented at several different spatial sizes. Figure 1.8(a) lists the size for each image, varying between 64×64 and 512×512. Each successively larger image was created by doubling each pixel in both directions. For the 512×512 sized image, the individual pixels from the original 64×64 sized image can easily be seen.

```
Integer x, Integer y
For y = 0 to N-1
  For x = 0 to M-1
    R(x, y) = K * G(x, y)
  next x
next y
```

Figure 1.7 An example of a pseudo program that multiplies each pixel in an image by a constant K.

Besides spatial sampling, the intensity level at each pixel must also be digitized into a finite set of numbers. The process of digitization converts an analog intensity value into a set of digital numbers that represent the intensity levels in the image. The quantity of numbers used to represent the intensities in a continuous tone image determines the final quality of the digitization process. We refer to this set of numbers as the *graylevels* or *grayscales* of an image. Since an image is the spatial distribution of light energy, the numbers assigned to graylevels of a digitized image can take on only positive values. Typically though, integer values are used to represent the graylevels present within a digitized image. Figure 1.9(a) gives a 4×4 sub-image taken from the lower right side of Figure 1.8(b), which highlights the boundary pixels between the background and the boy's shirt. Figure 1.9(b) gives the corresponding grayscale, with the value of 0 assigned to black and each grayscale value increasing in intensity until the value of 255 is reached, corresponding to white. Because of the low spatial sampling of this image the individual pixels are readily observable.

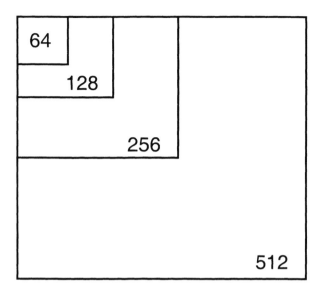

(a)

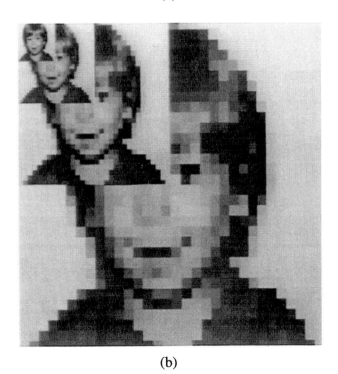

(b)

Figure 1.8 An example of a sampled image for various resolutions:
(a) the resolution size of each image and (b) the actual sampled images.

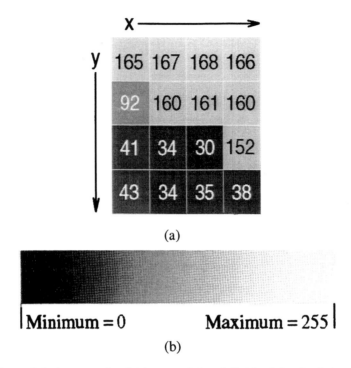

(a)

(b)

Figure 1.9 An example of (a) a sampled and digitized 4 × 4 sub-image and (b) its corresponding grayscale.

Consider a continuous tone image that has been digitized into four graylevels: black, dark gray, light gray, and white. Unfortunately, for most images, digitizing a continuous image into four graylevels does not produce an image of high quality. In fact, discrete contours are seen throughout the image at pixel boundaries between the different graylevels. These artificially introduced contours are commonly referred to as *false contours*. Figure 1.10(a) shows an image of a boy's face that has been digitized into 2 graylevels. The features of the face are observable, but this image is of poor image quality. Images that contain only two graylevels are typically called *binary images*. Binary images become important in image segmentation and binary morphological filters. Figure 1.10(b) is the same image of the boy's face but this image contains 4 graylevels. The image in Figure 1.10(b) is perceived as more acceptable than the binary image but is still of poor image quality. Present in this image are false contours, due to the pixel boundaries between the different graylevels. Figures 1.10(c) and (d) are 8 and 16 graylevel images of the boy's face, respectively. Notice the improvement in the image quality of the boy's face as the number of graylevels used to represent the image is increased.

The question comes down to how many graylevels are needed to digitize an image properly. The human eye can typically resolve between 40 to 60 different graylevels. Below this value the eye can detect false contours in an image. Figures 1.10(e) and (f) show the remarkable improvement in the image quality of the boy's face as the number of graylevels within the image is increased from 32 to 64 graylevels, respectively. In Figure 1.10(f), the false contours that were present in the smaller graylevel images have for the most part disappeared. Since most sampled and digitized images are stored as digital numbers and since digital numbers are stored using bytes of storage, one byte of digital storage is typically used to store one pixel within an image. Since a byte can represent 256 distinct values, one byte per pixel allows up to 256 graylevels to be stored per pixel. This number of graylevels is beyond the range in which the human eye can detect discontinuities in graylevels. A result of using 256 distinct graylevels for each pixel is a digitized image that is perceived as a continuous tone image, which is of high quality.

Many of the video digitizers that are presently available convert the *NTSC-RS170* video format into a two-dimensional array of digital numbers, which are then stored within the digital memory of a computer system using 256 graylevels for each pixel (one byte per pixel). For an RS170 video image, one frame of 640×480 video requires 307,200 *bytes* of storage. Even though 256 graylevels for each pixel satisfies the perceived image quality, there are many applications in which 256 graylevels do not provide enough graylevel dynamic range. Under these situations, specialized electronic cameras and image acquisition hardware are available that sample an image using 10 bits (1024 graylevels) or 12 bits (4096 graylevels) per pixel. It has been common practice to use one byte per pixel to store each of the red, blue, and green color components comprising a color image. This results in 3 bytes per pixel or 24 bits per pixel of image storage required for a color image. In fact, digitized and sampled color images are often referred to as 24-bit color images.

1.5 IMAGE NEIGHBORS AND DISTANCES

A very important concept in image processing is that of neighboring pixels. The use of neighboring pixels in an image processing algorithm provides a localized window of spatial information. Consider a pixel $p(x, y)$ within a sampled image at the coordinate x, y. The collection of the two surrounding vertical and the two surrounding horizontal pixels are defined as the 4-*neighbors* of the pixel $p(x, y)$. Figure 1.11(a) shows the relationship between the four neighboring pixels and $p(x, y)$. These four neighboring pixels define the vertical and horizontal spatial relationship to the pixel $p(x, y)$. Similarly, the four neighboring diagonal pixels form the set of *diagonal-neighbors* of the pixel $p(x, y)$, shown in Figure 1.11(b).

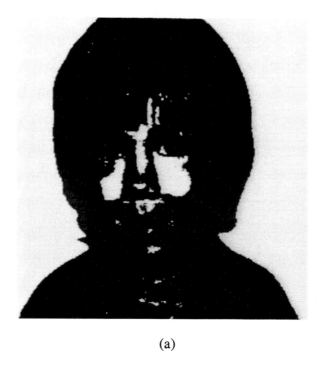

(a)

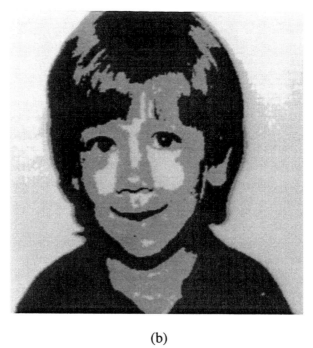

(b)

Figure 1.10 An example of image quantization: (a) 2 graylevels and (b) 4 graylevels.

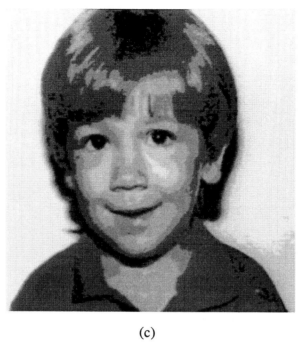

(c)

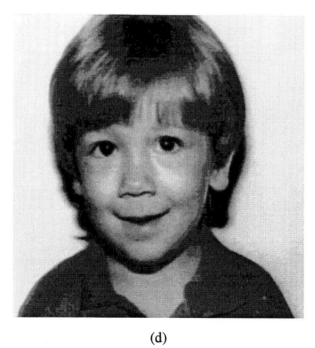

(d)

Figure 1.10 An example of image quantization: (c) 8 graylevels and (d) 16 graylevels.

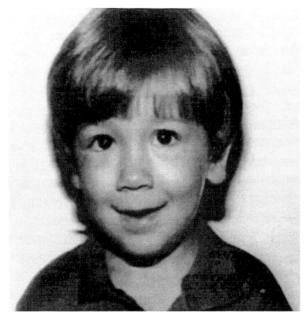

(e)

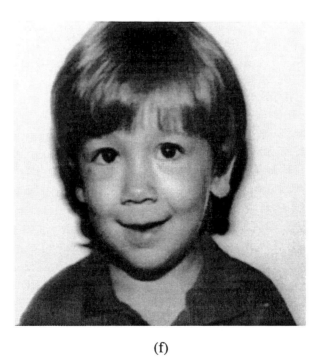

(f)

Figure 1.10 An example of image quantization: (e) 32 graylevels and (f) 64 graylevels.

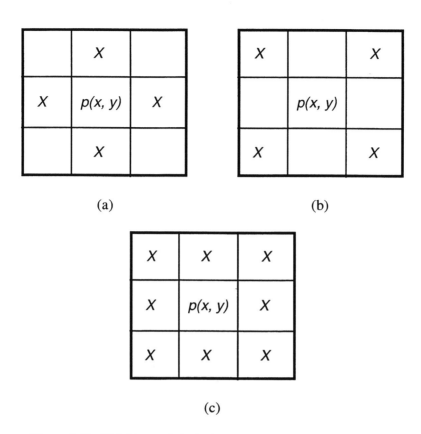

(a)   (b)

(c)

Figure 1.11   Neighbors of pixel $p(x, y)$: (a) 4-neighbors (b) diagonal-neighbors and (c) 8-neighbors.

The union of the set of 4-neighbors with the set of diagonal-neighbors form the set of 8-*neighbors*. Figure 1.11(c) shows the 8-neighbors relative to $p(x, y)$. The set of 8-neighbors is comprised of the 8 surrounding pixels of $p(x, y)$. The concept of neighboring pixels can be expanded to include as many neighboring pixels as desired. For example, it is common to use 24- or 48-neighbors (5×5 or 7×7) in many image processing techniques involving local neighbors. The use of local neighbors is key in applying spatial filtering methods to an image for image enhancement.

Another area important to electronic image processing is that of *connectivity*. Connectivity defines the relationship between neighboring pixels and is used to locate the borders between objects within an image. For example, connectivity defines the similarity between the graylevels of neighboring pixels.   Usually, neighboring pixels with large graylevel differences imply pixels that are separated by region boundaries. There are three types of connected pixels that describe how two neighboring pixels are related: 4-connected, 8-connected, and *m*-connected. Consider pixel *A*, which is a neighbor of pixel *B*. Two pixels *A* and *B* are 4-connected provided *A* is a 4-neighbor of *B* and their graylevels meet some predetermined criteria or predicate. One example criteria might be that all pixels must be between ±2 in graylevel to be considered part of the same region within an image. Figure 1.12(a) shows an example of 4-connectivity between pixels *A* and *B* for all pixels with a graylevel of 5. Four-connectivity allows only vertical or horizontal paths to be traced from pixel *A* to *B*.

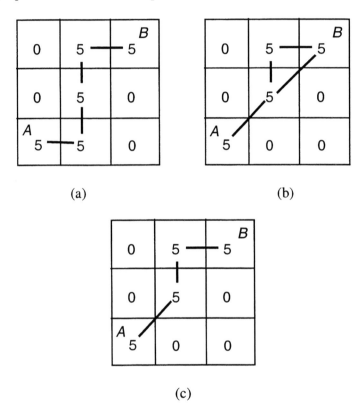

Figure 1.12 Examples of (a) 4-connectivity, (b) 8-connectivity, and (c) *m*-connectivity.

Similar to 4-connectivity is 8-connectivity, which includes the diagonal neighbors in determining a connected path. Two pixels *A* and *B* are 8-connected provided *A* is an 8-neighbor of *B* and their graylevels meet some predetermined criteria. Like 4-connectivity, 8-connectivity requires that a predetermined rule be set about graylevel similarity between pixels. The main difference between 4-connectivity and 8-connectivity is that 8-connectivity allows for diagonal paths between connected pixels. Figure 1.12(b) gives an example of 8-connectivity between pixels *A* and *B* for all pixels with a graylevel of 5. In this example, two paths exist between the two pixels *A* and B. In fact, 4-connectivity does not exist between pixels A and B, since there are no horizontal or vertical paths connecting pixel A and the center pixel of the 3×3 sub-image.

The main difficulty with 8-connectivity is that it produces two possible paths, as shown in the upper right hand corner of Figure 1.12(b). However, m-connectivity eliminates multiple paths by removing the diagonal path if 4-connectivity already exists between two pixels. This effectively leaves the horizontal and vertical paths and removes the diagonal path. Two pixels A and B are m-connected provided A is an 8-neighbor of B, their graylevels meet some predetermined criteria, and the 4-neighbor set of A with B does not intersect the 4-neighbor set of B with A. Figure 1.12(c) shows the use of m-connectivity and the elimination of the double path that was present in the 8-connected example given in Figure 1.12(b).

There are several distance measures that are used in image processing to measure the separation between pixels. The most common is the Euclidean distance. Given two pixels A and B at the coordinates (a, b) and (c, d), the Euclidean distance r_e is

$$r_e = \sqrt{(a - c)^2 + (b - d)^2} . \tag{1.3}$$

Similar to 4-connectivity and 4-neighbors, by allowing only vertical and horizontal paths in the calculation of the distance between two pixels, the 4-distance is defined as

$$r_4 = |a - c| + |b - d| . \tag{1.4}$$

The 4-distance is also referred to as the city block distance because this is the distance that an individual would travel ($r_4 = 2$) from one diagonal corner of a city block to the other diagonal corner if the path that was allowed to be traversed was along the sidewalks. Figure 1.13(a) gives the city block or 4-distance from the center pixel.

Another distance measure of interest is the 8-distance, defined as

$$r_8 = \text{maximum}(|a - c|, |b - d|) . \tag{1.5}$$

Figure 1.13(b) gives the 8-distance from the center pixel to neighboring pixels. Note that the contour of constant distances for the 4-distance measures is a diamond [Figure 1.13(a)], while the contour of constant distances for the 8-distance measures is a square [Figure 1.13(b)].

4	3	2	3	4
3	2	1	2	3
2	1	0	1	2
3	2	1	2	3
4	3	2	3	4

(a)

2	2	2	2	2
2	1	1	1	2
2	1	0	1	2
2	1	1	1	2
2	2	2	2	2

(b)

Figure 1.13 Examples of (a) 4-distance and (b) 8-distance.

REFERENCES

Abramson, Albert, *The History of Television, 1880 to 1941*, McFarland & Company, Inc., Jefferson, NC, 1987.

Baxes, Gregory A., *Digital Image Processing: A Practical Primer*, Prentice Hall, Inc., Englewood Cliffs, NJ, 1984.

Green, William B., *Digital Image Processing: A Systems Approach*, Van Nostrand Reinhold Company, New York, 1983.

Gregory, Richard L., *Eye and Brain: The Psychology of Seeing*, Princeton University Press, Princeton, NJ, 1990.

Goldman, Martin, *The Demon in the Aether*, Paul Harris Publishing, Edinburgh, Scotland, 1983.

Gonzalez, Rafael C., and Richard E. Woods, *Digital Image Processing*, Addison-Wesley, Reading, MA, 1992.

McFarlane, Maynard D., "Digital Pictures Fifty Years Ago," *Proc. of the IEEE*, Vol. 60, No. 7, pgs. 768-770, July 1972.

Mast, Gerald, *A Short History of the Movies*, Macmillan Publishing Company, New York, 1986.

Moses, Robert A., *Adler's Physiology of the Eye*, The C. V. Mosby Company, St. Louis, MO, 1970.

Myler, Harley R., and Arthur R. Weeks, *Computer Imaging Recipes in C*, PTR Prentice Hall, Englewood Cliffs, NJ, 1993.

North, Joseph H., *The Early Development of the Motion Picture (1987-1909)*, ARNO Press, New York, 1973.

Resnikoff, Howard L., *The Illusion of Reality*, Springer-Verlag, New York, 1989.

Tasman, William, ed., *Duane's Foundations of Clinical Ophthalmology*, Vol. 2, J. B. Lippincott Company, Philadelphia, PA, 1994.

Thorne, J. O., ed., *Chambers's Biographical Dictionary*, St. Martin's Press, New York, 1966.

Tissandier, Gaston, *A History and Handbook of Photography*, ARNO Press, New York, 1973.

2

VIDEO COMPRESSION STANDARDS

Ferran Marqués and Philippe Salembier
Universitat Politècnica de Catalunya
Barcelona, Spain

2.1 INTRODUCTION

A few years ago, the necessity of video compression algorithms able to reduce the bit rate requested for video transmission became clear. There were new emerging services such as high definition television, videotelephony, video databases, videoconferencing and video-on-demand with strong requirements in high quality and/or in high compression ratios for transmission and storage. These market interests drove the standardization activities that have lead to the H.261, MPEG-1, H.263 and MPEG-2 standards.

Nowadays, although the problems of the previous services have not been completely solved, the necessity has even increased with the explosion of multimedia services. Currently, the effort is not only to provide coding efficiency but new functionalities are included among the goals of the standardization bodies to address the new needs of multimedia services. Nevertheless, compression efficiency is still the main cornerstone of video standards.

Towards the goal of saving in transmission bit rate, redundant and irrelevant information can be removed from the data, prior to transmission. Redundant information can be derived from past data whilst irrelevant information is the part of the data produced by the source that is not necessary for the acceptable quality of the decoded data. This chapter will focus on techniques that mainly aim at reducing the redundancy of the sequence, that is, techniques that rely on the concepts of information theory [1]. Techniques addressing the removal of irrelevant information are based on the features of the human visual system (the so-called *Second Generation* techniques [2]). Nevertheless, current video compression algorithms and standards usually combine both strategies: redundant and irrelevant information removal. An analysis of current standards from the point of view of reduction of irrelevancy can be found in [3].

In video data, both spatial and temporal redundancy are present. This chapter will concentrate on those techniques aiming at reducing temporal redundancy, since this is specific of video compression. From the large number of techniques that exist in the literature, we have mainly selected those that are used in current video standards. Furthermore, some techniques that may help understanding the selected ones or that are extensions of those associated to the removal of spatial redundancy are also covered.

Given the scope of the chapter, several important video compression approaches are not included. In order to cover them, the reader is referred to the References section A basic reference for anyone interested in signal compression is [1]. Vector quantization techniques for still image and video compression are analyzed in [4]. The application of subband coding to the image domain is detailed in [5]. A good overview of still image and video coding techniques is presented in [6]. Finally, an updated summary of new advances in second generation video coding can be found in [7].

After this introduction, the chapter is structured as follows. In section 2.2, the basic algorithms that have been proposed in the literature for video compression are presented. Section 2.3 describes some of the current standards; namely, H.261, H.263, MPEG-1 and MPEG-2. Their descriptions rely on the algorithms that have been presented in section 2.2. The MPEG-4 standardization process is presented in section 2.4. Since MPEG-4 is currently under development, only the new concepts introduced by this standardization effort are discussed. Finally, section 2.5 summarizes the main notions addressed in this chapter and sketches the future lines of standardization.

2.2 ALGORITHMS FOR VIDEO COMPRESSION

Several techniques for still image coding have been proposed in the literature. The main idea of these techniques is to exploit the spatial redundancy that exists among neighbor pixels within an image. In the case of video sequences, redundancy is present not only among pixels belonging to the same image but also among pixels from consecutive images. This concept has led to a set of techniques that are a direct extension of those used in still image coding.

2.2.1   Predictive techniques for video compression

The basis of predictive coding is to estimate the value of a given pixel from the values of the neighboring ones. The prediction error is then quantified and transmitted. One of the main features of predictive image and video compression techniques is their low complexity. Fairly simple predictive schemes that use a few elements for the prediction have been proposed. In addition, simplicity usually translates into causal prediction, so that memory is saved. In the video framework, causal prediction is related to the order in which the frames are processed (natural temporal evolution or reordering the various frames) and the order in which the pixels in a frame are scanned (progressive or interlaced scanning). In the case of video compression, the causal prediction of a pixel value can rely on previous pixels in the current frame (spatial neighborhood) and on pixels in the previous frames (temporal neighborhood). For simplicity, spatial neighborhoods are usually constrained to a few pixels in the current and previous lines, whereas temporal neighborhoods are constrained to a few pixels in the same lines of the previous frame.

Several predictors have been proposed in the literature. In [8], a large set was analyzed in the context of videoconference data compression. A good predictor was found to have the following structure:

$$X_1 = \alpha X_0 + \beta(A_1 - \alpha A_0) + \delta(B_1 - \alpha B_0), \tag{2.1}$$

where X, A and B refer to pixels in the previous or current frame (see Fig. 2.1) and α, β and δ are weighting factors. The optimum values for these parameters, obtained through simulations, were reported to be α around 0.89 and β and δ approximately 0.5. Note that Eq. (2.1) estimates the temporal prediction error of the current pixel ($X_1 - \alpha X_0$) based on the temporal prediction errors of its spatial neighbors ($A_1 - \alpha A_0$ and $B_1 - \alpha B_0$).

This predictive algorithm is a combination of very simple intraframe and interframe predictions. In the context of video coding, intraframe prediction fails when dealing with contours, whereas interframe prediction is not able to handle moving objects. To increase the performance of this kind of predictors, adaptive intra/interframe prediction was proposed [9, 10]. In these techniques, a decision is made whether to use intraframe or interframe mode of prediction. The switch is based on the measurement of both prediction errors in a small causal neighborhood of the current pixel.

Predictive techniques can be improved by applying motion information in the prediction process. The estimation and use of motion information will be extensively described in section 2.2.3.

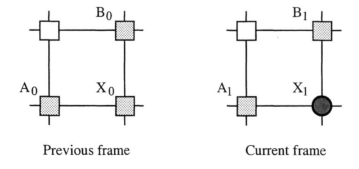

Previous frame   Current frame

Figure 2.1 Prediction approach for video compression.

2.2.2 Transform techniques for video compression

Mainly, two direct extensions of still image coding techniques applied to the case of video compression have been proposed in the literature. The first approach is that of coding separately each frame in a video sequence as if it was a set of nonrelated images. This idea is applied with the JPEG standard and has lead to the so-called MOTION JPEG [11].

The second direct extension deals with the use of transform coding on 3D blocks of data (two spatial and one temporal coordinate). This approach requires the storage of a large set of frames (typically, eight) in the encoder which results in a delay that, depending on the application, is not acceptable [12]. Nevertheless, some research is still active in the area and improvements have been presented that motion compensate the data prior to their transformation [13].

2.2.3 Hybrid predictive-transform techniques for video compression

In the previous sections, we have emphasized the use of motion information to improve the performance of video coding techniques. Actually, motion compensation is the most widely used prediction technique for video data. In the early seventies [14], the idea of combining motion information (for image prediction) and transform coding (for prediction error coding) was introduced. From this moment on, hybrid predictive-transform coding (for short, *hybrid coding*) has been a very active area of research and this approach has been adopted by the most important video compression international standards.

2.2.3.1 General scheme of a hybrid encoder-decoder

Basically, hybrid coding consists of the interleaving of a prediction loop in the temporal domain and a suitable decorrelation technique in the spatial domain. The general scheme of a hybrid codec is presented in Fig. 2.2. Motion is estimated between the current and the previous images. The current image is predicted by means of a motion compensated version of the previous coded image. The prediction error is compressed using intraframe techniques. Both the motion parameters and the prediction error information are encoded and transmitted. The coded information is used in the encoder in order to reproduce the information that will be available in the decoder. The decoded frame will be used as a reference image in the motion estimation and compensation steps.

The decoder receives the coded motion parameters and prediction error information. The previous decoded image is motion compensated using the motion parameters and the decoded prediction error information is added to the motion compensated image. Therefore, the receiver should store the previous decoded frame for motion compensation purposes.

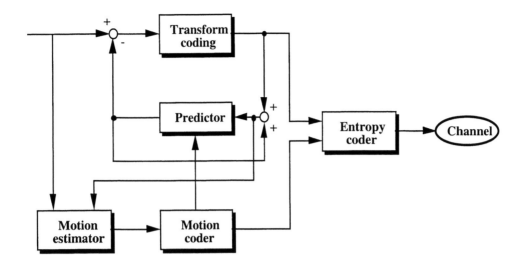

Figure 2.2 Block diagram of a hybrid predictive-transform coding.

The two basic blocks in a hybrid coding scheme are, therefore, the motion estimation and compensation and the prediction error coding. We mainly focus on the first one, given that this is a specific part of video compression. The prediction error image is handled as a still image.

2.2.3.2 Image prediction for video compression using motion information

Motion analysis [15] is a very wide field of research that has applications in several areas such as computer vision, industrial monitoring, biomedicine and video compression. The goal of motion estimation is to obtain the motion information of the objects present in the scene. However, the scene is a 4-D space (three spatial and one temporal coordinate) and video sequences are obtained by means of a projection and sampling of this 4-D space into a 3-D discrete space.

In practice, motion estimation algorithms try to reduce the complexity of the problem by assuming some of the following conditions, at least in a few nearby frames:

• Objects are rigid bodies and, therefore, object deformation can be neglected.

• The movement of an object can be approximated by a translational motion.

• Illumination is uniform and, therefore, the object intensities do not change.

• Occluded and uncovered areas are neglected.

In the context of motion analysis, an object is defined with respect to its motion. *Moving objects* can, therefore, be formed by several entities that share the same motion. For example, a person can be seen as a moving object as well as a static background if being still is considered a type of motion. The boundary of a moving object is very relevant information in order to obtain the object's motion. If all the pixels within a moving object have been grouped, the object's motion can be estimated in a more robust way. However, this is a typical chicken and egg problem since in order to have a precise moving object segmentation, accurate motion information is necessary.

Different solutions have been proposed to circumvent the aforementioned problem [15]. One solution is to look separately for the motion of each single pixel. However, this approach is not robust in the presence of noise. A second solution is to use a fixed partitioning of the image (typically, nonoverlapping blocks) and estimate the motion on each one of these regions. If the regions forming the fixed partition are small enough, a connected set of regions can approximate quite well the shape of the moving object. However, this assumption is not always fulfilled and regions may contain pixels from more than one moving object. Given that there is not a single approach that can solve the motion estimation problem, a specific technique has to be selected for each application.

Every application introduces different constraints in the motion analysis problem. Motion information in video compression applications is used to decrease the coding cost. Hence, it does not require the true motion parameters, but those motion parameters that reduce the total bit rate the most. Moreover, hardware complexity is, in real time applications, a very important parameter to take into account when designing a motion estimator.

Motion estimation techniques for video compression [16] can be roughly classified into three groups: pel-recursive [17], Fourier [18] and correlation methods [19], the last group being the most widely applied. Correlation methods divide the image into a set of nonoverlapping regions (partition). Each region is viewed as an independent moving object whose pixels follow a uniform motion. A given motion model is assumed, e.g., translation, zoom and/or rotation. For a set of motion model parameters, each region is motion compensated into the reference image and the matching between the current region and the pixels covered by the motion compensated region is computed. Motion model parameters are modified until the best match is found. Among this type of method, *block matching* is the most widely used.

2.2.3.3 Block matching approach for motion estimation and compensation

The block matching approach is a correlation technique. It uses a set of nonoverlapping blocks as regions for partitioning the image. The best match between every block in the current frame and the candidates of the reference frame is sought only in a confined zone (*searching area*). Finally, the approach merely assumes translational motion. This concept is illustrated in Fig 2.3. As

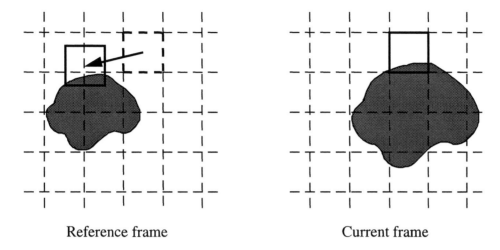

Reference frame   Current frame

Figure 2.3 Motion estimation using the block matching approach. (a) Reference frame. (b) current frame.

the region described by one set of motion parameters is usually small, a simple model, such as a translational one, can correctly describe its motion. The displacement of the center of the block from the current position to the position of the reference block yields the motion vector. In Fig. 2.3(a), the position of the current and the reference blocks are depicted using dark, dashed lines and dark, solid lines, respectively.

There are several reasons that have made the block matching approach very popular. It achieves reduced prediction error while implying acceptable computational load. In addition, as mentioned previously, the performance of video coding techniques can improve introducing the possibility of switching between interframe and intraframe coding approaches. The use of a motion compensation technique that works on a nonoverlapping block basis allows an easy combination of an interframe mode and the transform based intraframe mode that has been found to be the best approach for still image coding. In this case, a block-by-block decision can be taken on the type of technique to be used: if no suitable prediction has been obtained, the block is coded in intraframe mode.

In the example presented in Fig. 2.3, there are two parameters that have to be set: the size of the block and the size of the search area. Each of these parameters is related to a respective trade-off. Small blocks contain few pixels and, hence, the motion estimation is less robust against noise. In addition, the use of small blocks reduces the coding efficiency. When coding the motion information, a set of parameters is transmitted for each block. If small blocks are used, their number per image increases and so does the overhead information per frame. On the other hand, large blocks more likely contain pixels belonging to different moving objects and, therefore, the motion estimation is less accurate. In practice, block sizes of 8×8 or 16×16 pixels are adopted. The size of the search area is related to the largest displacement that can be detected. Therefore, a large search area will ensure that large movements are correctly detected. However, the larger the search area, the bigger the number of possible candidates that have to be analyzed. This leads to an increase of the computational complexity. Usually, the search area is chosen by experiment (application dependent) or due to hardware considerations.

Figure 2.4 presents an example that studies the influence of these parameters in the prediction results. Figure 2.4 (a) and (b) show two consecutive frames of the sequence *Table-Tennis*. Two different predictions of the image in Fig. 2.4 (b) using the block matching technique are presented in Fig 2.4 (c) and (d). The result in Fig. 2.4 (c) has been obtained using a block of 8×8 pixels and a search area of 17×17 pixels, whereas that in Fig. 2.4 (d) has been obtained using a block of 16×16 pixels and a search area of 33×33 pixels. The compensation errors of both predictions are presented in Fig. 2.4 (e) and (f), respectively.

The results obtained in both images illustrate several concepts. Note that the quality of the result presented in Fig. 2.4 (c) is quite good. This statement is further confirmed by the prediction error presented in Fig. 2.4 (e). However, the prediction has completely failed in tracking the motion of the ball. The ball, having a larger displacement than the largest one acceptable for a search area of 17×17 pixels, has not been correctly compensated. By contrast, the result presented in Fig. 2.4 (d) has been able to predict the motion of the ball, due to its larger search area.

The problems generated by the use of large block size can also be assessed in this example. Regardless of the zone related to the ball, the quality of the predicted image is better in Fig. 2.4 (c) and (e) than in Fig 2.4 (d) and (f). This is due to the problem of compensating large blocks that contain more than one moving object. The resulting effect can be observed in the wrong compensation of the tip of the racket in Fig. 2.4 (d).

Finally, let us comment on the problem that appears in the left-hand side of both predicted images. In both cases, the poster that appears on the left-hand side of the image is not correctly predicted. The reason is that there is global motion due to camera zoom-out. Since new objects appear in the scene (new areas of the board), they cannot be correctly predicted with a causal scheme.

(a)

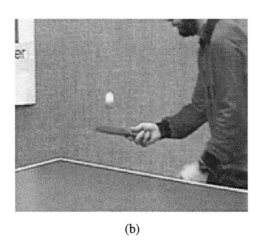

(b)

Figure 2.4 Example of motion compensation. (a) Original frame #41 of *Table-Tennis*, (b) Original frame #42 of *Table-Tennis*.

(c)

(d)

Figure 2.4 Example of motion compensation. (c) Motion compensated image (8×8), (d) Motion compensated image (16×16).

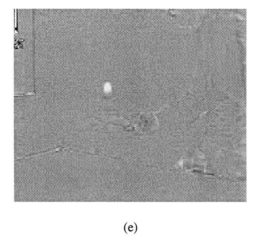

(e)

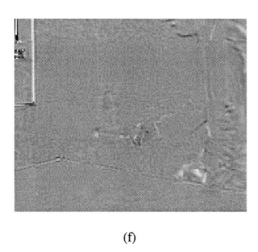

(f)

Figure 2.4 Example of motion compensation. (e) Motion compensation error (8×8), (f) Motion compensation error (16×16).

In addition to the value of the block and search area sizes, it is necessary to define the matching function to be used in the searching process. The selection of the matching function will be related to the computational complexity of the overall technique. Several matching functions have been proposed in the literature, the following being the most common:

$$MAD(d_x,d_y) = \frac{1}{N_x N_y} \sum_{n_x=0}^{N_x-1} \sum_{n_y=0}^{N_y-1} \left| f(n_x,n_y,t) - f(n_x - d_x, n_y - d_y, t-1) \right|$$

(2.2)

$$MSE(d_x,d_y) = \frac{1}{N_x N_y} \sum_{n_x=0}^{N_x-1} \sum_{n_y=0}^{N_y-1} \left[f(n_x,n_y,t) - f(n_x - d_x, n_y - d_y, t-1) \right]^2$$

(2.3)

where MAD and MSE stand for mean absolute difference and mean squared error, respectively; d_x and d_y are the displacement values in the x and y directions; and N_x and N_y are the number of pixels of the block in the x and y directions. MSE is the best matching criterion in the presence of white Gaussian noise. However, this noise model is not realistic in the case of coded images. The most commonly used matching function is the MAD: it leads to a simple hardware implementation and to relatively good performance.

Since the definition of the block matching approach, two main lines of research have been followed to improve its performance: reduction of the computational load and increase of the motion vector accuracy. Fast algorithms have been proposed to achieve the first goal (see section 2.2.3.4). To increase the motion vector accuracy, several techniques have been used such as hierarchical block matching [20] or fractional-pel accuracy [21]. From this set of techniques, only those which have been adopted by international standards will be described in section 2.3. For a recent and more complete review of these techniques, the reader is referred to [22]. In addition, [23] analyzes the trade-off between motion estimation accuracy and the amount of overhead information to be sent when the accuracy is increased.

2.2.3.4 Fast algorithms

Given a search area in the reference image, the exhaustive search of the best match for a given block of the current image requires a large amount of computations. In order to reduce this computational load, various fast algorithms have been proposed. These algorithms are based on the assumption that the matching function is monotone along any direction away from the optimal point; that is, the matching function has a unique optimum point within the search area. In this case, the search can be performed in several sequential steps. At every step, the matching function is evaluated in a few search points.

Search points are located around a central point and the distance between two nearby search points is called *step size*. The most promising search point is kept to initialize the procedure in the next iteration. Usually, the step size is reduced at each iteration and this procedure is repeated until the maximum of the matching function is obtained.

Fast algorithms reduce the number of positions in which the matching function is computed. However, in practice, the initial assumption of having a monotone matching function is not realistic. On the contrary, local maxima of the matching function may appear. Therefore, fast algorithms may get trapped on local optima, since they are based on local decisions.

Several fast algorithms have been proposed in the literature. In [24], a special case of the conjugate direction search was proposed, the so-called *once at a time* algorithm. This algorithm divides the 2-D search problem into two 1-D problems, which are separately solved. This way, the algorithm first finds the optimum point in one direction (e.g.: horizontal axis) and then in the orthogonal one. Two 1-D neighbors, and the central point are analyzed at each step. Although this technique usually requires the visiting of the lowest number of search points, it is considered to have the lowest performance in terms of prediction error reduction.

The *2D-log* search scheme was proposed in [25]. This algorithm analyzes at each step five different points: the central point and the four points located at a given distance (step size) from the central point in the vertical and horizontal directions. In the next step, the most promising point becomes the central point. If the most promising point is the central point of the previous search or it is located at the boundary of the search area, the step size is halved. Otherwise, the step size remains the same. In the last step (step size = 1), the nine points contained in a square of 3×3 pixels around the last best matching are evaluated. The complete procedure is illustrated in Fig. 2.5, where a search area of 15×15 pixels is used. All the visited points have been marked with a white circle and the initial and final points with a black one.

An algorithm that, for a given search area size, always uses the same amount of steps was presented in [26]. It is the so-called *three step* algorithm. At each step of this algorithm, nine points are analyzed: the central one and the eight points located in the corners and in the middle of each side of a square centered at the central point. The dimensions of the first square are half (or slightly larger than half) of the dimensions of the search area. As in the aforementioned techniques, the point leading to the best match becomes the central point of the following step. In this case, the step size is reduced by half at each step until it becomes equal to 1. This procedure is illustrated in Fig. 2.6, using a 15×15 search area and marking the visited points as in the previous example. This technique is often used in hardware implementations since it leads to good results while using the same amount of steps for every block and applying the same operation at every step.

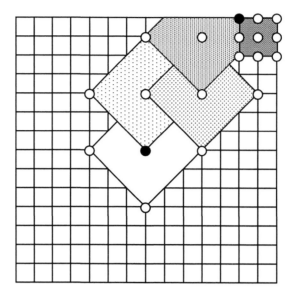

Figure 2.5 Illustration of the 2D-log algorithm.

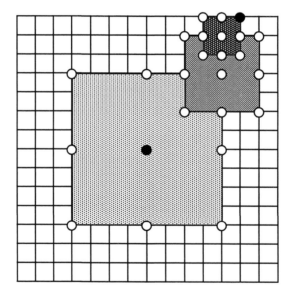

Figure 2.6 Illustration of the three step algorithm.

A comparison among these techniques in terms of number of visited search points and number of search steps is presented in Table 2.1. Data in Table 2.1 have been obtained assuming a square search area of 15×15 pixels. Therefore, the maximum displacement (horizontal and vertical) that can be detected is of 7 pixels. For a more complete comparison of these techniques as well as for the analysis of other fast algorithms, the reader is referred to [27].

Search Algorithm	Minimum number of search points	Maximum number of search points	Minimum number of search steps	Maximum number of search steps
Exhaustive	225	225	1	1
One-at-a-time	5	17	2	14
2D-log	13	26	2	8
Three step	25	25	3	3

Table 2.1 Comparison of performance of fast search algorithms (search area = 15×15 pixels).

2.2.3.5 Motion information coding

Once the motion vectors have been obtained, they have to be transmitted. In the case of the block matching algorithm, a predictive technique is usually applied. The motion vector components (horizontal and vertical, in the case of translational motion) of every block are separately coded using their spatial neighborhood. Typically, the candidate predictors are the motion vectors of three spatial neighbors of the current block. The prediction error is quantized and coded using variable length codes.

2.2.3.6 Prediction error coding

Once a block has been predicted using motion compensation, the prediction error is coded. Although the statistical characteristics of the prediction error images differ from those on which transform coding has proven very efficient, prediction error information is encoded by means of transform coding techniques. This approach has been largely criticized and this criticism has been the driving force of research in this area [6]. Nevertheless, none of the new approaches has outperformed the combination of motion compensation and transform coding.

In hybrid coding schemes, prediction error images present very little correlation due to the motion compensation step. They are rather noisy, and edges (high frequency components) appear emphasized. With these new statistical characteristics in mind, the basic quantization matrices used in intraframe coding are changed when used in interframe coding. These changes basically result in quantization matrices which are more flat than in the intraframe coding. Because the energy is not being concentrated in the low frequency coefficients, all coefficients are coded in a more homogeneous way.

2.2.3.7 Frame interpolation using motion information

Motion information can be used not only to predict the information in the current frame (causal prediction) but also to predict the information between two already known frames (noncausal prediction); that is, to interpolate frames [28].

The basic idea of motion-compensated frame interpolation is to use both past and future frames as references to predict the value of current frames. Frame interpolation is computed in the receiver based on the set of decoded frames. In this case the frame rate is increased in the receiver side. Once the reference and current frames have been decoded, motion vectors used to compensate the encoded reference frame are applied to generate intermediate frames. Forward compensation of the previous frame is combined with backward compensation of the current frame in order to interpolate intermediate frames.

2.2.3.8 Frame coding using bi-directional prediction

Alternatively, the interpolation can be carried out in the encoder side. This means that some of the frames to be coded are noncausally predicted and, therefore, they are coded using motion vectors from past and future reference frames. This procedure is illustrated in Fig. 2.7. In this case, a new set of motion parameters is computed from the past and future reference frames to the frame to be coded. These motion parameters are transmitted. In turn, the error prediction in interpolated frames may or may not be encoded.

This approach is more efficient than the basic motion prediction since:

- Noncausal prediction makes it possible to handle uncovered areas correctly
- The average of future and past information reduces the noise influence.

However, it presents as main drawbacks:

- An increasing overhead of motion vectors: two per block instead of one
- If bi-directional predicted images are to be interleaved, the search area for directly predicted images should be increased
- Noncausal prediction introduces an end-to-end delay
- Noncausal prediction increases the hardware complexity.

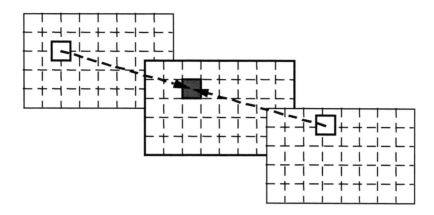

Figure 2.7 Frame interpolation for video compression.

2.3 CURRENT VIDEO COMPRESSION STANDARDS

Video compression is a key component of many commercial applications such as videophone, videoconference, digital TV or HDTV broadcasting, video-on-demand and services relying on video databases. Because of the very high industrial interest and motivation, the necessity of defining standards for video compression arose at the beginning of the 1980s. Video standards potentially offer two advantages: first, they allow the exchange of video information between equipment of different companies and, second, they reduce production costs since a very large number of key components can be massively manufactured. The goal of this section is to review four current video standards: H.261, MPEG-1, MPEG-2 and H.263. These standards have been selected because of their widespread use.

Historically, the first international committee that focused on video coding was the Study Group (SG) XV of CCITT, which started its activities in the early 1980s. In 1984, this SG agreed to work towards the definition of a worldwide standard for videophone and videoconference suitable for bit rates lower than 2 Mbits/s. The resulting Draft Recommendation H.261 was published in 1989. It defines codecs working at bit rates of p×64 kbits/s with p=1,...,30.

Meanwhile, in 1988, a Working Group (WG) was created in Subcommittee 2 (SC2) of ISO to define video coding algorithms for digital storage and bit rates lower than 1.5 Mbits/s. This group, known as MPEG (Moving Picture Experts Group), defined its first Draft International Standard (DIS) in 1991 and the corresponding International Standard, MPEG-1, was published in 1992. In contrast with H.261, MPEG-1 is a generic standard. This means that it is not devoted to any specific application (even if the original application was digital storage of video). In fact, only the syntax of the bit-stream is standardized, which mainly defines the decoder.

In 1990, MPEG-2 started its activities to define a standard for TV sequences (CCIR 601 resolution) and bit rates lower than 10 Mbits/s. This original goal was extended in 1992 to cope also with digital HDTV. As a result, MPEG-3, originally planned for HDTV, became useless. The Draft International Standard of MPEG-2 was issued in 1994. The main differences between MPEG-1 and MPEG-2 were two points: first, MPEG-2 deals with interlaced input format whereas MPEG-1 works with progressive format; second, MPEG-2 introduced the concept of various "Profiles," depending on the functionalities the user wants, and "Levels," depending on the resolution of the input sequences. This way of defining the standard is quite flexible and, for example, allows the creation of MPEG-2 compliant systems of limited complexity that only address one Profile and one Level.

Besides the activities in ISO, technical progress in video coding algorithms allowed ITU to define a new standard that significantly outperforms H.261 for very low bit rate application. This new standard called H.263 was defined in 1995 (Draft Recommendation).

Figure 2.8 summarizes the activity of both ISO and ITU for the four standards of interest in this section. On the technological side, it is interesting to note that all these standards are based on the same hybrid predictive-transform scheme described in section 2.2.3. Their basic strategy is to estimate the motion between frames, to perform a motion-oriented prediction of the frame to transmit and finally to code the prediction error using DCT, quantization and entropy coding. The main differences between the various standards result from the differences in input formats and from the constraints imposed by some applications. In the remaining part of this section, we will review the main objectives and features of each standard.

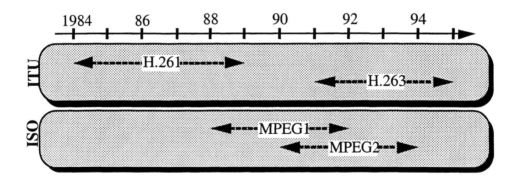

Figure 2.8 Summary of the main activities in ISO and ITU with respect to H.261, H.263, MPEG-1&2.

2.3.1 H.261: A standard for videophone applications [29]

The H.261 standard was designed for audiovisual applications over ISDN at p × 64 kbits/s with p ranging from 1 to 30. Typical applications include videophone and videoconference. For such applications, the encoding/decoding delay is of prime importance. As a result, H.261 was constrained to offer a low processing delay.

<u>Input formats and output bit rates:</u> The encoder input is assumed to be a noninterlaced video source of either CIF or QCIF format (please refer to Table 2.2 for the definition of these formats). In practice, if p=1 or 2, the QCIF format is used and only 10 (or less) frames per second are transmitted.

<u>Main features of the standard:</u> The basic strategy of the encoder is to code the first frame of the sequence in intraframe mode (I-frame). All remaining frames are coded by prediction (P-frames). The prediction is achieved by motion compensation of the previously transmitted frame (see Figure 2.9 (a)). As can be seen, the processing and transmission order corresponds to the natural order of the frame in the sequence. Moreover, since the encoding of the current frame only relies on the previous frame, the transmission delay remains low.

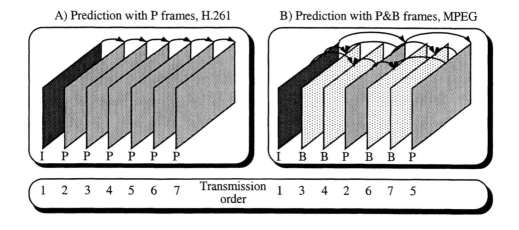

Figure 2.9 Prediction modes of H.261 and MPEG.

On the frame level, the algorithm relies on a partition of macroblocks of size 16×16 pixels of luminance. Each macroblock (MB) can be transmitted with one of the three following coding modes:

1) Not coded MB: if the difference between the current MB and the previous MB (without motion compensation) is small, the MB is not coded at all and the decoder displays the previous MB. Obviously, this transmission mode is quite efficient and in applications such as videophone or video-conference involving a rather high number of MB corresponding to a static background, this mode actually provides significant bit saving.

2) Intra MB: no prediction is used and the four 8×8 blocks of the MB are coded by using DCT as for the I-frame.

3) Inter MB: prediction is performed by motion compensation at the MB level and the compensation error is coded using 8×8 DCT.

Video Format	Spatial resolution (Hor.×Ver.)	Frame rate (Hz)	Chroma	Interlaced
QCIF	176×144	29.97	4:2:0	no
CIF	352×288	29.97	4:2:0	no
SIF-625	352×288	25	4:2:0	no
SIF-525	352×240	29.97	4:2:0	no
TV (CCIR 601)	720×576	50	4:2:2	yes
HDTV 1	1440×1152	50 / 60	4:2:2	yes
HDTV 2	1920×1152	50 / 60	4:2:2	yes

Table 2.2 Main video input formats. The figures in the Chroma column define the sampling relation between luminance and chrominance. 4:4:4 means equal sampling, 4:2:2 means half horizontal resolution for the chrominance and 4:2:0 means half horizontal and vertical resolutions.

2.3.2 MPEG-1: A generic standard for coding of moving pictures for media up to about 1.5 Mbits/s [30]

The MPEG-1 standard is devoted to the compression of progressive video and its associated audio. MPEG-1 was originally targeted for multimedia application on CD-ROM, however, it turned out to be a generic and flexible standard for a wide range of applications. Important features offered by MPEG-1 (and not by H.261) are the possibility to access randomly some frames of the sequence and the ability to search in fast forward or fast reverse mode directly in the compressed bit-stream.

Input formats and output bit rates: The main application of MPEG-1 is the representation of video sequences on digital storage such as CD-ROM. The algorithm has been particularly optimized for a SIF input format (see Table 2.2) and the corresponding output bit rate is equal to 1.15 Mbits/s. This defines the main application however it has to be mentioned that the user can modify to a large extent the input format (up to TV resolution) and the output bit rates (up to 1.86 Mbits/s). The standard itself is for progressive format, however, an Annex of the MPEG-1 International Standard [MPEG-1-93] proposes a set of pre-processing and post-processing steps suitable to deal with the representation of interlaced format. The strategy consists of converting the original interlaced sequence into a progressive sequence, to code this last sequence with MPEG-1 and to re-interlace the decoded sequence in the receiver side.

Main features of the standard: MPEG-1 is technically very similar to H.261. The main difference consists in the periodic use of I-frames and the introduction of bi-directional predicted frames (B frames). I-frames are coded without any reference to other frames. They are periodically introduced in order to allow random access and fast search. The use of P- and B-frames is illustrated in Fig. 2.9 (b). B-frames are frames that are predicted from the closest past and future I- or P-frames. The prediction is performed on a macroblock basis and can consist of selecting the motion compensated macroblock of either the past frame or the future frame or in computing the average of the motion compensated macroblocks of both frames. In this last case, two motion vectors per macroblock have to be transmitted to the receiver. B-frames allow efficient compression because uncovered areas of the image can be predicted from the future. B-frames themselves are never used for prediction. The main drawback of B-frames is the delay they introduce in the transmission. As can be seen in Fig. 2.9 (b), the transmission order of the frames does not correspond to the natural sequence order. In the example of Fig. 2.9 (b), one has to wait for the fourth frame of the original sequence before being able to process the second frame.

As mentioned previously, the user can define the pattern of the type of frames. For example, in an application such as videophone, the time delay is an important factor and B-frames may be undesirable. A possible solution could be IPPPPPPIPPPPPPI.... In a storage application, if the user wants to have a high degree of random access, the number of P-frames in between I-frames can be reduced. Finally, in broadcasting applications, where delay is not a crucial factor, a rather high number of B-frames can be introduced (i.e., IBBPBBPBBIBBP...).

2.3.3 MPEG-2: Generic coding of moving pictures [31]

MPEG-2 can be considered as an extension of MPEG-1. Initially, it was proposed to deal with interlaced sequences such as TV format. MPEG-2 is compatible with MPEG-1 in the sense that an MPEG-2 decoder should be able to decode an MPEG-1 bit-stream. Besides its ability to deal with interlaced format, MPEG-2 introduces a set of tools to allow scalable coding, providing functionalities such as embedded coding of TV and HDTV or graceful degradation in case of transmission errors. The syntax of MPEG-2 is fairly complex and, for some specific applications, it may be useless and even not practical to implement the full syntax. To solve this problem, MPEG-2 proposes the concept of Profiles and Levels. Profiles define a subset of functionalities the user wants to have and Levels define the type of input sequences output bit rates.

Input formats and output bit rates: The Level of the algorithm mainly defines the input format. A summary of the various profiles is given in Table 2.3. The Main Level is devoted to TV pictures, whereas High Levels are concerned with HDTV formats.

Main features of the standard: The main features of the algorithm are defined by the Profiles that are defined in a hierarchical way as described in Fig. 2.10. The core MPEG-2 algorithm is defined by the "Main" Profile that includes the coding of interlaced or progressive video (4:2:0 chrominance format) using I-, P- and B-frames without any scalable functionalities. The Main Profile of MPEG-2 is quite similar to MPEG-1: it is a hybrid predictive-transform scheme involving I-, P- and B-frames. To deal efficiently with interlaced sequences, MPEG-2 introduces the difference between frames and fields. Two fields may be coded either separately or together as a single frame. In practice, most of the MPEG-2 codecs will at least implement the Main Profile at the Main Level. In this case, MPEG-2 is able to encode TV sequences at PAL broadcast quality at about 4 Mbits/s and almost perfect quality at 9 Mbits/s.

Some MPEG-2 profiles also define tools allowing a scalable encoding of the sequence. The objective is to create several layers of representation either in spatial or in temporal quality and to separate physically the information in the bit-stream. There are two main objectives behind the implementation of scalable coding. The first one is to provide interoperability between different services that may share part of the information. A typical example is inter-working between regular TV and HDTV services. Instead of sending two independent bit-streams, it is more efficient to code the HDTV sequence in a scalable way such that the information corresponding to the TV sequence can be easily extracted from the HDTV bit-stream. The second objective of scalability is to allow the sharing of the information into various levels of priority. This idea is very attractive in the case of channel errors, where different transmission strategies can be used for the various levels of priority. For example, graceful degradation of the decoded image quality can be achieved.

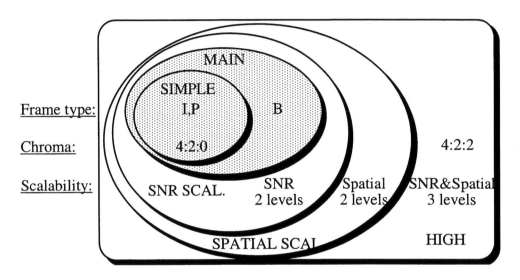

Figure 2.10 Summary of the MPEG-2 Profiles and their main features.

Level	Spatial resolution (Hor.xVer.)	Frame rate (Hz)	Output bit rate (Mbits/s)
Low	352×288	30	4
Main	720×576	30	15
High 1440	1440×1152	60	60
High	1920×1152	60	80

Table 2.3 Summary of upper bounds of the MPEG-2 levels.

2.3.4 H.263: Video codec for narrow channels (< 64 kbits/s) [32]

H.263 has mainly the same objectives as H.261 but takes advantage of the technical progress achieved in the field of video compression. The main improvements concern the temporal prediction used in the hybrid predictive-transform coding. The target service is to send video information with the ITU-T V.34 modem on telephone lines (28.8 kbits/s which leaves around 20 kbits/s for the video part).

Input formats and output bit rates: The input format can be CIF, QCIF or even sub-QCIF (see table 2.2). In any case, the standard is devoted to progressive sequences.

Main features of the standard: Roughly speaking, it can be considered that for very low bit rate applications (rates below 64 kbits/s) H.263 provides the same visual quality as H.261 but at half the bit rate. H.263 has several modes. In the default mode, the main difference between H.261 and H.263 is the accuracy of the motion estimation and compensation. In the case of H.263, this accuracy is equal to half a pixel (note that MPEG-2 already introduced this improvement). Bilinear interpolation is used when necessary. Half pixel accuracy provides a significant improvement: about 2 dB of PSNR over H.261 working at the same bit rates (between 30 and 64 kbits/s). The standard also defines four optional modes: Unrestricted Motion Vector, Advanced Prediction, PB-frames and Syntax-based Arithmetic modes. The mode that provides a significant improvement with respect to the default mode is the Advanced Prediction mode (gain of 1 dB over the default mode). The detailed description of this mode goes beyond the scope of this chapter and the reader is referred to [32] for more information.

The performance of this algorithm is illustrated in Fig. 2.11. The complete sequence *Foreman* has been coded at different bit rates and the result for a given frame is presented. This standard has been selected to present coding results given that it has been developed to cover low bit rates, where differences can better be noticed. Note the appearance of block artifacts with the decrease in bit rate as well as the blurring effect, due to the poor coding of the motion compensation error. These visual effects are typical of any standard based on the hybrid coding approach when the bit rate decreases.

To conclude this section, we present in Fig. 2.12 a schematic representation of the main range of applications of the standards that we have discussed. The picture assigns a location for each standard in the input/output bit rates plane. This location approximately defines the classical application of the standards and not their precise technical ability.

(a) Original QCIF frame of the Foreman sequence

(b) Decoded sequence at 128 kbits/s

(c) Decoded sequence at 64 kbits/s

(d) Decoded sequence at 32 kbits/s

(e) Decoded sequence at 24 kbits/s

Figure 2.11 Example of coding results with H.263 for low bit rate applications.

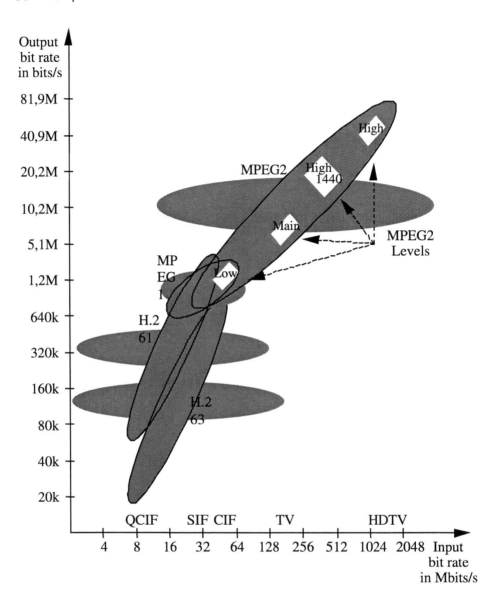

Figure 2.12 Summary of the range of use of the main video standards.

2.4 MPEG-4: A STANDARD UNDER DEVELOPMENT

The original target of MPEG-4 involved new audiovisual coding techniques for very low bit rates. However, due to the changes in the audiovisual world, MPEG adopted new objectives [33]. As mentioned in section 2.3, previous MPEG standards aimed to make the storage and transmission of video data more efficient. Although some additional functionalities were introduced, such as random video access and scalability, the user interaction is merely at video frame level.

The novelty and potential of these new objectives lead us to introduce in this chapter a section dealing with the MPEG-4 standard, although it is currently under development. Because the standardization process is in a very advanced phase (the Working Draft was issued in November 1996, the Committee Draft will be issued in November 1997 and the International Standard in November 1998), the main parts of the standard can already be described.

2.4.1   A new set of functionalities

The goal of MPEG-4 is not only to improve the coding efficiency, but also to support new functionalities that have not been addressed (or not well supported) in existing standards. There are 8 different functionalities that are said to be important for MPEG-4 [34]

1. Content-based multimedia data access tools

2. Content-based manipulation and bit-stream editing

3. Hybrid natural and synthetic data coding

4. Improved temporal random access

5. Improved coding efficiency

6. Coding of multiple concurrent data streams

7. Robustness in error-prone environments

8. Content-based scalability

These functionalities will be achieved by changing the description of the scene, representing it in terms of a composition of objects. Hence, *audiovisual objects* are the basic elements of the MPEG-4 standard, instead of video frames with associated audio as in the existing standards. In this chapter, since we are dealing with video compression, we will concentrate on video data and, therefore, we will use the concept of *video object*. A video object is a coded representation of an arbitrarily shaped object with certain relations in space and time. A special case of the previous definition is to take the complete frame as a single video object.

Nevertheless, in most applications a video object represents a semantically meaningful object in the scene. For instance the evolution of a person through a scene or a complete background. Figure 2.13 shows an example of video objects and their composition. In Fig. 2.13 (a), a frame of a background is shown. This background can be understood as a video object. Figure 2.13 (b) presents the first frame of a video object of the so-called *Akiyo* sequence. Finally, Fig. 2.13 (c) illustrates the composition of the two previous video objects.

The possibility of describing the scene by means of video objects opens the door to new concepts of interactivity. As in the example of Fig. 2.13, the user can select how to compose different video objects that are stored to create a new scene and how to synchronize their evolution in time. Therefore, the encoded data can be reused in a more flexible way than currently.

(a) VO background of *Akiyo*   (b) One frame of a VO of *Akiyo*

(c) Composition of the two previos Vos

Figure 2.13 Example of composition of different video objects.

The object-based representation adopted by MPEG-4 enables the combination of audiovisual objects of different natures, leading to the first true multimedia standard. This way, synthetic and natural video objects can be integrated together in a single sequence and bit-stream. Furthermore, different types of objects may be encoded with specific compression algorithms that allow their efficient representation. This capability of combining natural and computer-generated content will be very useful in various applications such as movies, computer-games, architecture, etc.

The ability to describe a scene as a combination of video objects allows us to address content-based functionalities. As the sequence can be handled as a composition of video objects, different actions can be applied to every object in the encoding process. For instance, a chosen object in the scene can be selectively coded with respect to its environment; that is, its quality or the error robustness of its description can be increased. Moreover, scalability can be implemented on an object basis, for instance, the temporal or spatial resolution of a selected object can be improved with respect to the other video objects in the scene.

Another important novelty in MPEG-4 is the concept of flexibility of the standard that will allow its adaptation to technological progress. This flexibility is mainly achieved at the system level and, therefore, it goes out of the scope of this chapter. However, let us comment that the object-oriented architecture adopted in MPEG-4 at the system level, the so-called MPEG-4 Systems and Description Languages (MSDL), already allows the flexible composition of audiovisual information. Moreover, it opens the door to other levels of flexibility such as the configuration of terminals and the downloading of tools [35].

2.4.2   MPEG-4 natural video

The new MPEG-4 standard has been named a *visual coding standard*. With this name, MPEG-4 highlights the fact that the standard should integrate synthetically generated visual objects as well as natural video objects. Synthetic objects may be represented in terms of their graphical parameters, if this representation is more efficient than that based on block motion vectors and transform coefficients of the prediction error. The Synthetic and Natural Hybrid Coding (SNHC) group in MPEG-4 is working towards the definition of new coding tools for this type of data and their integration in a visual coding standard [36]. This section concentrates on the description of natural video coding.

The MPEG-4 visual coding standard [37] should provide video compression at least as good as already existing standards. In addition to this main functionality, MPEG-4 video coding should enable functionalities such as object, spatial and temporal scalability or error resilience. The MPEG-4 video standard will provide a toolbox containing tools and algorithms bringing solutions to the previous functionalities.

As previously commented, the MPEG-4 video coding is based on the concept of video objects. Each temporal sample of a video object forms a two-dimensional spatial region that is called a *video object plane*. To enable content-based functionalities, each video object has to be coded separately. Therefore, for each frame, the various video object planes forming this frame have to be defined. The procedure for defining the shape of every video object plane is purely an encoder problem and, hence, it is not covered by the MPEG-4 standard. Figure 2.14 illustrates the MPEG-4 encoder structure. The MUX block depicted in Fig. 2.14 has to multiplex the various bit-streams that result from every VOP encoder, which is done at the syntax level of the standard.

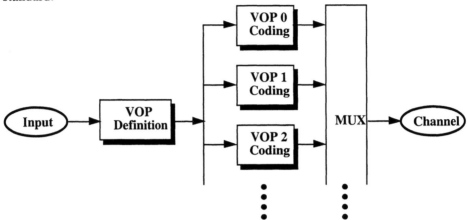

Figure 2.14 MPEG-4 encoder structure.

2.4.3 Video object plane information coding

Each one of the VOP encoders in Fig. 2.14 has to compress the data contained in an arbitrarily shaped region efficiently. The shape of 2-D objects is described by alpha maps. Binary alpha maps define which pixels belong to the video object plane at a given time. New algorithms to compress binary alpha maps
have been proposed mainly based on two approaches: block-based and vertex-based techniques. In both cases, the compression algorithms benefit from the temporal redundancy of the alpha map sequences by motion compensating the previously transmitted information. Due to the subjective relevance of shape information, both lossless and lossy coding techniques are being studied.

Multilevel (or gray scale) alpha maps are used to blend different video object planes to build the final scene. The gray scale represents the degree of transparency of a given pixel: the lowest level (usually 0) corresponds to a completely transparent pixel and the greatest level (usually 255) to a completely opaque pixel. This information is useful to create sequences where some video objects have transparencies. Shapes in multilevel alpha maps are coded using the same technique applied for binary alpha maps. The gray scale information is coded using the same techniques used for the pixel information, allowing lossy coding only.

The pixel data are compressed by means of motion compensated prediction coding and transform coding of prediction errors, as in the already available video standards. However, these techniques have to be extended to the case of arbitrarily shaped objects. For the motion information, this extension has resulted in the definition of a new tool, the so-called *polygon matching motion estimation*. This technique computes the matching error between the pixels of the current macroblock that are inside of the video object plane shape and those of the reference video object plane. As the matching may use some reference pixels that are outside the reference video object plane, a padding technique is applied to extrapolate their values from those of the pixels inside the video object plane.

In order to code the prediction error, a special procedure has to be applied to those blocks that do not completely belong to a video object plane and, therefore, do not have all their pixels defined. The same problem appears in the case of coding the pixel values in the intraframe coding. Two different approaches have been proposed to solve this problem. In the case of applying the traditional Discrete Cosine Transform (DCT), a padding is necessary to fill the missing pixels before transformation. An alternative approach is the so-called Shape Adaptive DCT (SA-DCT) [38].

2.5 DISCUSSION

This chapter has covered the main tools that are used in current video compression standards, as well as the standards themselves. The creation of international standards of video compression is a very important task. Standardization allows communication on an international scale. In addition, international standards enable the large-scale production of VLSI systems and devices. Such production makes final products cheaper and have a wider field of application.

International standards do not usually adopt the best technical solutions. In the creation of a standard, a compromise has to be achieved between the compression efficiency (or performance regarding other functionalities), the flexibility supported by the standard and the implementation complexity. Failing on this trade-off may lead to an unsuccessful standard. This was the case, for instance, of the H.120 standard (not covered in this chapter) whose cost was too high and capability for interoperability too low.

The success of the MPEG-1 and MPEG-2 standards seems clear, given their high level of flexibility. MPEG-1 has already taken the area of applications for which H.263 was initially designed. In addition, several vendors have already introduced in the market MPEG-1 encoder and decoder chips. In turn, the success of the MPEG-2 standard seems ensured by very strong commitment by a large amount of industries to use this standard. Note that currently, several decoder chips for MPEG-2 MAIN Profile at MAIN Level have already been developed.

The MPEG community is currently trying to create an even more challenging standard. The new standard should be capable of giving an answer to the future necessities of sectors such as telecommunication (communication services), information technology (interactive services) and entertainment (broadcasting services). This will be done by means of the first true multimedia standard, the so-called MPEG-4 standard.

REFERENCES

[1] N. S. Jayant and P. Noll, *Digital Coding of Waveforms*, Prentice-Hall, Englewood Cliffs, New Jersey, 1984.

[2] M. Kunt, A. Ikonomopoulos and M. Kocher, "Second generation image coding techniques," *IEEE Proc.*, vol. 73, pp. 549-574, 1985.

[3] N. García, F. Jaureguizar and José I. Ronda, "Pixel-based video compression schemes," in *Video Coding. The Second Generation Approach*, Kluwer Academic Publishers, Dordretch, The Netherlands, 1996.

[4] A. Gersho and R. M. Gray, *Vector Quantization and Signal Compression*, Kluwer Academic Publishers, Boston, 1990.

[5] T. A. Ramstad, S. O. Aase and J. H. Husoy, *Subband Compression of Images: Principles and Examples*, Elsevier Science B.V., Amsterdam, 1995.

[6] R. J. Clarke, *Digital Compression of Still Images and Video*, Academic Press, San Diego, 1995.

[7] L. Torres and M. Kunt, *Video Coding. The Second Generation Approach*, Kluwer Academic Publishers, Dordretch, The Netherlands, 1996.

[8] L. H. Zetterberg, S. Ericsson and H. Brusewitz, "Interframe DPCM with adaptive quantization and entropy coding," *IEEE Trans. Commun.*, vol. COM-30, pp. 1888-1899, 1981.

[9] P. Pirchs, "Adaptive intra-interframe DPCM coder," *Bell Syst. Tech. J.*, vol. 61, pp. 747-764, 1982.

[10] K. A. Prabhu, "A predictor switching scheme for DPCM coding of video signals," *IEEE Trans. Commun.*, vol. COM-33, pp. 373-379, 1985.

[11] ISO/IEC IS 10918 (JPEG), "Information technology - digital compression and coding of continuous-tone still images," 1994.

[12] R. J. Clarke, *Transform Coding of Images*, Academic Press, San Diego, 1985.

[13] T. Akiyama, T. Takahashi and K, Takahashi, "Adaptive three-dimensional transform coding for moving pictures," *Proc. Picture Coding Symp.*, paper 8.2, Cambridge, 1990.

[14] B. G. Haskell and J. O. Limb, "Predictive video coding using measured subjective velocity," U.S. Patent No. 3,362,865, Jan. 1972.

[15] M. I. Sezan and R. L. Lagendijk, *Motion Analysis and Image Sequence Processing*, Kluwer Academic Publishers, Dordretch, The Netherlands, 1993.

[16] G. Tziritas and C. Labit, *Motion Analysis for Image Sequence Coding*, Elsevier Science B.V., The Netherlands, 1994.

[17] C. Cafforio and F. Rocca, "Tracking moving objects in television images," *Signal Processing*, vol. 1, pp. 133-140, 1979.

[18] T. S, Huang, *Image Sequence Analysis*, Springer-Verlag, Berlin, 1981.

[19] H. G. Musmann, P. Pirsc and H. J. Graller, "Advances in picture coding," *Proc. IEEE*, vol. 73, pp. 523-548, April 1985.

[20] M. Bierling, "Displacement estimation by hierarchical blockmatching," *Proc. SPIE Visual Commun. Image Processing*, vol. 1001, pp. 942-951, Cambridge, MA, 1988.

[21] B. Girod, "Motion-compensating prediction with fractional-pel accuracy," *IEEE Trans. Commun.*, vol COM-41, pp. 604-612, April 1993.

[22] F. Dufaux and F. Moscheni, "Motion estimation techniques for digital TV: a review and a new contribution," *Proc. IEEE*, vol. 83, pp. 858-876, June 1995.

[23] B. Girod, "Rate-constrained motion estimation," *Proc. SPIE Visual Commun. Image Processing*, vol. 2094, pp. 235-242, Chicago, IL, 1994.

[24] R. Srinivasan and K. R. Rao, "Predictive coding based on efficient motion estimation," *IEEE Trans. Commun.*, vol. COM-33, pp. 888-896, Sept. 1985.

[25] J. R. Jain and A. K. Jain, "Displacement measures and its application in interframe image coding," *IEEE Trans. Commun.*, vol. COM-29, pp. 1799-1808, Dec. 1981.

[26] T. Koga, K. Iinuma, A. Hirano, Y. Iijima and T. Ishiguro, "Motion compensated interframe for video conferencing," *Proc. Nat. Telecommun. Conf.*, pp. G5.3.1-G5.3.2, New Orleans, Nov. 1981.

[27] H. M. Hang and Y. M. Chou, "Motion estimation for image sequence compression," Chap. 5 in *Handbook of Visual Communications*, Academic Press, San Diego, 1995.

[28] S. Tubaro and F. Rocca, "Motion field estimators and their application to image interpolation," in *Motion Analysis and Image Sequence Processing*, Kluwer Academic Publishers, Dordretch, The Netherlands, 1993.

[29] ITU-T Recommendation, H.261, "Video codec for audiovisual services at p×64 kbits/s," rev. 2, 1993.

[30] ISO/IEC IS 11172 (MPEG-1), "Information technology - coding of moving pictures and associated audio for digital storage media up to about 1.5 Mbits/s," 1993.

[31] ISO/IEC IS 13818 (MPEG-2), "Information technology - generic coding of moving pictures and associated audio," 1994.

[32] Draft ITU-T Recommendation H.263, "Video coding for narrow telecommunication channels at < 64 kbits/s," 1995.

[33] L. Chiariglione (Convenor), "MPEG-4 project description," Document ISO/IEC JTC1/SC29/WG11 N1177, Munich MPEG meeting, January 1996.

[34] MPEG AOE Group, "Proposal Package Description (PPD) - revision 3," Document ISO/IEC JTC1/SC29/WG11 N999, Tokyo MPEG meeting, July 1995.

[35] MPEG Systems Group, "Systems Working Draft, version 2.0," Document ISO/IEC JTC1/SC29/WG11 N1483, Maceió MPEG meeting, November 1996.

[36] SNHC Group, "Draft specification of SNHC verification model 2.0," Document ISO/IEC JTC1/SC29/WG11 N1454, Maceió MPEG meeting, November 1996.

[37] MPEG Video Group, "MPEG-4 video verification model 5.0," Document ISO/IEC JTC1/SC29/WG11 N1469, Maceió MPEG meeting, November 1996.

[38] T. Sikora, S. Bauer and B. Makai, "Efficiency of shape-adaptive 2-D transforms for coding of arbitrarily shaped image segments," *IEEE Trans. Circ. and Syst. Video Technol.*, vol. 5, No. 1, pp. 59-62, February 1995.

3

MULTIMEDIA SYSTEMS

Irek Defée
Tampere University of Technology
Tampere, Finland

Multimedia is a term often seen in the headlines of trends in information technology but its meaning is usually somewhat fuzzy [1], [2]. In a broad sense, multimedia can be understood as a new technology merging computer, communication and consumer electronics with content provided by the media industry. Traditionally, media industries like film, television and printing have been segmented along the lines of specific methods of content production and distribution. Developments in computers and communications bring opportunities for the creation of 'digital' media which will encompass and unify all the content into a single 'multimedia' platform running on universal multimedia devices. Multimedia can thus be viewed as synergy of technology and content into a new form of media.

To realize this synergy, multimedia demands sophisticated technology for visual, acoustical and graphical data processing and transmission. Multimedia content produced on the basis of this technology should provide users with enhanced perceptual experience and interaction compared to the presently available media like cinema and TV.

In this chapter we focus on multimedia systems from the point of view of requirements, system architecture, components and networking. This is a very broad range of topics tackling almost every aspect of digital technology and by necessity our presentation emphasizes system aspects without going into the details of specific solutions.

3.1 MULTIMEDIA SYSTEMS IN GENERAL

Human information processing is inherently multimodal since it relies on several sensory systems for gathering and producing information [3]. Visual, acoustical, haptic (that is, touch and gestures), smell and taste sensory modalities operate in a perfectly orchestrated manner and in a seemingly effortless way to provide rich perceptual experience of the world. This effortless way of operation is in fact hiding the enormous complexity of information processes which are involved, something which is observed everyday in humans at casual conversations, in lectures, in the cinema, and on TV.

Multimedia technology aims to devise methods and systems providing better match to the sophistication of human senses in the creation, processing and transfer of multimodal information. Multimedia systems become possible because any type of information can be represented and processed digitally. Digitization offers unlimited flexibility for the manipulation of data and we are at present moving towards increased complexity and sophistication of data processing required by multimedia.

On a general level, technology and its applications are much related in multimedia systems. From the users point of view, applications (i.e., content) are most important for multimedia, as in other media industry sectors. Multimedia systems are necessary for the production, distribution and use of the content and the form of content, depends on the technologies which are applied. From such a perspective one can see technology as a way of providing links between media content producers, media content providers/distributors and end users. This is similar to other such links in the current media industry. For example, the cinema industry and television industry link producers, distributors and consumers in a very specific way due to the properties of the systems they use.

Multimedia systems can offer in principle much greater flexibility than film and television media because of unlimited richness of digital content and variety of ways in which it can be produced, distributed and consumed. It is useful to think about multimedia systems in terms of a chain of content producers, distributors and users (Figure 3.1) since this helps in taxonomy of different types systems and possibilities offered by them. Adding complexity to this picture is the fact that contrary to the traditional media industry, the roles of producers and users are not strictly separated, almost everyone can become a producer and distributor of multimedia content due to the wide availability of content production and distribution tools.

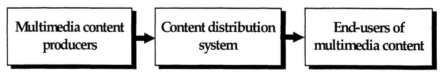

Figure 3.1 Multimedia system chain.

3.2 MULTIMEDIA APPLICATIONS

Multimedia makes information easier to access and more attractive to consume. This feature makes multimedia highly applicable to all areas of human activities which deal with the production, distribution and use of information. In places where computers and information processing are already in use, multimedia will broaden and enhance applications. Also, completely new ways of handling and dealing with information can appear. Business, education, science, technology, and the media industry will feel the impact of multimedia in several general ways. First, multimedia will change the way of presenting information from the current static form to an interactive and dynamic one based on video, sound and graphics. Most information accessed currently via computer terminals is in the text form requiring active reading by the user. Multimedia will make text an option to richer audiovisual forms of presentation, stimulating perception more actively and making a stronger impact.

The second area of multimedia use is enhanced communication. Provision of real time audiovisual communication by videotelephone or videoconferencing equipment integrated in desktop and portable devices will intensify remote contacts and will create conditions for distant cooperation and work appearing on a massive scale. This may have significant social impact by expanding telecommuting to areas where more personal contact is necessary than that provided currently by telephone and electronic mail. This might be relevant to business meetings and consulting but also to industrial activities like customer service, surveillance, and control and manipulation of equipment.

The third general area of multimedia applications is mass distribution of information, which is the domain of the traditional media industry. The media industry is based on the distribution of sound, images and printing but it will have to develop new forms of interactive, personalized and online accessible services. The World Wide Web gives us a taste of such services and its integration with TV, movies and printing industry is one potential scenario of the media evolution. Development in this direction will be very significant for the entertainment industry. However, new electronic distribution channels will enable almost everybody to become a media producer and distributor. One can expect new specialized media services appearing not only in entertainment but also in the areas of education, science and technology. Online access to unlimited information resources in multimedia form can make these areas less dependent on the physical location of providers and users. This may revolutionize access to high quality education, eliminate delays in the distribution of scientific results and spread rapidly new technology everywhere. Reduction of the costs of distribution of information will make it affordable much more widely.

To assess the full potential of multimedia, one has to consider the synergetic effect of the multimedia information presentation, multimedia communication and distribution (Figure 3.2). These application forms will be in simultaneous use creating an overall strong perceptual enhancement to human capabilities for producing and digesting information. At the same time ergonomic aspects of information processing tasks will be improved, resulting in reduced stress and fatigue.

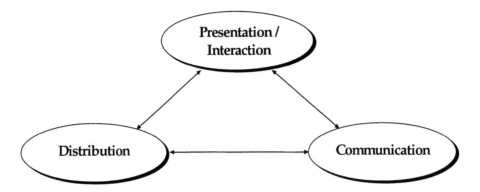

Figure 3.2   General categories of multimedia applications.

3.3 MULTIMEDIA TECHNOLOGY TAXONOMY

Creation of complete digital multimedia system chains shown in Figure 3.1 requires a large number of technologies, both existing and completely new. In fact it is quite difficult to list and systematize them all. The IEEE Signal Processing Society Technical Committee on Multimedia Signal Processing, in trying to understand what is the topic of its field, has defined [1] the following classification of multimedia technology for EDICS (Editors' Classification Information Scheme):

> Multimedia Signal Processing
> > Integration of Media
> > > Multimedia Compression
> > > Compressed Domain Processing
> > > Joint Audio-Video Processing
> > Multimedia Databases
> > > Indexing, Retrieval and Archiving
> > > Authoring and Editing
> > > Digital Library
> > Multimedia System Design and Implementation
> > > Parallel Architecture
> > > ASIC Design
> > > Software and Hardware Design
> > > System Integration
> > Human-Machine Interface and Perception
> > > Content Recognition/Analysis/Synthesis
> > > Multimodal Interaction
> > > Perception Quality and Human Factors
> > Multimedia Communications
> > > Equalization and Synchronization
> > > Transport Protocols

Quality of Service Control
Error Concealment and Loss Recovery
Rate Control and Hierarchical Coding
Multimedia Applications
WWW and Hypermedia
Videoconferencing and Collaboration Environment
Education and Distant Learning
Telemedicine
Home-Shopping Gaming and Virtual Reality
SDTV, HDTV, SHDTV and Video on Demand
Standards and Related Issues
ITU-T H-Series Standards for Audiovisual
Communication
MPEG-1, MPEG-2, MPEG-4, MPEG-7
MHEG, DAVIC
HTML, VRML and others
Other (specify)

The above classification shows how broad is the area of multimedia systems which makes even a general review of the field a nontrivial task. Three main parts can be differentiated in the above classification:

- systems hardware, software and user interfaces
- signal processing and information handling
- communication
- application environments

If such division is taken broadly, one can relate the general multimedia system of Figure 3.1 to its underlying technologies. For the content creation and delivery, one needs

- system infrastructure including multimedia hardware and software to make the systems run
- applications and tools for its development

Transferring multimedia content between different physical locations requires communications. This can be realized by

- physical media (CD-ROM, cartridges, discs)
- networking distribution

From the networking point of view one can differentiate between

- stand-alone multimedia systems (using physical media)
- networked multimedia systems

There is an obvious trend towards networked distribution with real time high-quality content delivery but physical storage media will long remain popular in specialized applications like games, movies and encyclopedias. In a similar way, one can divide the systems from the hardware and application point of view

- general purpose (multimedia computers)
- specialized (game consoles, set-top boxes)

Numerous options and solutions exist in all the categories listed above. In addition, as new products emerge very quickly, even qualitative technical descriptions become obsolete in a short time. For example, the once prevailing first generation CD-ROM for physical media distribution technology is being replaced by DVD-ROM (Digital Versatile Disc) which in turn may pave the way for HDVD-ROM (High Definition DVD). In a similar way, narrowband packet-oriented communication networks may be replaced by broadband connection-oriented networking. Hence, putting emphasis on basic aspects and system solutions is better than details of specific technologies which become obsolete very quickly.

3.4 MULTIMEDIA SYSTEM REQUIREMENTS

The difference between multimedia and traditional data processing stems from the richness and structure of media data types. Multimedia processing is based on data which are perceptually oriented, such as

- video
- audio
- three-dimensional graphics and animation
- still graphics and images
- text rendered in specific form

The basic multimedia paradigm is that all these data types are processed and presented simultaneously in a multitude of forms (Figure 3.3). Multimedia data are time and space dependent, a feature which has not been of particular concern in traditional data processing. Video, audio and animation differ from traditional computer data because they have a form of data streams ordered in time and space with very high precision. Besides this, digitized video and audio streams have huge sizes, virtually impossible to handle in their original forms. Thus, compression is necessary using algorithms which remove perceptual redundancy from the data. The algorithms are computationally very complex but reduce the data sizes to manageable amounts. The stream format of video and audio data implies also tight timing bounds on their processing and handling. When different types of multimedia data are handled, they usually need precise synchronization in space and time. Timing, synchronization and stream-oriented data handling and processing are fundamental system requirements for multimedia systems.

These requirements are difficult to satisfy in traditional digital systems which were originally not designed for them. Solutions developed to handle this problem tend to oscillate between special, limited-purpose hardware (e.g. game consoles) and upgrades to general-purpose machines (PC cards). Both these approaches have had serious limitations lacking either flexibility or performance.

For fully functional multimedia systems nothing less is required than complete revamping of almost every aspect of existing information processing technology including computer software and hardware, end-user devices, operating systems, servers and networks. This extremely demanding task is fascinating because it will lead to the creation of a new class of systems in which multimedia requirements will receive priority over the traditional data processing needs and eventually will make multimedia the main foundation of digital computing technology. We will concentrate on this trend and review ways how technology for the new multimedia systems is developing and replacing the present solutions.

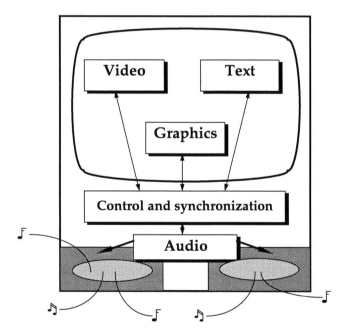

Figure 3.3 Multimedia system requirements.

3.5 PERSONAL COMPUTERS AND MULTIMEDIA

Microprocessors and personal computers were devised first as small cousins of mainframes used for data processing and numerical computations. When new applications demanded basic multimedia capabilities, i.e., sound, graphics and images, solutions were sought by adding cards with special processing hardware. Special cards solved basic processing needs but not the problem of multimedia integration that requires precise coordination and synchronization of different media types. Difficulties here result from the basic hardware and software architecture limitations. Internal PC buses were not designed for transfer of high bandwidth streams and operating systems are not prepared for strict timing. These problems can be solved only by radical changes in architecture. Below we describe such changes appearing in the processor design, mainboard architecture and on the operating system level.

3.5.1 Multimedia extensions to microprocessors

A radical way of providing multimedia processing capabilities is by complete redesign of the processor architecture. Traditional computing hardware technology emerged from the needs of numerical data processing and it has been optimized for these tasks. Multimedia processing in turn has its own unique requirements. Video and audio decompression, image processing and graphics rendering use algorithms which can be described by sets of specific, well-defined operations. These operations can be translated into processor instructions, which are needed to implement them, and incorporated into the microprocessor design.

Intensive studies have shown that multimedia processing dealing with visual and acoustical signals and graphics execute large numbers of simple repeating operations on massive amounts of short-format data. These operations are simple but different from standard processor instructions. There is thus a need for enhancing the microprocessor architecture with new multimedia processing instructions and execution units for those instructions. The architecture which emerged for this purpose is based on multiple simple processing units which operate in parallel on short data formats. Virtually every major processor optimized for multimedia processing incorporates such a solution.

This is exemplified by the Intel Pentium MMX processor [4]. In this processor, the standard 64-bit floating point arithmetic unit and its registers have been modified to perform a second role of a multimedia processor. Modification is done by splitting the floating point unit and its registers into several subunits which can operate in parallel. This can be done flexibly so the system can accept eight 8-bit data values or four 16-bit data values at the same time into its 64-bit registers. Executions units operating in parallel can perform one of the 57 new instructions which have been carefully selected to match the needs of multimedia data processing. Multimedia instructions operating in parallel on the data are called "packed." They include packed addition, packed subtraction, packed saturation arithmetic, packed logical operations, etc., which are efficiently executed by the processor logic. The overall processing speed is boosted by the new instructions and by the parallel execution. Depending on the application, the speed may be up to several times that of a conventional processor. The cost of doing this is very low due to the clever reuse of existing hardware on chip. For example, the Pentium MMX processor gate count is increased by only 3% comparing to the Pentium processor.

3.5.2 Architecture of the PC mainboard

In addition to changes to the processor itself, multimedia also requires changes in the system architecture. Especially the high-quality video and 3D graphics require very fast, continuous transfer of massive data streams to the display. This is difficult to achieve with the standard computer bus which is shared by many subsystems on a time basis. A general solution to this problem is by providing multiple buses for streams, but even a double-bus architecture with a dedicated graphics/video bus to the display provides significant improvement. This eliminates problems with bus congestion and simplifies integrating multiple visual sources, e.g., video overlayed on 3D graphics.

Following a long time of dominance by the single PCI (Personal Computer Interface) bus architecture, a dual bus system called Accelerated Graphics Port (AGP) was developed for standard personal computers [5]. AGP is also based on the PCI bus specification but since it deals only with dedicated graphics traffic and is localized on the mainboard it offers up to an order of magnitude increase in the data transfer from the processor to the display compared to the single PCI bus. Special graphics and video processing chips can be interfaced to the AGP bus for further increase of performance.

3.6 MULTIMEDIA TERMINALS

Multimedia terminals are specialized devices for interaction with multimedia content. Usually they are set-top boxes or consoles connected to the TV and have a higher degree of interactive control of information than the TV receiver (in this sense TV is not a multimedia terminal when receiving standard programming).

Ultimately, multimedia terminal devices are useable for presentation of a variety of high-quality media like video, audio, graphics, images and text with flexible and advanced interactive control (Figure 3.4). In practice this means playback and control of compressed video streams, playback of multichannel audio, and good quality 3D graphics rendering.

Multimedia terminal devices can have a variety of forms. Depending on the level of sensory involvement of the user interface, they can be classified as

- Presentation systems with low sensory involvement which rely on traditional displays, speakers, and keyboards or joysticks as interaction devices.

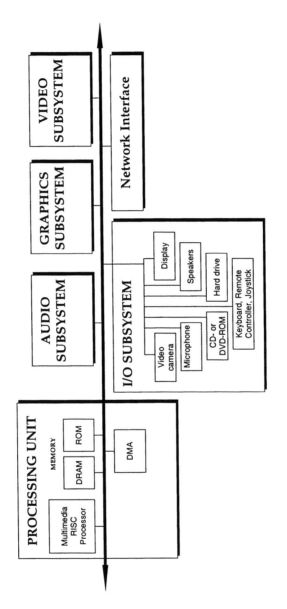

Figure 3.4 Architecture of multimedia terminal.

- Presentation systems with high sensory involvement, often called virtual reality devices. Enhanced perceptual sensations and interactions are achieved by the use of helmets with eye screen projection and special wear, e.g., gloves for interaction.

Multimedia terminal devices with low sensory involvement classified according to their capabilities are:

- set-top boxes connected to TV (WebTV)
- specialized consoles (games).

These devices differ in their type of use. A set-top box will be mostly oriented to general content with a relatively low level of interactivity. Specialized devices like game consoles are usually aiming for high-level interaction. They do not offer full flexibility and interaction such as that found with a personal computer which can be used with a large variety of content and also as a tool for content production. On the other hand, they are cheap and simple to operate.

3.7 OPERATING SYSTEMS FOR MULTIMEDIA

The requirements of media data processing make special demands on the operating system. High bandwidth streams with precise timing and synchronization can only be handled by a real time operating system. Widely used operating systems have far from real time performance and thus are unsuitable for multimedia. Operating systems are also the most valuable part of a computing platform, which makes a complete change rather unrealistic. The preferred way in adopting operating systems to multimedia is by extensions for real time media processing. This boils down in essence to providing real time kernel and services into a non-real time operating core and special interfaces for media data handling. The resulting system is a real-world compromise with performance limited by its non-media legacy.

Significant performance gain can be achieved if the operating system is designed from the beginning to handle media data. An example system of this kind known as BeOS [6] has many features which set it far apart from other operating systems for media processing. BeOS is designed to be multitasking, multithreading, and multiprocessing within a very compact kernel and performs fast task switching. This provides a high degree of processor load balancing even on systems with single processors by quick execution of a large number of small tasks and thus is a powerful tool for multiple stream processing. On true multiprocessing systems the real time performance will multiply robustly. This is important as one realizes that although the power of microprocessors seems to be increasing without end, the processing and coordination of real time multimedia streams will always be difficult. The burden of multiple media stream processing and coordination can be dealt with by multiprocessing architectures and real time operating systems like BeOS [6]. It is hard to tell if and when the mainstream operating systems will become truly media-oriented but at least the road for their potential long evolution, or for a quick revolutionary change, is now clear.

3.8 STORAGE DEVICES

Multimedia content storage requires very high capacity, even a single high-quality compressed full-feature movie stream has the size of several gigabytes.

CD-ROM has been the first generation device of choice for multimedia content storage because of its low cost, small size and ease of distribution. CD-ROM is in reality the same as compact discs used for audio recording. A version of compact disc called CD Video is also used for storing video encoded in the MPEG-1 standard. The capacity of a CD-ROM disc is about 680 megabytes which corresponds to 72 minutes of audio or MPEG-1 compressed video. While basic CD-ROM is a permanent medium like a CD, there are also recordable CD media, both write-once and reversible.

In the future one can expect a continuing increase in the capacity of CD-ROM type devices. The first in line is DVD (Digital Versatile Disc) and DVD-ROM [7]. They are greatly enhanced versions of CD and CD-ROM with density increased approximately 7 times, which corresponds to 4.7 gigabyte storage capacity. A single DVD disc can play up to a full-feature movie compressed in the MPEG-2 video compression standard [8]. DVD can also have two layers stacked on top of each other with the upper one transparent so the readout laser beam can focus on either of them. This boosts capacity to about 8.5 gigabytes. Furthermore, the DVD can also be made two-sided which increases the total capacity to 17 gigabytes.

3.9 MEDIA SERVERS

In networked computer systems, mass storage is concentrated in servers. Servers for multimedia differ from standard servers because of the need to deliver high-bandwidth streams to multiple users. This enforces strict timing bounds on the server data output operation. Data streams need to flow with a specified rate and they cannot break. With many simultaneous streams running at the same time with high bit rate, the server software and mass storage must be designed in a special way.

Mass storage with high output capacity is realized at present using RAID (Redundant Array of Inexpensive Discs) technology also called disc arrays [9]. A disc array is a system of hard discs synchronized and operating in parallel. Files are stripped throughout the discs and stored in parallel fashion. Readout from the discs is also done in parallel, this provides increased speed compared to a single hard disc. The speedup depends on the number of discs operating in parallel; disc arrays with output capacity of several hundred megabits per second are available. Operating systems and server software for stream generation and management needs to possess real time features. When a stream is started it must become part of the operating system process with guaranteed access to resources and protected status. This is because users expect high reliability in the delivery of multimedia streams like video and audio. Stream generation has to be precisely integrated with the networking interface.

Media servers with standard disk arrays can provide a sufficient number of streams for users in a local area network. Media servers can be realized by software and hardware modification to standard servers by providing real time protected processes and increased output capacity to the standard servers.

Servers for multimedia systems with hundreds or thousands of users need special machines. They can be parallel multiprocessor machines with very large RAM buffers, disk arrays, and archival storage on tapes. Such machines are currently rare and their architecture is still a research topic.

3.10 MULTIMEDIA NETWORKING

Networking is critically important for multimedia because it provides easy and instant access to content and its distribution. This is well illustrated by the success of the Internet, although the current Internet can be considered at best only a very limited multimedia system. For quality multimedia delivery a substantial upgrade to the Internet would be necessary to provide two absolutely essential elements necessary for networking of multimedia content: bandwidth and streaming. Streaming requires precise control and synchronization of packets (Figure 3.5) which puts very high demands on networking.

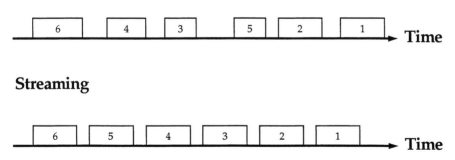

Figure 3.5 Packet vs. streaming networking.

Due to the physical constraints in the networks bandwidth and the structure of its protocols, the present Internet can provide neither the bandwidth nor the streaming necessary for quality multimedia. Audio and video require streaming bandwidth in the order of multiple megabits per second. Streaming is difficult to realize in traditional computer networks designed originally to carry bursty traffic with no time dependence. For high-bandwidth streaming, computer networks must be upgraded and/or redesigned both on the hardware and on the software (protocol) side. There is fast development in this area and one can expect significant upgrade to current networks in a near future. In local networks where the Ethernet has a dominant position, there are proposed extensions to it like Fast Ethernet, Gigabit Ethernet and Switched Ethernet. These extensions have speed ranging from 100 Mb/s to gigabits/s so much more data can be sent than over the standard 10 Mb/s Ethernet. Ethernet is a shared medium and with all users accessing the same bandwidth, streaming cannot be guaranteed in general. Switched Ethernet provides users with individual links which can be used for sending streams.

The network which can guarantee bandwidth, streaming and quality of service for multimedia is Asynchronous Transfer Mode or ATM [10]. ATM is a kind of broadband telephone system in which users can establish connection links with bandwidth flexibly allocated and with guaranteed streaming. The ATM system is in principle able to solve all problems of multimedia communication, but the cost of its wide deployment would be very high. ATM requires new infrastructure based on switches and links to terminals. Special software and hardware is needed for calling and link establishment. This means that networking for high-quality multimedia will most likely follow an evolutionary upgrade path from the current Internet infrastructure.

The Internet is in fact only a set of protocols running on top of every possible network type. An idea for Internet upgrading is first to increase the available network bandwidth by some cost-effective means and next to extend the current Internet protocols with streaming capability. One of the biggest obstacles in increasing the network bandwidth is end-user connections which usually run over telephone lines and provide transmission speeds in the range of a few kilobytes per second. Among the emerging technologies which can radically change this are new types of modems. ADSL (Asymmetric Digital Subscriber Loop) modems use sophisticated signal processing techniques to send up to 6 megabits per second of data over the local telephone loop between the subscriber and the nearest telephone exchange [11]. Cable modems [12] work over cable TV and provide users with their own digital modulated data channels. These new ADSL and cable modems may remove the local bandwidth barriers. As a follow-up to this development, the infrastructure of long-distance networks can be upgraded by introducing high-bandwidth Internet routes and ATM networks. Once these types of networks are in place, large-scale multimedia networking will happen (Figure 3.6).

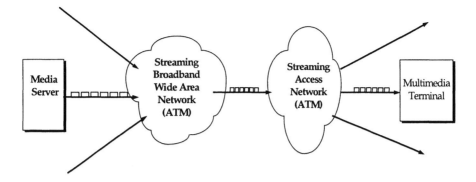

Figure 3.6 Networked multimedia system.

3.11 SOFTWARE AND STANDARDS

Software for multimedia is another broad area which ranges from basic instruction level in microprocessors through signal processing algorithms, system synchronization, control and management to networking and user interfaces. Standardization of software interfaces and data formats is of great help for system integration. This is especially important in networked systems where standards enable communication and distribution of content to everybody. Open systems are realized by software tools and nothing illustrates this idea better than the World Wide Web. Standardized communication interfaces (HTTP protocol) and scripting languages (HTML) help to build systems which can run on virtually any hardware platform.

Systems based on standards are often called open. There is no doubt that the concept of open systems and standardization is basic to multimedia since it provides universal access to content. Some of the most important standards developed for multimedia content and its delivery are:

- MPEG - a set of standards for video and audio compression including MPEG-1 for lower quality, MPEG-2 for high quality, adopted or digital television and DVD (Digital Versatile Disc) and MPEG-4 which covers very low bit rate streams, video manipulation and integration with 3D graphics.
- OpenGL [13], Open Inventor [14] - standard interfaces for 3D graphics objects description, and interaction.
- VRML [15] - Virtual Reality Modeling Language, standard for 3D networked virtual worlds
- JAVA [16] - language and virtual machine which allows downloading and remote execution of programs
- MHEG [17] - Multimedia and Hypermedia Expert Group, standard for producing synchronized multimedia presentations.

These standards together with complete system integration will enable creation of universal multimedia platforms with transparent access for everybody. Internet and World Wide Web can be seen as first steps in this direction but a lot remains to be done to build their full-scale multimedia equivalents.

3.12 SYSTEM INTEGRATION

The Internet is an example of an open integrated system composed of servers and clients tied by global communication networks. The Internet has at its core the universal TCP/IP data communication protocol on top of which there is a large number of standardized application-oriented protocols and tools which enable enormous applications like the World Wide Web to run. From the multimedia point of view, the Internet is very limited since it lacks the basic ability for high-bandwidth stream communications. But the Internet can serve as a model of integration, which is still missing for large scale multimedia systems.

DAVIC (Digital Audio Visual Council), a worldwide industrial body [18], has developed specifications for an integrated multimedia system aimed at home applications. DAVIC's vision is of a set-top box connected to a TV retrieving multimedia content from remote servers via a broadband network. DAVIC specifications are a set of interfaces and communication protocols which enable a complete system for high-quality multimedia content retrieval and playback. The underlying idea is to ensure high quality of content presentation and system operation reliability. Because of this, the DAVIC communication system is based on ATM networks for transferring video and other data using the MPEG-2 standard. An elaborate scheme is used for establishment of communications links between a server and client advanced ATM signaling protocols and video stream control commands using the Digital Storage Media Command and Control part of the MPEG-2 standard. For synchronized application playback, the system uses the MHEG standard. As a result, the DAVIC system provides tightly integrated architecture for multimedia content delivery with guaranteed quality of service (Fig. 3.7).

While ATM networks are not deployed widely and this limits the DAVIC system potential in practice, the DAVIC specifications provide excellent insight into the problems of building integrated multimedia systems. The systems are conceptually complicated and their technical realization is difficult and will require time to master. However, potential benefits of this technology are great. A vision which comes to mind is that of a "Multimedia World Wide Web" enabling instant global access, real-time retrieval, and communication of high quality video, grahics, images, and sound. This would obviously bring a huge change to the current media industries.

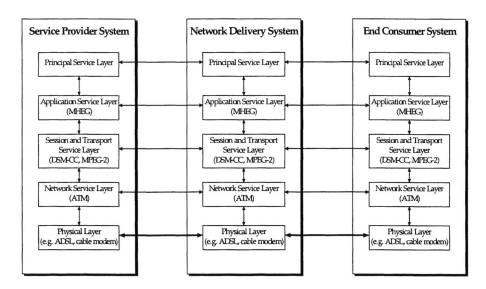

Figure 3.7 Standardized interfaces of the DAVIC multimedia system.

3.13 FUTURE DEVELOPMENTS

Viewed from the current perspective there is a clear trend towards high-quality networked multimedia systems based on broadband networks. Such systems may evolve from the current Internet and by gradual merging with other media, most notably with TV, may become a universal 'digital media' platform. Key technologies needed for this evolution are high-quality multimedia processing and above all increased network bandwidth and streaming.

More futuristic developments are related to the progress in wireless cellular networking which may lead to new types of mobile personal multimedia systems, unifying mobile phones and personal digital assistants with graphics and video communication.

At the same time significant developments may occur in user interfaces for accessing and manipulation of content. Voice and visual input may at least partially eliminate the need for keyboard inputs.

REFERENCES

[1]   "The Past, Present and Future of Multimedia Signal Processing," IEEE Signal Processing Magazine, July 1997, vol. 15, No. 3, pp. 28-51.

[2]   R. Steinmetz and K. Nahrstedt, Multimedia: Computing, Communications and Applications, Prentice Hall, 1995.

[3]   B.E. Stein and M. A. Meredith, The Merging of the Senses, MIT Press, 1993.

[4]   Pentium MMX Processor:
M. Miilind, A. Peleg, U. Weiser, MMX Technology Architecture
Overview, Intel Technology Review, Q3 1997,
`http://developer.intel.com/technology/itj/q31997.htm`

[5]   Accelerated Graphics Port (AGP): `http://www.agpforum.org/`

[6]   BeOS Operating System: `http://www.be.com/`

[7]   Digital Versatile Disc (DVD):
`http://www.mpeg.org/~tristan/MPEG/dvd.html/`

[8]   MPEG video compression standard: `http://www.mpeg.org/`

[9]   P.M. Chen et al., "RAID: High Performance, Reliable Secondary
Storage," ACM Computing Surveys, vol. 26, No. 2, June 1994, pp.
145-185.

[10]   D.E. McDyson and D.L. Spohn, ATM Theory and Applications,
McGraw-Hill, 1994.

[11]   Asymmetric Digital Subscriber Loop (ADSL):
`http://www.adsl.com/`

[12]   Cable modems: `http://www.cablemodem.com/`

[13]   J. Neider, Tom Davis, and M. Woo, OpenGL Programming Guide,
Addison-Wesley, 1993.

[14]   J. Wernecke, The Inventor Mentor, Addison-Wesley, 1994.

[15]   Virtual Reality Modelling Language (VRML):
`http://www.vrml.org`

[16]   A. van Hoff, S. Shaio, and O. Starbuck, Hooked on Java, Addison-
Wesley, 1996.

[17]   Multimedia and Hypermedia Expert Group (MHEG):
`http://www.fokus.gmd.de/ovma/mug/`

[18]   Digital Audio Visual Council (DAVIC):
`http://www.davic.org/`

4

VISION AND VISUALIZATION

Robert Moorhead
Mississippi State University
Mississippi State, Mississippi

Penny Rheingans
University of Maryland, Baltimore County
Baltimore, Maryland

This chapter addresses techniques used to represent information visually, called *visualization,* and the mechanisms and characteristics of the human visual system. Visualization can serve as an effective tool in the exploration and communication of the structure of complex data by harnessing the substantial processing power of the human visual system. Knowledge of the mechanisms of perception can be used to improve visualization effectiveness by ensuring that the most striking features of the image are also the most important, by minimizing the effects of interactions and illusions, and by designing representations which take advantage of the strengths of the visual system.

4.1 VISION

This section gives a brief overview of the anatomy, physiology, characteristics, and idiosyncrasies of the human visual system. An understanding of these factors will yield valuable insights into the process through which visualizations are perceived and the ways in which visualizations can be made more effective.

4.1.1 Physiology of the human visual system

Processing of visual stimuli by the human visual system begins in the eye and continues in the brain. Although the anatomical structures of the eye and the brain are physically separate, linked only by the optic nerve, the characteristics of the two structures are similar enough that it would be reasonable to consider the eyes to be satellite portions of the brain.

4.1.1.1 The eye

Light enters the eye, is focused by the cornea, passes through the variable-diameter iris, is further focused by the lens, and strikes the light-sensitive receptors of the retina in the back of the eye [Figure 4.1]. The photoreceptors of the retina can be divided into two basic categories: rods and cones. Rods function at very low light levels, for instance, a moonless night, and have a single light-sensitive pigment with a maximum response to light of 507 nm (green). Vision under these conditions, called *scotopic vision*, does not convey differences in the wavelength of emitted or reflected light; all objects appear to be shades of slightly greenish gray. Because most perception takes place in normal light situations, the contribution of rods to human vision will not be emphasized in the following discussion.

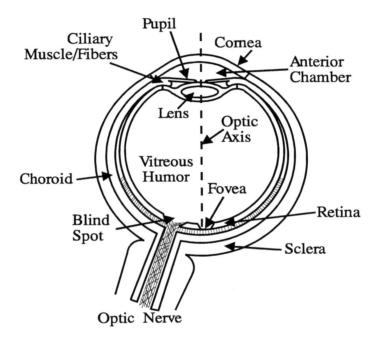

Figure 4.1 Cross-section of human eye.

At normal light levels, cones dominate the initial response to light, resulting in *photopic vision*. Three different kinds of cones are active in the human visual system: short-wavelength (S) cones with a maximum response at 450 nm (violet-blue), medium-wavelength (M) cones with a maximum response at 530 nm (slightly yellowish-green), and long-wavelength (L) cones with a maximum response at 560 nm (slightly greenish-yellow). Figure 4.2 shows the spectral sensitivity function of each cone type. The three types are sometimes called the blue, green, and red cones, respectively, because of the general wavelengths to which they respond most strongly. Notice that the maximum response of the S cone is much smaller than either of the other two, limiting our sensitivity to small blue objects (such as blue letters on a black background). The response of cones to incoming light has a logarithmic relationship with the intensity of light, making small differences in intensity more apparent at low light levels than at high levels.

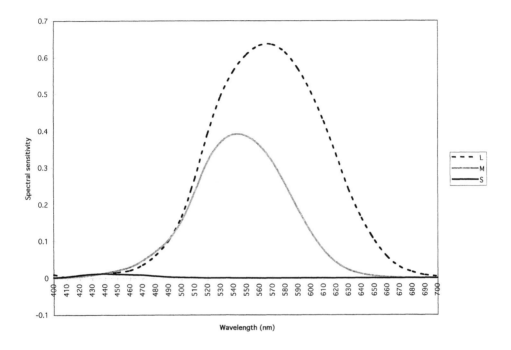

Figure 4.2 Cone spectral sensitivity functions.[1]

Judgments about the dominant wavelength, or *hue*, of light can be made by comparing the responses of the three cone types. For instance, Figure 4.3a shows the response of the S, M, and L cones to light of 610 nm (orange). For light of 480 nm (blue), the response of the M cones would be similar, but that of the S and L cones would be different (Figure 4.3b). This mechanism gives rise to the Tristimulus Theory of Color, which states that any color sensation can be matched by an appropriate mixture of any three other colors.

Cones are most densely packed into the central region of the retina, or *fovea*, while rods are located almost exclusively in the outer regions of the retina, or *periphery*. When we look directly at an object, it will be imaged on the fovea where the densely packed receptors provide high spatial acuity. In the center-most section of the fovea, the *foveola*, S cones are virtually absent, further reducing our sensitivity to small blue objects.

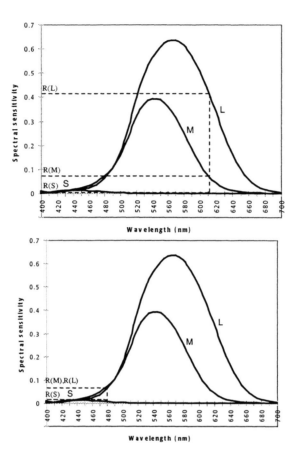

Figure 4.3 Cone responses. (a) response to light of 610 nm, (b) response to light of 480 nm.

From the photoreceptors, neural responses pass through a series of linking cells, called bipolar, horizontal, and amacrine cells, which combine and compare the responses from individual photoreceptors before transmitting the signals to the retinal ganglia. These linkages between neighboring cells provide a mechanism for *lateral inhibition*, whereby the response of a cell is attenuated by a large response in nearby cells. This mechanism facilitates relative, rather than absolute, judgments of intensity, emphasizing edges and other areas of change in the visual field.

Retinal ganglia cells have concentric receptive fields with a center-surround organization. The *receptive field* (RF) of a ganglion cell is the area of the retina where a light stimulus will give rise to a response in that particular cell. In general, receptive fields are small in the fovea and grow larger with increasing eccentricity (distance into the periphery). The receptive field of a ganglion cell has two distinct regions: a circular center region and an annular surrounding region. See Figure 4.4. Most ganglia cells are excited by stimulus

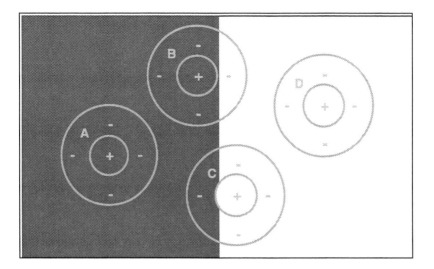

Figure 4.4 Enhanced perception of edges.

in the center region and inhibited by stimulus in the surrounding region, called *on-center, off-surround organization*, though some cells have off-center, on-surround organization. Other ganglia cells respond to color differences in the center and surrounding region. Most wavelength sensitive ganglia cells are excited by red light in the center and inhibited by green light in the surround, or vice versa, though some cells respond to differences between blue and yellow light.

This center-surround organization serves to enhance the perception of edges further, by producing more extreme responses at the edge of an illuminated region. Figure 4.4 shows this situation. Cell A, located in a dark area, produces only a small response. Cell B receives the same amount of stimulation to its center region but is more inhibited by the bright light falling on part of its surround, resulting in an even smaller response than that of Cell A. Cell D, located in a bright region, produces a large response. Cell C receives the same amount of stimulation to its center region, but is less inhibited since part of its surround is located in the dark region. Accordingly, the response of C is even larger than that of Cell D.

This center–surround organization can also explain the mechanism of lateral inhibition. Figure 4.5a shows an illusion resulting from lateral inhibition. Apparent dark spots occur at the intersections of bright lines due to a greater inhibitory effect at intersections than at other bright areas. Figure 4.5b shows representative center-surround receptive fields at various positions in the image. Receptive fields located between the sides of two dark squares receive inhibitory input from two areas of the annular surround. Receptive fields at the intersection of two bright lines receive inhibitory input from four areas of the surround, resulting in a smaller response than from fields along the square sides. This smaller response results in the impression of dark spots where none really exist.

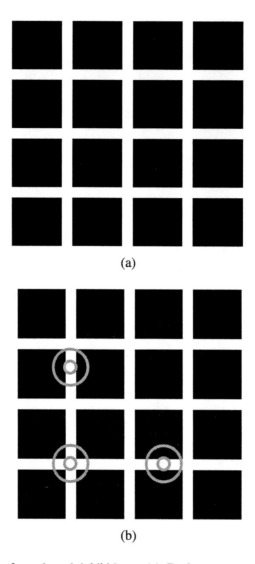

(a)

(b)

Figure 4.5 Illusion from lateral inhibition. (a) Dark spots seem to appear at the intersections of the white lines, (b) differing inhibitory stimuli based on RF location. Receptive fields at and between the intersections receive the same amount of positive stimulation in the circular center region, while fields at the intersections receive more inhibition from light falling on the annular surround.

In the retinal ganglia cells, incoming signals indicating the responses of S, M, and L cones are re-encoded in terms of the *opponent color channels*.[2] These channels describe light sensations in terms of their achromatic, R-G, and Y-B components. The *achromatic channel* represents the intensity of light, without regard to its wavelength characteristics. It is computed as the sum of the signals from the M and L cones. The contribution of the S cones is ignored, due to its relatively small magnitude. The *R-G channel* measures the chromatic content of the light on a scale from red to green. It is computed as the difference of the L and M signals. The *Y-B channel* encodes the chromatic content of the light on a scale from yellow to blue. Some reflection will reveal that the content of this channel is orthogonal to that of the R-G channel. It is computed as the difference between the sum of the M and L signals and the S signal. This process is illustrated in Figure 4.6.

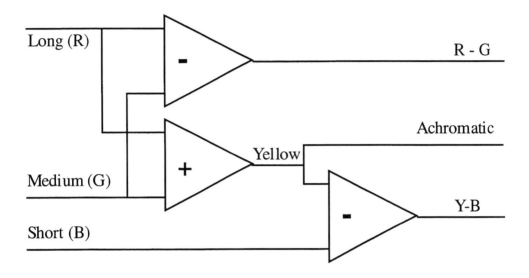

Figure 4.6 Opponent color channel encloding.

In the retinal ganglia cells the visual system splits into two independent pathways.[3] The *magnocellular pathway* detects objects and their boundaries as well as provides a basis for the perception of depth and motion. This pathway begins in a subset of the retinal ganglia cells with large receptive fields and receives input from the achromatic opponent color channel. All processing along this pathway is independent of the wavelength of light received, depending only on its intensity. The *parvocellular pathway* is responsible for the perception of color and fine detail. This pathway receives input from all three opponent color channels. These two pathways function independently and the judgments of each are reconciled at a much later stage in visual processing.

4.1.1.2 The brain

Neural responses leave the eye via the optic nerve, passing through the optic chiasm where visual responses from each side of the visual field are sent to the opposite side of the brain, and to the *lateral geniculate nucleus (LGN)* located deep in that side of the brain. See Figure 4.7. Each LGN (one for each side of the visual field) is structured into six layers. Two layers belong to the magnocellular pathway, each receiving input from the achromatic opponent color channel originating in one retina. The remaining four layers belong to the parvocellular pathway and receive input from all three opponent color channels. The LGN also receives a large amount of input (approximately 60 percent) from later stages of the visual pathways in the cortex. These upstream connections appear to provide a mechanism for expectations and previous perception to influence even the early stages of visual processing. From the LGN, visual signals proceed to the primary visual cortex, visual area 2, and then to various areas of higher visual processing.

The primary visual cortex (V1), located at the back of the brain, is composed of six different layers with differences in type and density of neurons and their interconnections. Layers 1, 2, and 3, collectively called the *superficial layers* of the cortex, appear to receive input from the LGN. Layer 4 receives input from the superficial layers. Layer 5 outputs primarily to the superior colliculus, which is responsible for driving eye movements. Layer 6 sends a substantial portion of its output back to the LGN.

After V1, most visual signals appear to pass to area V2, located just forward of V1. Area V2 is segregated into stripes, with separate stripes specialized for processing color, form, and stereo. Specific functions of later stages of the visual pipeline, called the higher visual areas, are not completely understood. Visual area 4, located further forward, is believed to specialize in color perception. The area in the middle temporal lobe (MT), located along the sides of the brain, appears responsible for the perception of motion and stereo.

Parvocellular Pathway. The parvocellular pathway provides information about the color and small detail of objects in a scene. The anatomical components of this pathway appear to include one type of retinal ganglia cell, parvocellular layers of the LGN, some layers of the primary visual cortex, the stripes in visual area 2 responsible for color (and to some extent form) perception, visual area 4 which seems to be responsible for the higher level processing of color information, and perhaps the temporal-occipital region responsible for the identification of objects. Later stages of the pathway appear to be split into parts which are sensitive to wavelength differences, presumably for color perception, and parts which are not, presumably for perception of small detail. Characteristics of this pathway include small receptive fields, sensitivity to differences of both wavelength and brightness, and relatively slow response times.

Magnocellular Pathway. The magnocellular pathway determines the locations and boundaries of objects in a scene. The anatomy of this pathway appears to include the achromatic opponent color channel, the magnocellular layers of the LGN, some layers of the primary visual cortex, the stripes in visual area 2 responsible for stereo and form perception, the middle temporal lobe (MT), and perhaps the parieto-occipital region responsible for tasks involving the positions of objects. The characteristics of processing done by this pathway closely match the inherent requirements of its tasks. All judgments are made

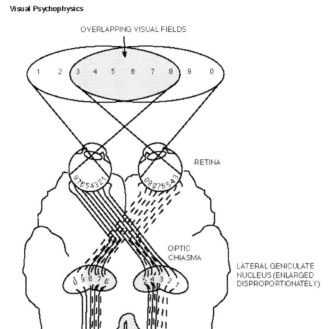

Visual Psychophysics

OVERLAPPING VISUAL FIELDS

1 2 3 4 5 6 7 8 9 0

RETINA

OPTIC
CHIASMA

LATERAL GENICULATE
NUCLEUS (ENLARGED
DISPROPORTIONATELY)

VISUAL CORTEX

Figure 4.7 Visual pathways.

without regard to difference in wavelength; only brightness differences are considered. Compared to the parvocellular pathway, receptive fields are large, responses are fast and transient, and only small contrast differences are required for discrimination.

Depth Perception. Since the picture of the world imaged onto the retina is a two-dimensional projection of the three-dimensional world, other mechanisms must be used to infer depth information from the information available. Cues to depth come from a variety of sources; most seem to be processed by the magnocellular pathway. *Monocular depth cues* are derived from the information available in a single retinal image. These cues include occlusion, shading, and perspective. *Binocular depth cues* are available from the stereopsis provided by the overlapping visual fields of the two eyes. Starting in visual area 2, the magnocellular pathway contains cells sensitive to *ocular disparity*. Such cells compare the retinal location of an object in each eye, using differences to deduce depth information. Small retinal disparities indicate objects which are far away; large retinal disparities indicate objects which are near.

Motion Perception. Later stages of the magnocellular pathway, particularly MT, have a large number of cells which are selective to movement and direction, suggesting that this pathway is also responsible for motion perception. Motion sensitive cells respond most strongly to stimuli which are moving across their receptive field. Additionally, most motion sensitive cells are selective to the direction and velocity of motion. This motion perception pathway functions independently of other visual pathways, as demonstrated by a few people who have no motion perception but have otherwise normal vision. Motion plays several roles in human vision. At the most basic level, motion makes pattern vision possible by ensuring that the stimulus on an area of the retina is constantly changing. Motion gives valuable clues to our relationship with our environment: where to direct our eye movements, the time to collision with objects around us, and information about the location of parts of our body relative to other objects (*exproprioceptive* information). Motion also provides information about the relative depths, 3D structure, and grouping of objects that we see.

Interaction. Interaction with the environment is also important for accurate perception. *Interaction* is the state of being able to change the contents of, or at least control the view of, what one is viewing. Control over visual experience, rather than just the visual experience itself, is necessary for the normal development of the visual system of cats.[4] Kittens who passively receive visual stimulation never develop the ability to perform visually guided behaviors. Interaction is also important to the performance of tasks by humans. Interaction has been shown to improve perception of abstract graphs,[5] perception of the shapes of dot distributions,[6] performance of a spatial assembly task,[7] and accuracy and confidence in quantitative and qualitative judgements.[8,9]

Equiluminance. Judgments made primarily by the magnocellular pathway, those of object boundaries, stereopsis, or motion, break down under conditions with no brightness differences between objects. The term *equiluminance* is used to describe the condition where the appearance of objects in the scene differs only in hue and saturation, not in brightness. Visual relationships that break down or degrade under equiluminance include perspective depth cues, depth cues from relative motion, linking by common movement or collinearity, illusory borders, and illusions of size.

4.1.2 Characteristics of human visual perception

Several mechanisms of the human visual system enable the perception of stimuli over an enormous dynamic range of light level and stimulus magnitude.

These include the contraction and dilation of the iris, the logarithmic response of photoreceptors, the inhibitory surround of retinal ganglia receptive fields, contrast effects, the phenomenon of adaptation, and the constancy of visual qualities. In general, such mechanisms optimize the judgment of relative quantities at the expense of absolute judgments, facilitating the detection of spatial and temporal change.

Contrast effects. The perceived intensity or hue of an area can be significantly affected by nearby colors. This phenomenon is called *simultaneous contrast.* For example, a gray patch on a red background will seem slightly green, while the same patch on a green background will seem slightly red. A similar effect occurs for achromatic contrast in situations where only luminance differs. In Figure 4.8, the inner square at the far left appears to be darker than the inner square at the far right. In actuality, the squares have the same intensity. Simultaneous contrast seems to occur independently on each of the opponent channels and have effects of comparable magnitude.[10] Ware observes that these contrast effects are strongest where smooth color gradients are present, i.e., where adjacent colors and color changes are similar.

Adaptation refers to the reduction in response to a constant stimulus over time. For example, a person walking from inside a dark building out into the bright sunlight will first experience the sunlight as extremely bright, reducing the world to an almost featureless field of glare. After a few minutes outside, the perceived brightness has diminished, allowing the world to be seen much more clearly. Through adaptation the visual system becomes more sensitive to small differences in stimulus value around the value of the adapting stimulus and less sensitive to small differences far from the value of the adapting stimulus. Essentially, the visual system dynamically adapts for optimum performance under the current conditions. This phenomenon of stable perception under differing absolute conditions is called *constancy.*

In addition to adaptation to intensity level, the human visual system adapts to the wavelength of light, the spatial frequency of a pattern, and the direction and velocity of movement. A number of interesting after-effect illusions occur after a strong adapting stimulus is replaced by a neutral stimulus. In such situations, the viewer sees the opposite of the adapting stimulus, not the actual neutral stimulus.

Relative size and shape judgments. Just as our judgments of relative brightness are more accurate than our judgments of absolute brightness, so are other relative judgments more accurate than the corresponding absolute judgments. For example, we have a much finer discrimination between the relative size of objects than we do of absolute size. As with contrast effects, this optimization for relative judgments can result in perceptual illusions such as misjudged sizes [Figure 4.9a], orientations [Figure 4.9b], and curvatures [Figure 4.9c]. In 4.9a, the upper inner circle appears to be larger than the lower one, while the two are

Figure 4.8 Simultaneous contrast. The inner squares on the far right and far left have the same intensity.

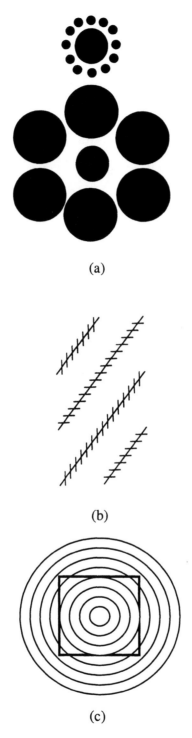

(a)

(b)

(c)

Figure 4.9 Relative judgment illusions. (a) the inner circles are the same size, (b) the long lines are parallel, (c) the sides of the square are actually straight.

actually the same size. In 4.9b, the long lines are parallel, but appear not to be. In 4.9c, the sides of the square appear to bow in, but are actually straight.

4.1.3 Cognitive influences on perception

Our experiences with our environment, particularly with the three-dimensional nature of our world, heavily influence our perception of ambiguous, and not so ambiguous, stimuli. We use the expectations that have been built through our experiences to interpret new situations. While this process can lead to incorrect interpretations, it can also be exploited to improve the perception of three-dimensional relationships from two-dimensional views.

Size Judgments. Judgments about the size of an object are influenced by the viewer's estimate of that object's depth. Our experiences with the effects of perspective in our three-dimensional environment have led us to expect that if two objects with the same size retinal images are located at different distances from us, the object which is farther away must be larger. In situations where these 3D expectations are applied to 2D drawings, size illusions can result. Figure 4.10 illustrates this phenomenon. Three identical size rectangles appear to be of different sizes, because they are expected to lie at different depths.

Shape Judgments. In a similar way, our judgments about the shapes of objects can be influenced by our unconscious attempts to place them in a familiar three-dimensional world. This process can lead us to misjudge angles and primitive shapes. In Figure 4.11a, most people would claim to see 12 right angles; the drawing only contains 4. The other 8 angles are only present in 3D, not in the 2D drawing. In Figure 4.11b, most people would claim to see three circles, rather than the one circle and two crescents which are actually present.

Texture. Texture can help disambiguate obscuration cues by clarifying how different parts of the surface meet and hide one another. The effects of perspective, and from them additional depth cues, are visible in the texture gradient. Specifically, the apparent scale of the projected texture decreases as the surface recedes from the observer. The texture in distant patches appears to be compressed, a denser pattern with smaller features. These perspective effects are likely to be more obvious on the texture than on the object itself. This is because texture shape in one area can be compared to texture shape in another, with shape differences indicating depth differences. Similar comparisons among geometric features of the object would not be as enlightening because object shape in different areas cannot necessarily be expected to be similar. The texture gradient also provides information about the orientation of surface segments, since textures converge on surfaces that drop away from the observer. More information about surface perception from texture can be found in the perception literature, Refs. 11 through 14.

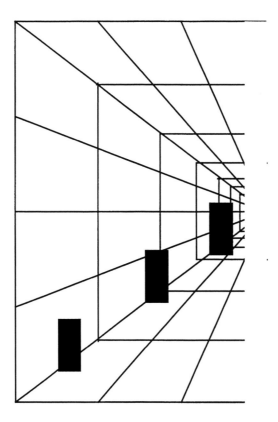

Figure 4.10 Size illusion. The three dark rectangles are the same size.

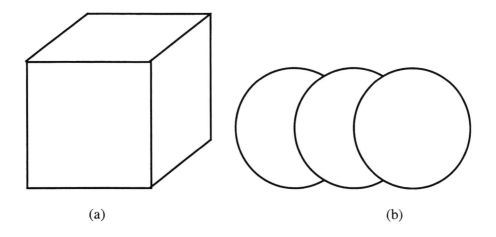

(a)   (b)

Figure 4.11 Shape judgment illusions. (a) How many right angles? (b) How many circles?

4.2 VISUALIZATION

Visualization is the mapping of data to imagery to obtain insight and understanding about the data or an associated phenomenon. Both computed and measured data can be visualized. The goal of visualization is to acquire a deeper understanding of the data being investigated and to foster new insight into the underlying phenomena by exploiting the widest information path to the brain, namely the human visual system. To achieve this goal, visualization utilizes aspects of computer graphics and image processing as well as techniques from other fields.

Visualization allows one to organize the data or information in a form more amenable to the pattern recognition and pattern matching skills of the human visual system. By visualizing the data we are able to see patterns or relationships that might not be detected by simply looking at a series or matrix of numbers. Representing data using imagery rather than numerical characters allows us to represent more information in one view.

Visualization is currently used in a wide range of applications areas. In the medical field it is used to create arbitrary views of volumetric data for surgical planning or diagnostic purposes. It is used in industry to represent the output of computational models of fluid flow around objects such as cars, airplanes, missiles, and bicycle helmets and inside objects such as ocean basins, lakes, engines, and compressors. It is used in the financial industry to detect economic trends. It is used by sports enthusiasts to determine where to catch fish, how to putt a golf ball, and what play to run against a particular defense.

The visualization process [Figure 4.12] often involves a filter at the first stage either to select a region of interest in time or space or to reduce the resolution in time or space so that the data can be more easily manipulated or the major aspects of the phenomenon understood. If the output of this filtering—more data—is defined on a surface and has enough spatial density, each data point can be mapped to one pixel value to form an image. Otherwise the filtered data are mapped into geometric primitives, which are then rendered into images. This is the more common occurrence since the data are usually defined within a volume and the spatial resolution is usually much less than the pixel density on a monitor. These geometric primitives can be formed in a number of ways. The most common method is to use the data points as the endpoints of lines or vertices of polygons. In 2D, the term *cell* is often used to define a quadrilateral created in such a manner. In 3D, we use the term *cube* to denote a hexahedron whose eight vertices are data points. See Figure 4.13. The rendering process converts a set of geometric primitives into one or more images. This process can be simple or very complex. The degree of realism sought in the rendering process is usually determined by the intended use of the resulting visualization. The images are analyzed to understand phenomena or to examine features such as extrema, gradients, flow rates, range of values, and spatial patterns. This may cause the analyst to change the filtering parameters to focus on a region or a data range or to expand the view to see even more of the data. This examination may also lead to more data collection or more simulation runs.

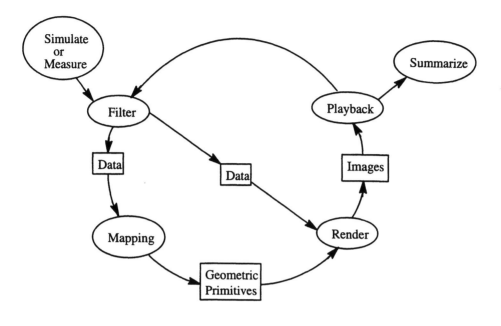

Figure 4.12 Diagram of the visualization process.[15]

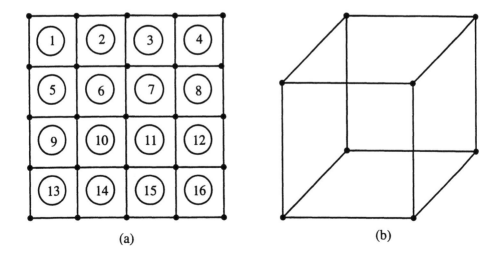

Figure 4.13 Data cells and cubes. (a) Cells are formed by connecting adjacent data points in 2D. The cells are indicated by circled numbers; the data points are represented by heavy dots. (b) Cells are formed by connecting adjacent data points in 3D. The data points are represented by heavy dots.

4.2.1 Types of visualization

Visualization can be classified into three categories: I-see, We-see, and They-see. *I-see* visualization is visualization for self-edification. Usually in I-see visualization a softcopy image is analyzed—usually on the monitor of the computer used to create the image. This is exploratory visualization. Contextual information is not very important if the person who created the image knows the context of the data. In this type of visualization one is likely to use more simplistic rendering to make the visualization process more interactive. *We-see* visualization is visualization done to show a peer, colleague, or collaborator. The researcher or the one who created the visualization is there to explain the visualization. In this case, the image that is analyzed may be softcopy or hardcopy, depending on a number of factors. Here more contextual information is not only useful for documentation, but often is necessary to understand the imagery. Contextual information includes color legends, axis labels, context indicators, contour values, etc. *They-see* visualization is visualization intended for others, from a technical audience to a lay audience. This is a rather broad category, but they-see visuals should include enough contextual information for the image to be self-explanatory. They-see visuals are almost always hardcopy, such as a photographic slide, a photographic print, a dye-sublimation print, a videotape, or similar hardcopy display.

4.2.2 Types of data

The data we visualize take many forms. Since we use digital computers, visualization data must be discrete in value, space, and time. To recreate a continuous function, an underlying relationship is assumed and an appropriate interpolation function is used to regenerate in-between values. Data quality is always an issue. Are some data missing? Are there redundant values at some locations, as might occur in multi-zone computational fluid dynamic simulations? Do simulation models provide an error estimate?

Other issues in visualizing data are the dimension of the domain (the independent variables) and the range (the dependent variables). The data which we visualize usually come from 2D or 3D domains, which is to say there are usually two or three independent (spatial) variables. Exceptions to this include financial and socio-economic data, which often have many independent values and may not have a spatial context. Range data are usually either scalar or vector data. Scalar data are single-valued at each sample point; vector data have a direction and a magnitude at each sample point. Data sets which have multiple scalar values at each sample point are also common.

Examples of common 2D data sets include:

- topography data sets for which the dependent variable is a scalar, such as height.

- surface data sets in which the dependent variable(s) may be scalars like pressure, temperature, viscosity, surface coverage or vectors like flow velocity, wind speed/direction, or a gradient.
- census data sets in which the scalars may be age, education, income, race, sex, religion, etc.

Examples of common 3D data sets include:

- medical data sets for which the dependent variable may be density (scalar), blood velocity (vector), or air velocity (vector). Figure 4.14 contains two visualizations of density data.
- oceanographic data sets for which the dependent variables might include temperature, salinity, 2D or 3D currents, or 2D sea surface height fields. Figure 4.15 contains five visualizations of 2D currents.
- atmospheric data sets for which the dependent variables might include wind velocity, wind stress, surface pressure, accumulated precipitation, terrain height, solar radiation, heat flux, temperature, and potential temperature.

If the data are defined on a structure and the sampling is at a fixed interval, we are able to store the data in a more compact form. The relationship between data points, i.e., the structure of the data, also affects interpolation schemes and error analysis. The data must be structured—or a structure created—to use visualization algorithms which operate on cells or cubes.

4.2.3 Mappings

There are a number of mappings from data to images. In fact, we often go through a number of stages of mapping. The choice of mapping is influenced by the data, the problem or physical phenomenon being studied, the question or issue being addressed, the audience, and the purpose of the visualization (to demonstrate something or to convince someone).

4.2.3.1 Color mapping or pseudo coloring

Color mapping maps data to colors. Most often the mapping is applied to scalar data, but schemes exist to map multiple variables to different components of color.

For scalar color mapping, the mapping is implemented by using the scalar values as indices into a list of colors. This is called a *color lookup table* or LUT. We may choose to map only part of the range of scalar values or we may map all of the scalar values into only part of the color lookup table. This allows the user to trade-off precision and range, since the

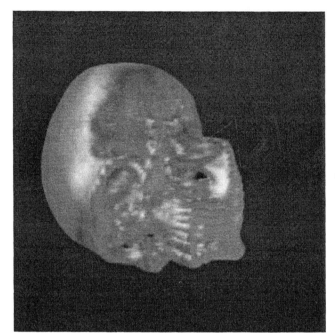

(a)

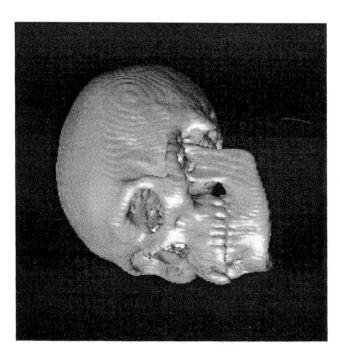

(b)

Figure 4.14 Comparison of density data rendered as a) an isosurface and b) a volume rendering.

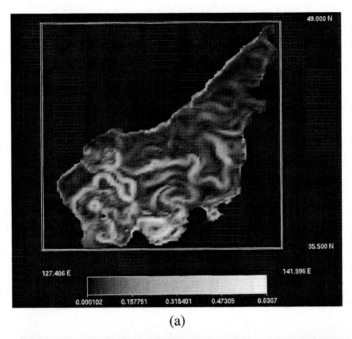

(a)

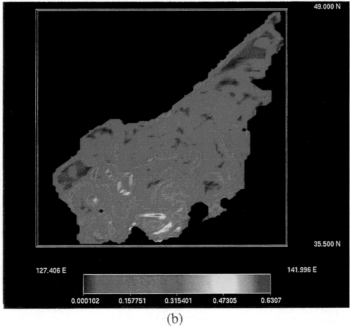

(b)

Figure 4.15 Modeled sea surface height in the Sea of Japan on January 17, 1981, a) grayscale, b) rainbow (blue to red). Data courtesy of the Naval Research Lab, Stennis Space Center, MS.

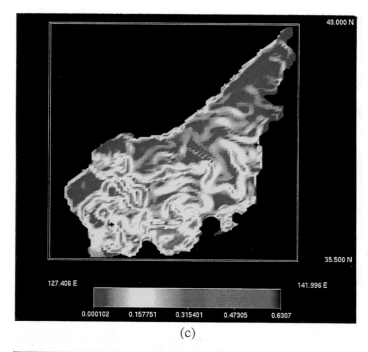

(c)

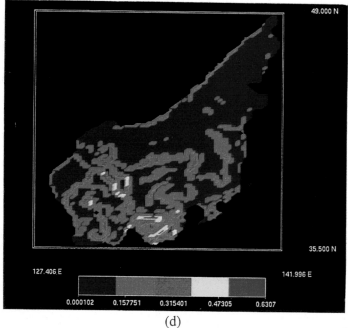

(d)

Figure 4.15 Modeled sea surface height in the Sea of Japan on January 17, 1981, c) rainbow (red to blue), d) banded. Data courtesy of the Naval Research Lab, Stennis Space Center, MS.

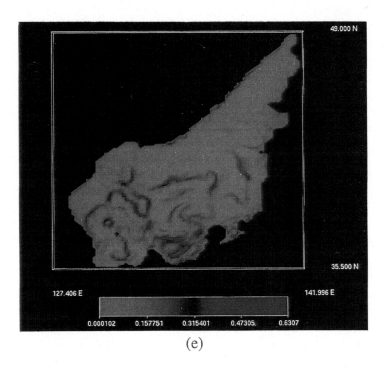

(e)

Figure 4.15 Modeled sea surface height in the Sea of Japan on January 17, 1981, e) double-ended. Data courtesy of the Naval Research Lab, Stennis Space Center, MS.

human visual system can only distinguish 20,000 to 30,000 colors simultaneously. Thus, it may be desirable to only map a subrange of scalar values into the color lookup table, map the values below the subrange to one constant color, and map the values above the subrange to another constant color. This allows the viewer to examine a range of scalar data with greater precision. This may be useful when there are a number of outliers in the data set or when one particular range of values is of particular interest. Other alternatives include allowing color re-use, i.e., colors may not represent unique values or value ranges, but may be context dependent.

Figure 4.15 shows five different lookup tables (colormaps) used to indicate flow speed in the Sea of Japan. Figure 4.15a is a grayscale map, in which the slowest flow is black, the mean flow is mid-level gray, and the fastest flow is white. In 4.15b the hues mimic a rainbow going from blue to red, while in 4.15c the hues once again mimic a rainbow, but go from red to blue. In 4.15d the banded color map displays ranges of scalar values as a constant color. Many oceanographers claim this allows them to see the structure better due to the well-defined breaks. In 4.15e the double-ended colormap shows flow slower than the median speed in red and faster than the median speed in blue. There is a smooth (linear) transition from black to saturated blue or red. This mapping is often used for data sets whose range includes zero.

4.2.3.2 Contouring

Contours are lines (in 2D) of constant functional value. Contouring follows naturally from color mapping. The image in Figure 4-15d can be constructed by first contouring the data and then filling in between contour levels with a constant color. As we increase the number of contour levels, we approximate colormaps like those in Figures 4-15a, 4-15b, 4-15c, and 4-15e. Examples of 2D contour displays include weather maps with lines of constant temperature (isotherms) and topological maps with lines of constant elevation. Many people consider contouring a more analytical visualization technique than color mapping, since the location of particular values are more visible. The human visual system has a very hard time picking a particular color from a smoothly varying range of colors.

Contours can be found by a number of techniques. The two most prevalent are often called *edge tracking* and *marching squares*. Both start with the selection of a contour value. The location along every cell edge where this contour value occurs is found using an interpolation technique. Once the intersection points are found, the two techniques vary. Edge tracking detects an edge intersection and then "tracks" the contour as it moves across cell boundaries, i.e., what goes in must come out. In marching squares, the cells are treated independently. The premise is that there are only a finite number of ways that a contour can pass through a cell. A case table is created that contains all the possible topologies. For more details, see Ref. 16. For both techniques, ambiguity arises when the contour intersects all four edges of one cell. A choice can be made arbitrarily, resulting in either breaking or extending the current contours as shown in Figure 4.16.

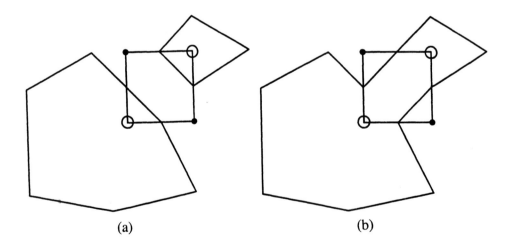

(a)   (b)

Figure 4.16 Choosing a particular contour case will break (a) or join (b) the current contour.

4.2.3.3 Isosurfaces

Contouring can be extended to three dimensions. Analogous to finding a curve in 2D, a surface—an isosurface—is found in 3D. Accordingly, an *isosurface* is a surface of constant value. Alternatively, the isosurface can be thought to divide the volume into two regions, one above and one below a given threshold. The two most prevalent techniques are called *marching cubes*[17] and *marching tetrahedra*.[18] Marching cubes is an extension of marching squares. The technique is applicable to volumes of data sampled at regular intervals in all three dimensions, i.e., volumes in which hexahedra or cubes are formed when adjacent data samples are connected. The major difficulties in the extension to 3D are the additional number of topological cases and the more difficult ambiguity resolution. The 256 possible topologies can be reduced to 15 by exploiting symmetry, using rotation and mirroring. The ambiguity resolution must now consider the abutting cube or else the surface will have holes. Several approaches have been proposed. Nielson and Hamann[19] developed a technique called the asymptotic decider, which is based on an analysis of the variation of the scalar variable across an ambiguous face to determine how the intersection points should be connected. The marching tetrahedra approach divides each cube into five or six tetrahedra. The advantage of this approach is that there are no ambiguous cases. The disadvantage is that the isosurface that is generated has more triangles and these extra triangles may actually make the surface less smooth.[16]

Other subsetting techniques exist. A simple technique is *cutting planes*, in which only the data on (or interpolated onto) one plane in shown. More complicated subsetting schemes are usually based on a feature detection algorithm. These techniques allow the viewer to see particular features, values, or locations in the data field.

4.2.3.4 Volume rendering

There are two basic ways to render data whose domain is 3D: surface rendering (such as isosurfaces) and volume rendering. In applying *surface rendering* techniques, we use surface primitives like points, lines, and triangles to represent a 3D surface. In visualizing a 3D data set using surface rendering, we have to generate isosurfaces or other surfaces consisting of surface primitives. *Volume rendering* is a technique that allows us to render 3D data sets directly, without having to generate intermediate surface primitives. An example of each is shown in Figure 4.14. Figure 14.14a shows an isosurface created from a 3D data set of density values with 64 samples in each dimension. The isosurface or threshold value was picked to show bony structure. Figure 4.14b shows a volume rendering of the same data set. Volume rendering lets us see inside or see through our data. It can be argued that this lets us see more than does surface rendering. To see inside objects, we map each scalar value into an opacity value and a color value. This allows us to cast light and attenuate it in one of two ways. We can either cast light from behind the volume and accumulate what passes through the volume on an image plane or we can cast light into the front of the volume and see how the light is attenuated in the volume. Figure 4.17 shows how light rays would be projected in parallel through a volume of data or how light could be cast from the image plane into the volume. For more information on volume rendering see Ref. 20.

The advantage of volume rendering, from a data field understanding perspective, is that you can see all the data in the data field in one image. Thus volume rendering is useful for visualizing data fields like pollutant distributions if one is interested in the variations throughout the whole volume. The disadvantages of volume rendering include the computational complexity, the memory required, and the necessity of adjusting values for various transfer functions to obtain the best image.

The advantage of surface rendering is that you can see exactly where a particular value occurs within the data field. Surface rendering yields a hard surface which is more satisfying to an engineer or scientist looking for a precise answer.

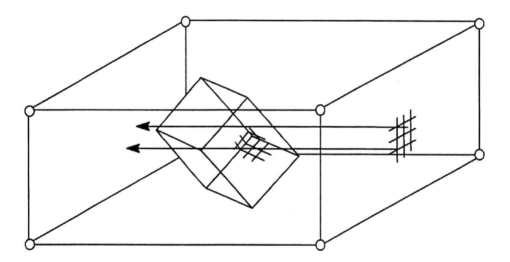

Figure 4.17 Conceptual diagram of direct volume rendering process.

4.2.3.5 Vector visualization

Vector data result from numerous scientific investigations, such as those of weather, fluid flow, electromagnetic fields, or mechanics. Vector data are usually represented by components in two or three dimensions, but can be represented by a direction (in two-space or three-space) and a magnitude. The most common visualization technique for vector data is to draw a line scaled by the magnitude, oriented by the direction, and anchored at each data point. This technique is sometimes referred to as a *hedgehog* due to the bristly appearance of the images it produces.[16] The technique is also referred to as *arrows* or *tufts*. A visualization of the computed flow in the Sea of Japan using tufts is shown in Figure 4.18. This basic technique can be modified in numerous ways to reduce ambiguities and provide more information. For example, arrowheads can be added to indicate direction and the lines can be colored to indicate vector magnitude or some other scalar quantity.

A tuft display is not without its limitations though. Adding arrowheads can clutter the image. In 3D perspective views, it is hard to tell if the length of the lines is determined by the vector magnitude or the projection. When one line falls on top of another it is difficult to tell which is in front, thus a key depth cue—obscuration—is missing. In Figure 4.19 the same data set is shown in a perspective view.

One vector visualization technique which solves the projection problems for 2D vector field visualization is called *colorwheel*.[21] In this mapping no intermediate geometric primitives are used. Instead, the mapping uses the HSV color space. The direction of the vector is represented by a hue and the magnitude of the vector is redundantly mapped to both value and saturation since the human visual system has a larger perceived dynamic range in hue than value or saturation.[22]

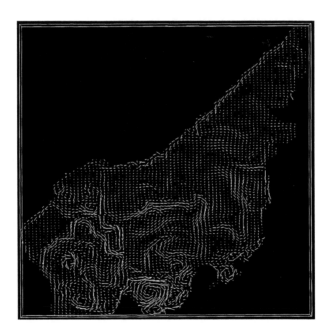

Figure 4.18 Surface flow in the Sea of Japan visualized using tufts.

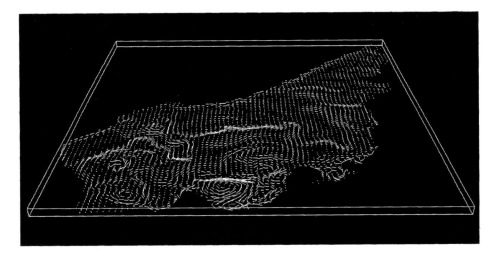

Figure 4.19 Surface flow in the Sea of Japan shown in a perspective view. The location of the tufts and their size is ambiguous due to the lack of obscuration.

In this mapping, each pixel can represent a different vector value. It has been found that a log-scale mapping of vector magnitude to saturation/value works well, given the human visual system's nonlinear response. The functional value resolution, i.e., the ability to see the variances over space as well as over the range of functional values on displays like 24-bit CRTs and most color printers, should be better with the colorwheel technique than with tufts. In Figure 4.20 the same data set is visualized using the colorwheel technique.

The major problem in visualizing vector data is the difficulty of showing global structure and local values. For example, a vortex at one scale is a current or a wave at a more highly resolved sampling density. Projecting magnitude and direction from 3D space into one uncluttered 2D image is a difficult problem. Techniques that advect smoke, clouds, and flow volumes[23-25] do a good job of showing the flow at a global level, but not at a detailed level. Particle-based techniques[26,27] show local flow properties well, but are inferior at presenting the big picture.

Hall[28] has developed a technique called *colorsphere* which visualizes 3D vectors using perceptually based color spaces, similar to the colorwheel. The biggest advantages over tuft diagrams are that (1) color is invariant under projection and (2) high sampling rates can be used. The principal drawbacks of the color-based representations are the lack of a standard mapping from physical-space to color-space and user inexperience. The first drawback can be mitigated by showing the color space used; the second can only be overcome with use. Hall proposed using a perceptually linear color space such as the Munsell space[29,30] which is based on the opponent theory of color vision. The color-space can be quantized to produce abrupt edges which may help a

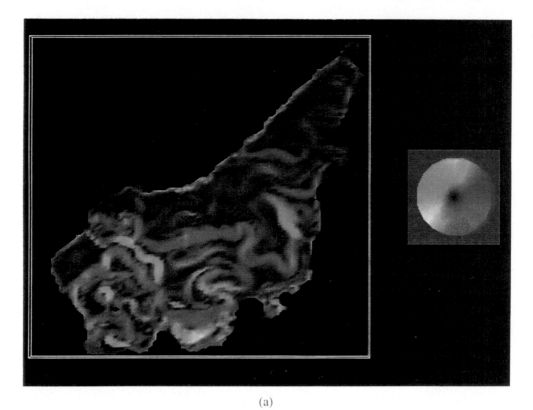

(a)

(b)

Figure 4.20 Surface flow in the Sea of Japan shown using the colorwheel technique: a) plan view, b) perspective view.

scientist see global variations better. Color is a way to represent a 3D vector quantity in a single sample (pixel) and thus provide a dense display of multivariate information. However, in the end, many scientists still use tufts to visualize vector data due to the intuitive nature of the visualization method.

4.2.4 Viewing

4.2.4.1 Softcopy viewing

Most exploratory visualization is performed using softcopy displays, specifically a cathode ray tube (CRT) or monitor. Exploratory visualization is the mode in which the scientist is trying to analyze or explore his or her data. To explore the data effectively, it is useful to have an understanding of how the data are mapped into pixels and of the potential artifacts and limitations. We have discussed the first step in most visualization schemes, namely, converting data to graphical primitives. The second step is called rendering and is the process of converting graphical primitives to pixels.

Although the number of samples per unit area (pixel density) on a softcopy display is usually much less than on a hardcopy display, this limitation is often mitigated by the ability to alter the image by changing various mapping parameters dynamically. In other words, the data set can quickly be viewed from a different viewpoint or with a different perspective, the color mapping can be altered to emphasize a different aspect of the data, or less of a data volume can be visualized to allow smaller details to be seen.

The color range (gamut) and precision of the CRT raises another issue in softcopy viewing. For example, how different are two neighboring colors in the displayable gamut? Fortunately, CRTs can display a wide range of colors as well as a large number of distinct colors. Although the eye can discriminate on the order of 100,000 colors pairwise, it can only discriminate about 30,000 colors simultaneously. For exploratory visualization, softcopy is a very functional medium. A problem can occur when it is necessary to show others the softcopy image on another monitor or via a hardcopy display. The problem is the gamut mismatch between various displays—both softcopy and hardcopy. Often an image that has sufficient contrast on one monitor is too dark or too light on another monitor. Colors may shift so a data value that was purple on one monitor may now be blue or red. This can be overcome by using device independent colors, but few scientists take the time to calibrate their monitors. Also most softcopy images look much different when printed, once again due to the gamut mismatch.

4.2.4.2 Hardcopy viewing

Most presentation visualizations are seen in hardcopy form. Single images are usually printed, whereas sequences of images are generally written to video media. Many types of printers are used to create hardcopy: laser, dye-sublimation, wax-transfer, offset press, photographic, etc. Most printers can produce positives (reflect light) or negatives (absorb light). Each printer has its strengths and weaknesses. For example, dye-sublimation printers have a very large gamut, but the marking engines and the paper/transparency material are relatively expensive. Photographic slides and prints have high spatial (on the order of 3000×4000 pixels) and intensity resolution and the production process is relatively cheap, but the process is analog and requires the use of chemicals, both of which limit the reproducibility of results. For more information see Chapter 7 (Digital Halftoning for Printing and Display of Electronic Technology) or Ref. 30.

Video is a technology in the midst of a transition. There are three basic methods to present a sequence of images today: (1) NTSC/PAL, (2) HDTV/MPEG-2, and (3) online. NTSC and PAL are the television standards in most of the world today. Both were developed over 40 years ago when technology was more limited. Neither have the image quality of more recent standards, such as HDTV and MPEG-2. In particular they have limited intensity, spatial, and temporal resolution. NTSC has a spatial resolution of about 440×484 shown as 30 interlaced frames/sec.[29] PAL has about a 20% higher spatial resolution, but a lower temporal resolution (25 interlaced frames/sec).[31] Both standards introduce[29] numerous artifacts which have become more noticeable as technology has improved. However, NTSC and PAL recorders are relatively cheap and players are readily available, so most scientists use those formats to show image sequences. Even worse most scientists use the lowest quality video format—VHS—which uses a compressed and modified form of either the NTSC or PAL format. The VHS format compresses the luminance bandwidth of the NTSC signal from 4.2 Mhz to 2.2 MHz and multiplexes and downsamples[31] the chrominance bandwidth to 500 kHz. This gives a horizontal resolution of about 230 lines, i.e., only 115 cycles can be resolved in one scan line. The small chrominance bandwidth limits the displayable gamut.

Over the past 20 years, the television, communication, and computing technical communities have tried in various standards bodies to address the inadequacies of NTSC and PAL. Recently the MPEG-2 standard[32] (see Chapter 2 on Video Compression Standards) has been accepted as the basis of a new U.S. television standard, called Advanced TV (ATV) or High Definition TV (HDTV). MPEG-2 will allow better utilization of the available bandwidth and allow scientists to trade off spatial, temporal, and intensity resolution. In fact, with the inherent compression defined in the standard, one is usually able to obtain improvement in all three resolutions. The standard has multiple formats at a 16:9 aspect ratio, as opposed to the existing 4:3 in both NTSC and PAL. For applications needing high temporal resolution, 60 progressively scanned frames with 1280×720 pixels/frame can be shown per second. For applications requiring high spatial resolution, frames with 1920x1080 pixels can be shown 30 times per second.[32]

The present impediment to hardcopy use (videotape or videodisk) of HDTV is the lack of video production standards. However, this does not affect online presentations. MPEG, Quicktime, and other video file formats have been standardized either officially or de facto. Although there are some incompatibilities, MPEG players exist for most computers and operating systems. At many conferences today, presenters are able to show scientific animations directly from a computer, even a laptop, that are far superior in quality to animations shown from VHS tape.

4.3 PERCEPTUAL PRINCIPLES FOR VISUALIZATION DESIGN

Careful attention to the mechanisms and characteristics of human perception can yield more effective visualizations by exploiting the strengths of the visual system and avoiding its weaknesses. Additionally, visualization design should ensure that the most striking aspects of a visualization are also the most important. Representations which draw the viewer's eye to unimportant features may cause more interesting features to be overlooked. Consideration of the characteristics of human perception can be a valuable guide in predicting which aspects of a visualization will draw the attention of the average viewer. Features likely to catch the eye are those that are brightly colored, highly saturated, well-defined, moving, or changing.

4.3.1 Avoiding unwanted interactions

The visual characteristics judged by the human visual system are not orthogonal quantities; that is, the perception of one characteristic may be influenced by the value of another.

Color-size interactions. Some visual experiments have suggested that the color of an object can influence the perceived size of that object. Tedford, Berguist, and Flynn[33] conducted observer experiments under precisely controlled conditions and found a significant color-size effect. Specifically, rectangles of the same size, saturation, and brightness appeared to have different sizes when colored red-purple, yellow-red, purple-blue, or green (in order of decreasing apparent size). At high saturations, this effect was statistically significant for all color pairs except yellow-red and purple-blue. At low saturations, only the difference between yellow-red and green rectangles was significant. In trials where hue was held constant and saturation varied, rectangles with higher saturations were consistently judged to be smaller than less saturated rectangles.

Cleveland and McGill[34] investigated the implications of the color-size illusion for statistical maps. Subjects were shown a map of Nevada in which counties were colored either red or green with the total area of red and green nearly equal. Subjects were asked to judge which color, if any, represented the larger land area. Each subject was shown 10 maps. On the average, subjects judged that the red areas were larger more often than they judged the areas the same or the green areas larger. When the experiment was repeated using low-saturation tones of red and green (formed by adding yellow), no such bias was observed. Their results suggest that the color of a region influences the perceived size of the region and that the effect is strongest for very saturated colors.

This interaction has obvious ramifications on visualization effectiveness. If perceptions about object size are important, the visualization designer should take care in assigning distinct colors to objects, either explicitly or through a color mapping process. If color coding of objects is desirable, desaturated colors are likely to cause less pronounced interaction effects.

Interactions between color components. A common strategy of multivariate color schemes is to map each variable to a different component of color. These components could be intuitive (hue, saturation, brightness), physiological (opponent color channels), or device-derived (red, green, blue). One would expect that these color model components would be perceptually orthogonal. Perceptual studies suggest that this is not entirely the case. Interactions have been observed between hue and brightness and between saturation and brightness. This explains in part the advantage of colorwheel over colorsphere visualization for vector data. A saturated color is perceived as brighter than a desaturated color when the two are related in brightness (Helmholtz-Kohlrausch effect).[35] Yaguchi and Ikedo hypothesized that a cancellation of hues in the chromatic channels was resulting in decreased perceived brightness. The Bezold-Brucke Phenomenon, describing the changes in perceived hue with increasing illumination levels, has been observed in experiments where subjects are asked to match the hues of patches with differing luminances.[2] As the luminance of the brighter patch was increased, perceived hue shifted away from green and toward blue and yellow. While these effects may not be strong enough to make color schemes based on color components impractical, they can be expected to create slightly distorted perceptions.

4.3.2 Designing effective colormaps

Designing colormaps is as much an art as a science. Although well-chosen color maps can emphasize important features of a data set, color maps can also exaggerate unimportant details or create visual artifacts due to unforeseen interactions between the choice of colors, the expectations of the viewer, the data, the purpose of the visualization, and human physiology.

Ideally, a colormap should be intuitive. Unfortunately, a colormap, which is intuitive for one purpose or one audience, may not be intuitive in another situation. Consider a rainbow scale which increases from blue to red. This matches most people's concept of blue being cold and red hot, and thus is a good colormap to show temperature. However, a physicist would think of blue being hotter than red, since hotter objects emit more blue light than red.[16]

A colormap should be appropriate to the characteristics of the data, calling attention to the most important features. The common rainbow color scale maps the middle values to yellow, a particularly striking color. In applications where the location of middle values is of particular interest, this is appropriate.

Such applications are not very common, however. More often, the high or low values are of greatest interest, and middle values are of least interest. In such situations, a double-ended colormap might be appropriate. Such a colormap would map the middle values to some unobtrusive color, such as gray, while mapping the high and low values to distinct and more prominent colors.

An effective colormap should also be well-suited to the primary purpose of the visualization. One fundamental division in visual analysis tasks arises from the dichotomy between quantitative and qualitative information display. For *quantitative* information, a color scale which is monotonically increasing in any of the opponent channels will tend to cause contrast effects and can encourage errors in mapping from a displayed color back to the represented value. In *qualitative* display, where judgments about the shape or surface properties of a structure are of primary importance, simultaneous contrast does not seem to pose a difficulty. The human visual system is experienced at identifying surface tendencies from luminance gradients in the presence of contrast effects. This suggests that for tasks which require reading metric values from a representation, a color sequence which does not vary monotonically with any opponent channel (such as the rainbow scale) is superior to one which does (such as the gray scale). Cues about surface properties, however, are best judged from lightness differences presented by scales such as the gray scale. This design guideline has been supported by experimental evidence.[10] Ware's findings also suggest that a color scale which varies in both luminance and hue can be used to accurately represent both metric and surface properties by minimizing the effects of simultaneous contrast.

Clearly, no one map is universally acceptable for all purposes. For more information on colormaps see Refs. 10 and 36 through 38.

4.4 SUMMARY

An understanding of both how humans see[40] (vision) and of how images are created (electronic imaging) are crucial in developing effective visualizations of data. An awareness of the mechanisms and characteristics of human visual perception can improve visualization research by guiding the selection of new techniques and by improving the results obtained. Since the information contained in visualizations must pass through the perceptual system, careful attention to the system's characteristics can greatly improve the effectiveness of visualizations. Perceptually based visualization strives to avoid distortions caused by perceptual anomalies, to exploit cognitive and cultural expectations, and to convey the features of the data displayed accurately.

REFERENCES

1. G. Wyszecki and W. S. Stiles, *Color Science: Concepts and Methods, Quantitative Data and Formulae, 2nd Edition,* John Wiley & Sons, New York, 1982.
2. L. Hurvich, *Color Vision*, Sinauer Associates, Inc., Sunderland MA, 1981.

3. M. Livingstone and D. Hubel, "Segregation of Form, Color, Movement, and Depth: Anatomy, Physiology, and Perception," *Science*, vol. 240, 1988, pp. 740-749.

4. R. Held and A. Hein, "Movement-Produced Stimulation in the Development of Visually Guided Behavior," *Journal of Comparative and Physiological Psychology*, vol. 56, no. 5, 1963, pp. 872-876.

5. K. W. Arthur, K. S. Booth, and C. Ware, "Evaluating 3D Task Performance for Fish Tank Virtual Worlds," *ACM Transactions on Information Systems*, vol. 11, no. 3, 1993, pp. 239-265.

6. W. van Damme, *Active Vision: Exploration of Three-Dimensional Structure*, PhD dissertation, University of Utrecht, ISBN 90-393-0803-0, 1994.

7. G. J. F. Smets and K. J. Overbeeke, "Trade-Off Between Resolution and Interactivity in Spatial Task Performance," *IEEE Computer Graphics and Applications*, vol. 15, no. 5, 1995, pp. 46-51

8. P. Rheingans, "Color, Change, and Control for Quantitative Data Display," *IEEE Visualization '92 Proceedings*, Boston, MA, Oct. 1992, pp. 252-258.

9. P. Rheingans and C. Landreth, "Perceptual Principles for Effective Visualizations," *Perceptual Issues in Visualization*, G. Grinstein and H. Levkowitz (eds.) Springer-Verlag, New York, 1995, pp. 59-73.

10. C. Ware, "Color Sequences for Univariate Maps: Theory, Experiments, and Principles," *IEEE Computer Graphics and Applications*, Sept. 1988, pp. 41-49.

11. J. Gibson, *The Perception of the Visual World,* Houghton Mifflin, 1950.

12. J. Cutting and R. Millard, "Three Gradients and the Perception and Flat of Curved Surfaces," *Journal of Experimental Psychology: General*, vol. 113, no. 2, 1984, pp. 198-216.

13. J. Todd and R. Akerstrom, "Perception of Three-Dimensional Form from Patterns of Optical Texture," *Journal of Experimental Psychology: Human Perception and Performance,* vol. 13, no. 2, 1987, pp. 242-255.

14. B. Cumming, E. Johnston, and A. Parker, "Effects of Different Texture Cues on Curved Surfaces Viewed Stereoscopically," *Vision Research*, vol. 33, no. 5/6, 1993, pp. 827-838.

15. C. Upson, "2D and 3D Visual Workshop," Shortcourse #13, SIGGRAPH '89, 1989.

16. W. Schroeder, K. Martin, and B. Lorensen, *The Visualization Toolkit, 2nd edition*, Prentice Hall, 1997.

17. W. E. Lorensen and H. E. Cline, "Marching Cubes: A High Resolution Surface Construction Algorithm," ACM SIGGRAPH, *Computer Graphics*, vol. 21, no. 4, July 1987, pp. 163-169.

18. J. Bloomenthal, "Polygonization of Implicit Surfaces," *Computer-Aided Geometric Design*, Vol. 5, no. 4, 1988, pp. 341-355.

19. G. M. Nielson and B. Hamann, "The Asymptotic Decider: Resolving the Ambiguity in Marching Cubes," *IEEE Visualization '91 Proceedings*, San Diego, CA, Oct. 1991, pp. 83-91.

20. A. Kaufman (ed.), *Volume Visualization*, IEEE Computer Society Press, 1990.

21. A. Johannsen and R. Moorhead, "AGP: Ocean Model Flow Visualization," *IEEE Computer Graphics and Applications,* vol. 15, no. 4, July 1995.

22. S. M. Pizer, J. B. Zimmerman, and R. E. Johnston, "Contrast Transmission in Medical Image Display," *Proceedings of ISMIII '82* (International Symposium on Medical Imaging and Image Interpretation), IEEE Computer Society, vol. 2, no. 9, 1982.

23. K.-L. Ma and P. Smith, "Virtual Smoke: An Interactive 3D Flow Visualization Technique," *IEEE Visualization '92 Proceedings*, Boston, MA, Oct. 1992, pp. 46-53.

24. N. Max, B. Becker, and R. Crawfis, "Flow Volumes for Interactive Vector Field Visualization," *IEEE Visualization '93 Proceedings*, San Jose, CA, Oct. 1993, pp. 19-

25. N. Max, R. Crawfis, and D. Williams, "Visualizing Wind Velocities by Advecting Cloud Textures," *IEEE Visualization '92 Proceedings*, Boston, MA, Oct. 1992, pp. 179-184.

26. A. Hin and F. Post, "Visualization of Turbulent Flow with Particles," *IEEE Visualization '93 Proceedings*, San Jose, CA, Oct. 1993, pp. 46-52.

27. J. van Wijk, "Rendering Surface-Particles," *IEEE Visualization '92 Proceedings*, Boston, MA, Oct. 1992, pp. 54-61.

28. P. Hall, "Volume Rendering for Vector Fields," *The Visual Computer*, vol. 10, no. 2, 1993, pp. 69-78.

29. R. Hall, *Illumination and Color in Computer Generated Imagery*, Springer-Verlag, 1988.

30. J. Foley, A. van Dam, S. Fiener, and J. Hughes, *Computer Graphics Principles and Practices*, Second Edition, Addison-Wesley, 1990.

31. A. M. Noll, *Television Technology*, Artech House, 1988, pp. 98-101.

32. B. Haskell, A. Puri, and A. Netravali, *Digital Video: An Introduction to MPEG-2*, Chapman and Hall, 1996.

33. W. H. Tedford, Jr, S. L. Berquist, and W. E. Flynn, "The Size-Color Illusion," *The Journal of General Psychology*, vol. 97, 1977, pp. 145-140.

34. Cleveland, W. S. and R. McGill, "A Color-Caused Optical Illusion on a Statistical Graph, The American Statistician, vol. 37, no. 2, pp. 101-105 (1983).

35. Yaguchi, H. and M. Ikeda, *Contribution of Opponent-Color Channels to Brightness, Colour Vision*, J.D. Mollon and L.T. Sharpe, eds., Academic Press, Inc., pp. 353-360 1993.

36. H. Levkowitz, "Linearized Optimal Color Scale Compared with Linearized Gray Scale for Medical Image Data," *IEEE Computer Graphics and Applications*, vol. 12, no. 1, Jan. 1992, pp. 72-80.

37. L. Bergman, B. Rogowitz, and L. Treinish, "A Rule-based Tool for Assisting Colormap Selection," *IEEE Visualization '95 Proceedings*, Atlanta, GA, Oct. 1995, pp. 118-125.

38. B. Trumbo, "Theory for Coloring Bivariate Statistical Maps," *The American Statistician*, vol. 34, no. 2, 1981, pp. 81-93.
39. P. K. Robertson and J. F. O'Callaghan, "The Generation of Color Sequences for Univariate and Bivariate Mapping," *IEEE Computer Graphics and Applications*, 1996, pp. 24-32.
40. B. Wandell, *Foundations of Vision*, Sinauer Associates, Inc., Sunderland, MA, 1995.

5

COLOR IMAGE PROCESSING

Arthur Robert Weeks
University of Central Florida
Orlando, Florida

Color image processing is one of the newest and most exciting areas of electronic image processing. Until recently, the computing hardware required to store and manipulate color images was limited to a special few. Today, it is not uncommon to find a standard desktop computer system with a true-color 24-bit display, at least 32 million bytes of memory, and 2 to 4 gigabytes of hard disk storage. The use of desktop color scanners and color printers attached to a desktop computer system are also becoming commonplace. The demand has increased to integrate both color and black and white images into presentations and documents. The new operating systems that are being used with these desktop computer systems are integrating to make it easier to transfer color images from one application to another. Several completely electronic photography systems have appeared in recent years that totally eliminate the use of film and chemicals. An image is acquired by an electronic camera, digitized, and then stored digitally within the camera. These images are then transferred at a later time to a desktop computer for processing and inclusion in documents. Even though the systems presently do not have the resolution of film, they are becoming quite popular because of the speed with which images can be acquired and placed within a document or presentation.

This chapter first presents the fundamentals of color followed by a discussion of the several commonly used color models. Next, several examples of using color image processing will be given, including the correction of the tint and saturation, the spatial filtering, and the color white balancing of color images. The final section discusses a technique known as pseudocoloring and the different type of displays that are used to highlight specific graylevels in a grayscale image in color to enhance important features so that they are clearly observable.

5.1 COLOR FUNDAMENTALS

One of the initial studies of color was done by Sir Isaac Newton in the eighteenth century and is contained in his treatise *Opticks*. Newton showed that white light was made from a combination of colors. Using an optical prism, he was able to separate sunlight into a rainbow of colors ranging from blue to red. Newton was also able to combine colors together to form other colors. For example, he combined green with red to produce yellow. From his research, Newton concluded that seven colors were needed to represent all the combinations of visual colors. It was Thomas Young, James Forbes, and James Clerk Maxwell in the nineteenth century who showed that only three primary colors are needed in different combinations to represent the visible spectrum of light. It was Maxwell who confirmed this when he showed that three primary colors were all that were needed to generate a color image in his now famous experiment performed during one of the Royal Society meetings in 1861. Using three separate black and white photographic plates, representing the filtered red, green, and blue color components of an image of a ribbon, Maxwell recreated the image by illuminating each photographic plate with the color light source corresponding to the color component of the plate. He then imaged the three color images from each photographic plate onto each other, producing a color image of the ribbon.

Figure 5.1 shows a diagram of the electromagnetic spectrum ranging from 0.001 nanometers to 1000 meters in wavelength. In the center of the spectrum is the visible spectrum, ranging in colors from violet to red and in wavelengths from 0.4 μm to 0.7 μm. Table 5.1 gives the seven colors that will be of interest in this chapter, with their approximate wavelengths. In the center of the visible spectrum is the color green, defined over the approximate wavelengths of 0.49 to 0.56 μm.

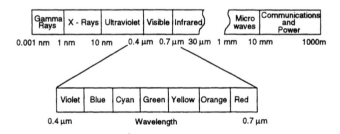

Figure 5.1 The electromagnetic spectrum.

Adjacent to the visible spectrum, are the ultraviolet and the infrared spectra. The ultraviolet spectrum is not used much in image processing because it is difficult to manufacture optical elements and detectors for use at these short wavelengths. As an example, to manufacture lenses for use at these short wavelengths, they must be extremely well polished to reduce the amount of light scattering off them, making them very expensive to build. The infrared spectrum is of interest to electronic image processing because of the enormous number of images that have been obtained with infrared camera systems. All objects at a temperature above absolute zero radiate electromagnetic energy, and if the object is at room temperature it radiates this energy in the infrared part of the electromagnetic spectrum. By the *Stefan-Boltzmann law*, a blackbody radiator radiates energy proportional to the fourth power of its temperature:

$$I = 5.67 \times 10^{-12} T^4 \quad \text{watts/cm}^2 , \quad (5.1)$$

where T is absolute temperature in degrees K.

Table 5.1 Wavelengths for the seven visible colors.

Color	Wavelength
Violet	0.38 - 0.45 μm
Blue	0.45 - 0.48 μm
Cyan	0.48 - 0.49 μm
Green	0.49 - 0.56 μm
Yellow	0.56 - 0.58 μm
Orange	0.58 - 0.60 μm
Red	0.60 - 0.70 μm

This radiated energy is distributed across the visible and infrared spectrums with the peak of this radiating energy occurring at

$$\lambda_{peak} = \frac{2898}{T} , \quad (5.2)$$

where T is the absolute temperature given in degrees Kelvin. For objects at a typical room temperature of 27°C, which corresponds to an absolute temperature of 27 + 273°K, the peak wavelength of the radiated energy from a room temperature object is 9.65 μm. Temperature differences are easily observable with camera systems that are sensitive to infrared wavelengths.

A common infrared camera system is the *forward looking infrared* (FLIR), which is sensitive to 10 μm. The FLIR camera system uses a linear array of detectors that scans a scene, producing a two-dimensional image that is converted to a standard video format for viewing. Other camera systems have been developed that use a two-dimensional array of sensors that are sensitive to wavelengths in either the 3 to 5 μm band or the 8 to 10 μm band. In the absence of any visible radiation (complete darkness to the human eye), these camera systems can detect and produce a video image of objects that would otherwise be undetected. Initial applications of these cameras were for military purposes, but they have found a use in remote sensing of the earth, criminal surveillance, emergency rescue, and security.

Other electronic cameras have been developed that also produce images in the microwave part of the electromagnetic spectrum. The combining of images of the same scene collected from a set of imaging sensors that are each sensitive to different parts of the electromagnetic spectrum is known as *multispectral image processing.* Color image processing is multispectral image processing in which the multispectral images are limited to the visible spectrum. The three primary colors of light as used by Maxwell in his color imaging experiment were red, green, and blue. Figure 5.2 shows the relationship between these three primary colors and the colors produced by combinations of them. The new color produced by the combination of two primary colors is referred to as a *secondary color.* For example, the addition of the primary colors red and green produces the secondary color yellow. The other two secondary colors are cyan, created by the addition of green and blue, and magenta, created by combining equal amounts of red and blue. In the center of the three circles is the color white formed, from the addition of equal amounts of the three primary colors.

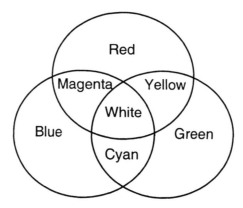

Figure 5.2 The primary and secondary colors of light (additive colors).

The mixture of the primary colors of light plays an important role in the generation of color images that are viewed on an electronic video monitor such as a television or a computer monitor. Each of these display devices uses a three gun cathode ray tube to generate three electron beams that are focused onto red, green, and blue phosphors, producing the primary colors of light. As the electron beams are scanned across the surface of the CRT, each beam is modulated, changing the amount of light that is generated proportional to the three primary colors contained within the displayed image.

The three primary colors of light are different from the primary colors of paint, or subtractive colors, as shown in Figure 5.3. The primary colors of paint are defined as the absorption of a primary color of light and the reflection of the other two primary colors of light. For example, consider a white light source incident on an object that absorbs the color green and reflects the colors red and blue. As a result, this object takes on a magenta color. The primary subtractive colors given in Figure 5.3 are created by absorption of a primary color of light, which is not included among the subtractive primary colors. For example, the subtractive primary color cyan, which is defined as the combination of green and blue, is created by an object that absorbs the color red. The secondary subtractive colors are created by the absorption of a secondary color of light by an object. An object illuminated with a white light source that absorbs the color yellow can only reflect the color blue. From Figure 5.3, the mixture of the subtractive primary colors magenta and cyan produces the color blue, while the mixture of magenta with yellow produces the color red.

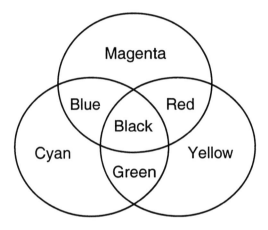

Figure 5.3 The primary and secondary colors of paint (subtractive colors).

The subtractive colors are primarily used to predict the color outputs from color photographic film and color printers. Both of these technologies use paper, a white light source, and the absorption and reflection of colors to produce a color image. Even though equal combinations of the three subtractive colors magenta, cyan, and yellow produce the color black, most high quality color printers add the color black as a fourth print color in order to produce a better perceived black in the final color print. When a black color is to be printed, this fourth print color is used instead of the equal mixtures of yellow, magenta, and cyan.

The use of three coefficients to describe a color is consistent with our understanding of how the human eye perceives color. The sensory section of the human eye is composed of rods and cones. Located centrally about the fovea are the cones, which are sensitive to color. Figure 5.4 shows a typical response of the eye as a function of wavelength for both (low light level) scotopic and (bright light level) photopic vision. For scotopic vision, the peak response occurs in the green region of the visible spectrum at a wavelength of approximately 0.51 μm. On the other hand, for photopic vision, the peak of the response shifts to the right to 0.56 µm, which corresponds to the yellow-green part of the visible spectrum. For both types of vision, the response of the eye decreases toward violet and red. This is why red and blue colors seem to have a perceived brightness which is lower than green colors. For many years, this was one of the reasons manufacturers chose green color display terminals for computers, believing that this color would produce less strain on the eyes under continuous hours of work.

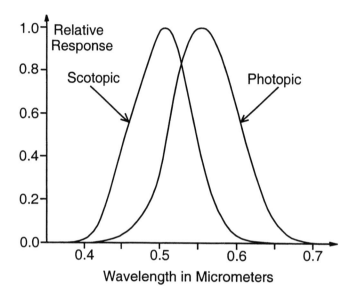

Figure 5.4 The visual response of the human eye as a function of wavelength (adapted from Carterette and Friedman, 1975).

In 1931, the Commission Internationale de L'Eclairage (CIE) devised a method of standardizing the definition of a color. A number of visual color experiments were performed in which observers were asked to match a composite color created from the combination of a set of known colors at a given wavelength. From these observations, it was determined that three color components were all that were needed to match an observed color accurately. Figure 5.5 shows, as defined by the CIE, the response of the eye to each color component as a function of wavelength. Essentially, this figure illustrates that the human visual system can be modeled using three color sensors each having a peak sensitivity corresponding to the colors red, blue, and green. Unlike the blue and green sensors, which only have one spectral maximum, the red sensor has two maximums, at approximately 0.6 µm and 0.45 µm. As it will be shown in the next section, it is these two peaks that will make the hue component in several of the different color models a modulus 2π.

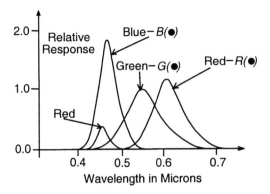

Figure 5.5 The three color components as defined by the CIE (adapted from Martin, 1962).

Consider a light source with a color distribution $L(\lambda)$ that depends on the wavelength λ. The perceived color by the eye is then defined by its *tristimulus values X, Y,* and *Z*:

$$X = C \int_{-\infty}^{\infty} R(\lambda)\, L(\lambda)\, d\lambda \quad , \tag{5.3}$$

$$Y = C \int_{-\infty}^{\infty} G(\lambda)\, L(\lambda)\, d\lambda \quad , \tag{5.4}$$

and

$$Z = C \int_{-\infty}^{\infty} B(\lambda)\, L(\lambda)\, d\lambda \quad , \tag{5.5}$$

where C is a constant that defines the total overall brightness response of the eye. Each of the three tristimulus values X, Y, and Z are directly related to the three primary colors of red, green, and blue. The use of Equations (5.3) through (5.5) includes the visual response of the human eye in determining the tristimulus values that defines the perceived color. Equations (5.3) through (5.5) give a better match of a perceived color than using the mixture of the three primary colors as proposed by Maxwell, Forbes, and Young. The sum of the tristimulus values gives the total luminance (brightness) perceived by the eye:

$$L = X + Y + Z \ . \tag{5.6}$$

Equation (5.6) produces a more accurate perceived brightness than adding the red, green, and blue components together equally by taking into account the different spectral responses of the eye for each of the three primary colors. Each of the three tristimulus values can be normalized by the luminance value L to give the percentage of each color:

$$x = \frac{X}{L} = \frac{X}{(X + Y + Z)} \ , \tag{5.7}$$

$$y = \frac{Y}{L} = \frac{Y}{(X + Y + Z)} \ , \tag{5.8}$$

and

$$z = \frac{Z}{L} = \frac{Z}{(X + Y + Z)} \ . \tag{5.9}$$

Equations (5.7) through (5.9) are known as the *trichromatic coefficients* of a color. The trichromatic coefficients do not depend on the total brightness of a color but only on the percentage of red, blue, and green color components.

A color can also be described by its luminance, hue, and saturation. The luminance defines the brightness of a color, for example, a dark blue sport suit versus a bright blue shirt. The hue (also referred to as tint) of a color defines its wavelength, and a color's saturation defines the percentage of white in it. Pastel colors, which contain large amounts of white, are low saturated colors. A color's saturation value varies between 0 and 100%. The hue, saturation, and luminance color coefficients describe a color similar to the human visual system and are better than the description of a color using the tristimulus values X, Y, and Z or the red, blue, and green color coefficients. For example, consider a color containing 40% red, 30% green, and 30% blue. The reader must think about this combination of colors before realizing it contains 90% white and 10% red. But if this color was described as the color pink, the reader could immediately visualize it without thinking about the combination of primary colors.

The hue and saturation of a color define its *chromaticity*, while the luminance describes its overall brightness. The luminance component will be used later in this chapter to generate grayscale images from color images. Often the luminance image is referred to as the grayscale image of a color image. Equations (5.7) and (5.8) can also describe the chromaticity of a color by defining its hue and saturation in terms of x and y. The luminance dependence has been removed from the trichromatic coefficients by the normalization process. In particular, only the x and y trichromatic coefficients are needed to describe a color's chromaticity, since the third coefficient z can be written in terms of the other two coefficients:

$$z = 1 - x - y . \tag{5.10}$$

Plate 1 gives the distribution of color as a function of the x and y trichromatic coefficients. This chart was defined by the CIE in 1931 and is known as the CIE *chromaticity diagram*. Plotted along the horizontal axis is the x trichromatic coefficient which represents the amount of red in a color, and along the vertical axis is the y trichromatic coefficient, giving the percentage of green in a color. The amount of blue is defined using Equation (5.10). Located at the center of the chart is the color white, given by equal amounts of x, y, and z. Along the outer edge of the chart are the wavelengths of the spectral colors (in micrometers) describing the various color hues as defined by the CIE. The saturation of a color increases radially from the center of the chart to the outer edge, which describes 100% saturated colors. Along the bottom of the chart, the line connecting the wavelengths from 0.380 µm to 0.780 µm gives the nonspectral color of violet. The color violet is a perceived color due to both the blue and red color receptors of the eye having a sensitivity to the visible wavelength of approximately 0.43 µm. The color violet is created as the mixture of the perceived color blue with the perceived color red.

The three primary colors of red, green, and blue are also shown on the chart, but the CIE definitions show that they are not completely monochromatic in color as one would expect. The color red, for example, contains approximately 73% red, 27% green, and no blue. The color blue contains approximately 18% red, 18% green, and 64% blue. By the CIE definition, the primary color blue is only 46% saturated.

The CIE chromaticity diagram is useful in determining the wavelength (hue), the percentage of white (saturation), and the mixture of two colors. Given the location of two colors on the CIE chromaticity diagram, the line connecting them gives all of the colors that can be created by different amounts of the two colors. Also, a line drawn from any color to the color white, located at the point of equal amounts of the trichromatic coefficients, gives all possible saturations of that color. For example, drawing a line from the color white to the color red generates colors that change from white at the center, to pink near the midpoint, and to red at the outer edge of the chart. Given a color on one side of the CIE diagram and a line drawn from this color through the color white, the point at which the line intersects the outer edge of the diagram on the other side of the color white is defined as the *complementary hue* of the selected color. For example, drawing a line from 0.7 µm through the color white gives a complementary hue of 0.493 µm.

The mixture of two colors can be extended to include the mixture of three colors. The three colors that are to be mixed together represent the vertices of a triangle. Enclosed within this triangle are all of the possible colors that can be created from different combinations of the three colors. Colors outside the triangle are not realizable by any mixture of the three colors. An interesting point is that the selection of no three colors within the CIE chromaticity diagram will produce a triangle that encloses all of the colors defined by the diagram. No matter what three colors are chosen, some colors will not be included within the triangle and hence are not realizable. A combination of more than three colors is needed to produce every color defined within the CIE chart. The realization of colors is an important part of color display and printer technology. Due to the limited color phosphors that are available, only some of the colors as defined in the CIE chromaticity diagram are realized by a color CRT in a color display. The color blue is the most difficult to produce accurately. In addition, since most color printers are also based upon a three color model, this limits the range of colors that are printable.

5.2 COLOR MODELS

Because of the limited understanding of the human visual system, several different color models have been proposed to model the characteristics of a color. All of these color models are based on using at least three components to describe a color, similar to the X, Y, Z components used to describe a color given by the CIE chromaticity diagram. Many of these color models start with the red, green, and blue color components and perform a transformation into a new color coordinate system. For example, the transformation from the red, green, and blue coordinate system to the hue, saturation, and intensity color system yields the *HSI color model*. Before any color models can be given, the definition of the red, green, and blue color components must be given, and there are two basic standards defining these. The first standard is given by the National Television Standards Committee (NTSC) for use in color television and the other standard is defined by the CIE.

The CIE red, green, and blue color components are related to the NTSC color components using the following linear transformation:

$$\begin{bmatrix} R_{CIE} \\ G_{CIE} \\ B_{CIE} \end{bmatrix} = \begin{bmatrix} 1.167 & -0.146 & -0.151 \\ 0.114 & 0.753 & 0.159 \\ -0.001 & 0.059 & 1.128 \end{bmatrix} \begin{bmatrix} R_n \\ G_n \\ B_n \end{bmatrix}, \quad (5.11)$$

where R_{CIE}, G_{CIE}, and B_{CIE} are the red, green, and blue CIE color components and R_n, G_n, and B_n are the corresponding NTSC color components. The inverse transformation is found by finding the inverse of the 3×3 transformation matrix given in Equation (5.11) and is defined as

$$\begin{bmatrix} R_n \\ G_n \\ B_n \end{bmatrix} = \begin{bmatrix} 0.842 & 0.156 & 0.091 \\ -0.129 & 1.319 & -0.203 \\ 0.006 & -0.069 & 0.897 \end{bmatrix} \begin{bmatrix} R_{CIE} \\ G_{CIE} \\ B_{CIE} \end{bmatrix} . \tag{5.12}$$

Equations (5.11) and (5.12) give the spectral relationships between the two color standards. A close inspection of Equation (5.11) shows that a single nonzero NTSC color component such as $R_n = 0$, $G_n = 1$, and $B_n = 0$ produces the nonrealizable CIE color $R_{CIE} = -0.146$, $G_{CIE} = 0.753$, and $B_{CIE} = 0.059$. The CIE color components must be nonnegative values, and this requires at least two nonzero NTSC color components. The CIE chromaticity diagram in Plate 1 shows that the CIE color components of red, green, and blue are not monochromatic in color but contain a composition of several colors. The requirement of at least two nonzero NTSC color components is consistent with this observation. Unless otherwise stated, the color models that will be presented in this section are based upon the NTSC RGB color components. The subscript n will be dropped to reduce the notation complexity of the equations in the rest of this chapter.

The simplest of the color models is the *RGB color model*. This model uses the three NTSC primary colors to describe a color within a color image. Each color component represents an orthogonal axis in a three-dimensional Euclidean space as shown in Figure 5.6. The RGB color model is a normalized color space and is defined by the color components

$$r_o = \frac{R}{R_{max}} , \tag{5.13}$$

$$g_o = \frac{G}{G_{max}} , \tag{5.14}$$

and

$$b_o = \frac{B}{B_{max}} , \tag{5.15}$$

where R_{max}, G_{max}, and B_{max} are the maximum color intensities for each of the corresponding color components and r_o, g_o, and b_o are the normalized RGB color components. For a 24-bit color system, $R_{max} = 255$, $G_{max} = 255$, and $B_{max} = 255$. The color components as defined by Equations (5.13) through (5.15) have been normalized between 0 and 1, guaranteeing that the lengths of the eight edges of the cube in Figure 5.6 will always be equal to one. All possible RGB colors are contained within this cube.

Six of the eight corners of the color cube in Figure 5.6 describe the three primary colors red, green, and blue and the three secondary colors yellow, magenta, and cyan. Figure 5.6 shows that the secondary color yellow is formed by equal amounts of red and green. The two additional corners describe the color white, created from equal amounts of r_o, g_o, and b_o, and the color black, defined as all three color components equal to zero. The dotted line between white and black corresponds to all of the possible combinations of equal values of all three color components. All the different graylevels of a grayscale image reside along this line.

The RGB color model treats a color image as a set of three independent grayscale images, each of which represents one of the red, green, and blue components of a color image. Typically, 256 graylevels are used to represent a grayscale image to meet the computer storage requirement of one byte per pixel of storage. RGB color images need one byte per pixel for each of the color planes, or three times the storage of a grayscale image of the same spatial dimensions. Each pixel in a color image requires 3 bytes, or 24 bits, to represent all of the possible colors. Color images of this type are now standard and are often referred to as 24-bit or *true-color* images. A 24-bit color image can contain up to 16.7 million different colors, 256^3. Thirty bit color images have recently appeared using 10 bits per primary color. These color images can contain as many as 107.3 million colors.

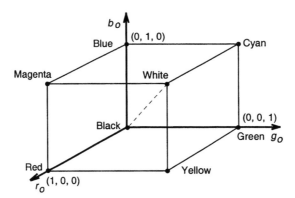

Figure 5.6 The RGB color model.

Plate 2 is an example 24-bit color image showing a cube containing several different color squares. The colors of the squares were chosen to include the three primary colors as well as the secondary color yellow located in the center bottom square. The white squares were chosen to show that combining approximately equal amounts of the three primary color components produces the color white. The three grayscale images given in Figure 5.7 corresponds to the red, green, and blue color images of Plate 2. The graylevel of each of these

COLOR PLATES

Chapter Five

Plate 1

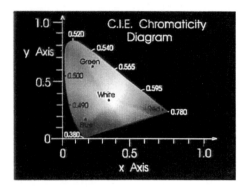

Plate 1: The CIE chromacity diagram.

Plate 2

Plate 2: A color image showing several different color squares.

Plate 3

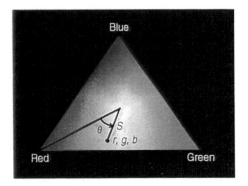

Plate 3: The HSI color space.

Plate 4

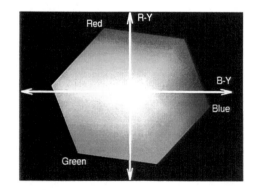

Plate 4: The C-Y color space.

Plate 5

Plate 5: The original low color
saturation image.

Plate 6

Plate 6: The enhanced image of Plate 5
with the saturation increased
by 1.8.

Plate 7

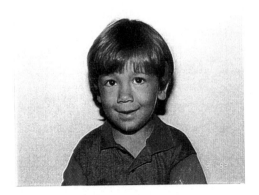

Plate 7: The original image containing
the incorrect hue.

Plate 8

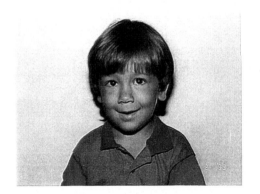

Plate 8: The enhanced image of Plate
7 with the hue rotated by 30°.

Plate 9

Plate 9: The original color image.

Plate 10

Plate 10: The selective color removal based upon the hue component.

Plate 11

Plate 11: The original Gaussian noise degraded image.

Plate 12

Plate 12: The enhanced image obtained using a 5×5 mean filter.

Plate 13

Plate 13: The original poorly white
balanced image.

Plate 14

Plate 14: The enhanced white balanced
image of Plate 13.

Plate 15

Plate 15:   The pseudocolored image of
Figure 5.13 highlighting the keys in red.

(a)

(b)

Figure 5.7 The three color component images of Plate 2: (a) the red and (b) the blue color images.

(c)

Figure 5.7 The three color component images of Plate 2: (c) green images.

images represents the amount of red, green, or blue contained in the original color image, with white indicating a higher intensity. The squares in the three grayscale images that correspond to the green square in the original color image have a medium graylevel in the green image and are black in the other two grayscale images. This is as expected, since the color green should produce a bright graylevel in the green component image and approximately a black graylevel in the red and blue color images. Consider the two white squares in the upper left-hand corner of the original color image. Comparing the two corresponding squares in the three color component images shows that the graylevel in each is about the same. This shows that approximately equal amounts of red, green, and blue are needed to generate the color white.

The yellow and orange squares produce bright white squares in the red color component image, implying that large amounts of red are required to produce the colors yellow and orange. The blue color component image shows that both of these colors contain very little blue. The green component image shows that the difference in yellow and orange is the amount of green added to red. The yellow color square has more green than the orange color square. This is also confirmed by the CIE chromaticity diagram given in Plate 1. As stated earlier, drawing a line from the color red to the color green gives all of the possible combinations of colors that can be created by mixing these two colors. The CIE diagram shows that adding less green to the color red produces the color orange, while adding more green makes the color more yellow.

The difficulty with the RGB color model is that it produces color components that do not closely follow those of the human visual system. A better set of color models to produce color components that follow the human understanding of color is that of hue, saturation, and intensity (HSI), or luminance. Of these three components, the hue is considered a key component in the human recognition process. For example, consider an image of a light blue body of water. Changing the luminance of the water would lighten or darken it, while changing its saturation would change its color from a pale blue to a deep blue. Neither of these changes in the saturation or luminance components would produce an image that seems greatly out of the ordinary. On the other hand, changing the hue of the water from blue to red produces an image of the body of water that seems unnatural and would be objectionable to most observers.

Hence, it is very important when applying electronic imaging techniques to a color image that the hue component be maintained and unaltered. The requirement that the hue component be maintained during electronic image processing of a color image is one of the major difficulties associated with using the RGB color model to generate three color images that are then processed individually as grayscale images. A linear image processing operation applied equally to each of the RGB component images does not change the percentage of red, green, or blue, maintaining the hue present in the original image. A nonlinear filtering operator can change the percentage of each of the RGB components, changing the hue of a pixel. Using the HSI color model to represent a color image allows for its processing using only its luminance and saturation components, leaving the hue component unchanged.

There are several color models that are used to represent the hue and saturation components of a color image. The first of these models, the *HSI color model*, is based upon Maxwell's triangle, derived from the RGB color cube shown in Figure 5.6. Figure 5.8 shows the Maxwell or HSI triangle as a plane that intercepts the r_o, g_o, and b_o coordinates of (1, 0, 0), (0, 0, 1), and (0, 1, 0). The HSI model collapses the three-dimensional RGB cube into a two-dimensional triangle by separating the luminance component from the chromatic components of a color. Going through the center of this triangle is a line connecting the colors white and black. Since this line gives all the possible shades of gray of a color, the center of the HSI triangle corresponds to zero saturated colors. Plate 3 shows the color version of the HSI triangle, with the low saturated color at the center. Located at the vertices are the three primary colors. The color saturation S is measured as the length of the vector from a given r, g, b color to the center of the triangle. The outer edge of the triangle defines fully 100% saturated colors. The hue θ is defined as an angle between 0 and 360° and is measured from the reference line drawn from the center of the triangle to the red vertex.

To derive the equations that give the RGB to HSI transformation and its inverse, the derivation requires that the H component be divided into three regions, which essentially divides the triangle into three regions along each of its vertices. When the blue b component is the minimum of the three, the color is located in the bottom region of the triangle and the hue is in the range of 0 to 120°. Likewise when the red r component is the minimum, the r, g, b color is located on the right side of the triangle and the hue is in the range of 120° to 240°. Finally, when the green g component is the minimum, the r, g, b color is located on the left side of the triangle and the hue is in the range of 240° to 360°. Three normalized color components,

$$r = \frac{R}{R+G+B} \ , \tag{5.16}$$

$$g = \frac{G}{R+G+B} \ , \tag{5.17}$$

and

$$b = \frac{B}{R+G+B} \tag{5.18}$$

are used to define the normalized red, green, and blue color components within the HSI color space. For equal values of R, G, and B, all three normalized color components given in the above three equations become equal to 1/3, which is the location intercept of the line connecting the colors of black and white with the HSI triangle in the RGB color space.

The intensity, saturation, and hue components in terms of the RGB color components are defined as

$$I = \frac{R+G+B}{3} \ , \tag{5.19}$$

$$S = 1 - 3 \cdot \min[r, g, b] \ , \tag{5.20}$$

and

$$\theta = \cos^{-1}\left[\frac{\frac{2}{3}\left(r - \frac{1}{3}\right) - \frac{1}{3}\left(b - \frac{1}{3}\right) - \frac{1}{3}\left(g - \frac{1}{3}\right)}{\sqrt{\left(\frac{2}{3}\right)\left[\left(r - \frac{1}{3}\right)^2 + \left(b - \frac{1}{3}\right)^2 + \left(g - \frac{1}{3}\right)^2\right]}}\right].$$ (5.21)

Whenever $b > g$, the hue θ will be greater than 180°. For this case, since the inverse cosine is defined only over the range of 0 to 180°, θ is replaced by $360° - \theta$. Three grayscale images that represent the hue, saturation, and intensity components can be generated by scanning the RGB color image pixel by pixel and then using the values computed from Equations (5.19) through (5.21) as the graylevels for each of these grayscale images. Since most grayscale images are represented by 256 graylevels requiring one byte per pixel, the hue and saturation components are typically scaled between 0 and 255. If the original color image is a 24-bit color image, the intensity component will already be in the range of 0 to 255.

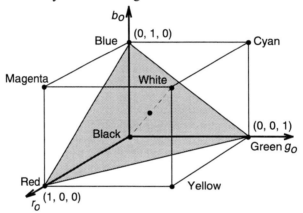

Figure 5.8 The HSI triangle shown in the RGB color cube.

The inverse transformation from HSI color space to the RGB color space requires the use of three separate equations, depending on which one of the three regions the color is located, within the HSI triangle.

For $0° \leq \theta < 120°$

$$b = \frac{1}{3}(1 - S) \ ,$$ (5.22)

$$r = \frac{1}{3}\left[1 + \frac{S \cos \theta}{\cos(60° - \theta)}\right],$$ (5.23)

$$g = 1 - r - b \ . \tag{5.24}$$

For $120° \le \theta < 240°$

$$r = \frac{1}{3}(1 - S) \ , \tag{5.25}$$

$$g = \frac{1}{3}\left[1 + \frac{S\cos(\theta - 120°)}{\cos(180° - \theta)}\right] \ , \tag{5.26}$$

$$b = 1 - r - g \ . \tag{5.27}$$

For $240° \le \theta < 360°$

$$g = \frac{1}{3}(1 - S) \ , \tag{5.28}$$

$$b = \frac{1}{3}\left[1 + \frac{S\cos(\theta - 240°)}{\cos(300° - \theta)}\right] \ , \tag{5.29}$$

$$r = 1 - b - g \ . \tag{5.30}$$

The final unnormalized RGB color components are obtained by multiplying the normalized r, g, and b components by $3 \cdot I$.

Electronic image processing using the HSI color space usually entails the manipulation of the intensity and saturation components while leaving the hue component unchanged. Consider an image of a red square located against a blue background. This image can be divided into two different regions, containing the red square and the blue background. Applying a mean filter to the luminance component blurs the sharp edges associated with the red square but leaves the chromatic part unchanged. Applying a mean filter to the saturation component produces a gradual change in saturation between the two color regions. Since the human visual system is very insensitive to changes in color saturation, these changes are usually unnoticed in the processed color image.

The difficulty of processing the hue component can easily be explained with the mean filtering of it. In the hue space, the red and blue color regions generate two different hue regions with four edges located at the discontinuity between the two regions. Application of a mean filter changes the sharp discontinuities to gradual edges that linearly change from one hue value (blue) to the other hue value (red). This results in a rainbow of colors at the edges of the red square. The rainbow of colors that are displayed are the colors that the hue transitions through as it changes gradually from blue to red, which is easily observable and objectionable in the processed color image. Except for changing the overall hue of an image, similar to the tint control on a color television, the hue color component is normally ignored in the processing of color images. There has been some recent work in using the hue space to detect edges. For a further discussion on using the hue space to process color images, the interested reader is referred to the *Fundamentals of Electronic Image Processing* given in the bibliography.

Care must be taken when using any color model that transforms the RGB color components into intensity and chromatic components. Modification of a color's hue, saturation, and intensity can produce nonrealizable colors when transforming back from the HSI space to the RGB space. The RGB to HSI transformation implies that the hue and saturation components are independent of the intensity. This is true only if there are no limits on the maximum allowable R, G, B values. Dealing with digitized color images, there will always be a minimum and maximum limit on the allowable values for the R, G, B components. The RGB cube given in Figure 5.6 contains all the possible combinations of allowable colors for a given set of R_{max}, G_{max}, and B_{max}. A better way to look at the HSI color model is that for a given intensity there is a corresponding HSI triangle. As the intensity I of a color changes, the size and shape of the HSI triangle must change to guarantee that all colors as defined by the HSI color model are still contained within the RGB color cube.

For low intensity values the HSI triangle shape remains the same, but its size decreases as the intensity decreases. At the point of black, the size of the HSI triangle reduces to a single point. This makes sense in that the color black, which corresponds to all three RGB components being equal to zero, must also have a saturation value of zero. The size of the triangle is determined by the maximum saturation, with the outer edge of the triangle defining the maximum saturated colors allowed for a given intensity. Figure 5.9 shows the HSI color triangle for several different intensities. As the intensity increases, the size of the HSI triangle increases until the vertices of the triangle meet the outer edge of the RGB cube. At this point, the vertices of the triangle become clipped. For a 24-bit color image this corresponds to intensity values greater than 85. Eventually the triangle's shape becomes inverted and reduces to a single point at the coordinate R_{max}, G_{max}, B_{max} (white).

The HSI color model has several limitations. The first is that it gives equal weighting to the RGB components when computing the intensity, or luminance, of an image. As shown previously in this chapter, the sensitivity of the eye varies for each of the RGB color components. A better mapping for the luminance or intensity is

$$Y = 0.299 \cdot R + 0.587 \cdot G + 0.114 \cdot B \ . \tag{5.31}$$

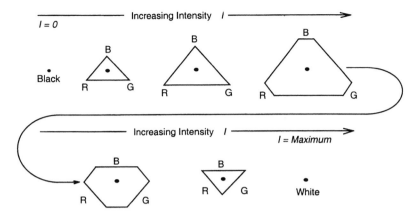

Figure 5.9 The shape of the HSI triangle as a function of intensity I.

Equation (5.31) gives a better correspondence with the perceived brightness of a color than using equal weightings of the colors of red, green, and blue. The second difficulty with the HSI color model is that the length of the maximum saturation vector varies depending on the hue of the color. As shown in Plate 3, the length of the saturation vector is longest at the vertices and is shortest halfway between the vertices, compressing the same saturation range into a smaller geometrical length. Since the saturation component many times is digitized to a finite dynamic range either for memory size restrictions or for real-time hardware processing of color images, there are fewer values of saturation available for hues that are halfway between the vertices of the HSI triangle as compared to colors with hues located near the vertices of the triangle. The third difficulty with the HSI color model is the complexity of the equations, in that different equations must be used depending on the location of a color within the HSI triangle.

The *C-Y color model* as proposed by the NTSC for use in color television solves these three limitations of the HSI model. The C-Y color model has three color components *B-Y*, *R-Y*, and *G-Y* and one luminance component *Y*. Only two of three color components are needed to define a color. The third component can be derived in terms of the other two. Typically, the *R-Y* and the *B-Y* color components are used and the *G-Y* component is derived from them. The luminance component *Y* in terms of the RGB components is given by Equation (5.31). The C-Y color components can be computed in two different ways. The first method is to subtract the luminance *Y* from each of the *R*, *G*, and *B* components to produce the *R-Y*, *G-Y*, and *B-Y* components. The other method is to use the following 3×3 transformation

$$\begin{bmatrix} Y \\ R\text{-}Y \\ B\text{-}Y \end{bmatrix} = \begin{bmatrix} 0.299 & 0.587 & 0.114 \\ 0.701 & -0.587 & -0.114 \\ -0.299 & -0.587 & 0.886 \end{bmatrix} \begin{bmatrix} R \\ G \\ B \end{bmatrix}. \qquad (5.32)$$

The *G-Y* color component is then given in terms of the *R-Y* and *B-Y* components as

$$G\text{-}Y = -0.509 \cdot R\text{-}Y - 0.194 \cdot B\text{-}Y .$$ (5.33)

The *B-Y* and *R-Y* components form two orthogonal components that represent the chromaticity of a color. Plate 4 shows the C-Y color space plotted with the *B-Y* component along the horizontal axis and the *R-Y* component along the vertical axis. At the center of this space are the zero saturated colors. Radially from the center, the saturation increases until it reaches its maximum at the outer edges of the color space, shown as a hexagon shape. The angle relative to the *B-Y* axis gives the hue of the color. The hue θ and saturation *S* are derived from the *B-Y* and *R-Y* components as

$$S = \sqrt{B\text{-}Y^2 + R\text{-}Y^2}$$ (5.34)

and

$$\theta = \begin{cases} \tan^{-1}\left[\dfrac{R\text{-}Y}{B\text{-}Y}\right] & \text{for } S \neq 0 \\ \text{undefined} & \text{for } S = 0 \end{cases} .$$ (5.35)

Equation (5.35) shows that the hue is undefined for zero saturation. Colors with zero saturation contain no color information and are grayscales ranging from black to white. Also shown in Plate 4 are the three primary colors red, green, and blue each separated by 120°. A close observation of Plate 4 shows that the C-Y color space is a two-dimensional perspective image of the RGB color cube looking down the axis of the line from white to black in Figure 5.6. Even though the C-Y color space is not perfectly circular, this space is better at providing a more uniform maximum saturation than the HSI color model.

The inverse transformation to go from the C-Y color components to the RGB components again can be computed in two ways. The first is to add the *Y* component to the *B-Y* and *R-Y* components to yield the *B* and *R* components. The *Y* component is then added to Equation (5.33) to yield the *G* component. The other method is to use

$$\begin{bmatrix} R \\ G \\ B \end{bmatrix} = \begin{bmatrix} 1.0 & 1.0 & 0.0 \\ 1.0 & -0.509 & -0.194 \\ 1.0 & 0.0 & 1.0 \end{bmatrix} \begin{bmatrix} Y \\ R\text{-}Y \\ B\text{-}Y \end{bmatrix} .$$ (5.36)

The processing of color images using the C-Y color space is basically the same as was explained for the HSI color space. For every pixel in the color image, the *Y*, *B-Y*, and *R-Y* color components are computed. Next, Equations (5.34) and (5.35) are used to find the saturation and hue. The hue, saturation, and luminance components can then be processed to enhance the color image. The inverse transformation of C-Y to RGB yields the enhanced RGB color image. For example, multiplying the saturation component by a gain term for every pixel in the color image changes the saturation level of the processed color image. This is the process that is used to adjust the saturation level in a color television when the color level control is changed. The luminance component given in Equation (5.31) can also be used to convert a color image to a grayscale image. For each pixel in the color image, only the luminance component is computed to create the grayscale image of the original color image.

As with the HSI space, the shape and size of the maximum allowable saturation changes as a function of luminance, resulting in a change of the size and shape of the C-Y space. Figure 5.10 shows the allowable values for the *R-Y* and *B-Y* components of a 24-bit color image for several different luminance levels. The merging of each of these planes into one plane to form the total C-Y space as shown in Plate 4 defines the hexagon shape for the allowable colors. As with the HSI model, the area of the allowable colors decreases for very low and high luminance values. At the extreme luminance levels of white (255, 255, 255) and black (0, 0, 0), the allowable area for the *B-Y* and *R-Y* components reduces to a single point where the maximum possible saturation must be zero. The size and shape limitations of the C-Y color space are not due to the definition of the color model itself but to limiting the C-Y color space to within the finite volume of the RGB cube. As stated earlier, this is a practical limitation that has been imposed by using only a finite number of values to represent the original RGB color components.

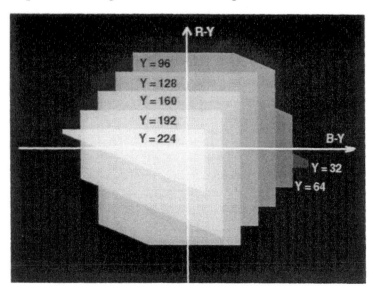

Figure 5.10 The shape of the C-Y space as a function of luminance
(From Weeks et al., 1995).

For the color cube given in Plate 2, a set of 256 graylevel images corresponding to the luminance, saturation, and hue images can be generated using Equations (5.31), (5.34), and (5.35). As can be seen from Figure 5.11(a), the luminance image is simply the grayscale image of the original color image. The graylevels in the saturation image given in Figure 5.11(b) correspond to the color saturation level of a color. The black graylevels represent low saturated colors, while the high saturated colors are given in white. This saturation image was scaled so that the maximum saturation within the color image is scaled to graylevel 255. Inspection of the saturation image shows that seven out of the nine squares contain large saturation values. A comparison of Figure 5.11(b) with Plate 2 shows that the zero saturated squares located in the upper left-hand corner correspond to the white color squares in the color image. This is as expected since the color white has zero saturation.

The hue component as defined by Equation (5.35) ranges from 0 to 360°. So that the hue component could be displayed as a grayscale image, Equation (5.35) was also scaled between 0 and 255. Figure 5.11(c) gives the hue image, showing the angle of the hue for each of the seven color squares. Since the hue is undefined for pixels containing zero saturation, the hue of all pixels containing zero saturation was set to zero. Any other value could have been chosen, but setting the undefined hue to zero also sets the white background surrounding the nine color squares to zero, emphasizing the hues of the seven color squares. A comparison of Plates 2 and 4 with the hue image shows that the degree values do match the colors represented. For example, the hue value of 113° in the lower left-hand square is the angle this color makes with the *B-Y* axis in Plate 4. This places the square's color slightly to the left of the *R-Y* axis in the second quadrant. At this location, Plate 4 gives the color red, which matches the color of the square shown in the original color image in Plate 2. The hue of the lower right-hand corner square is 135°. This places the color at the center of quadrant 2, 135° from the *B-Y* axis. Inspection of Plate 4 yields the color orange, which is in agreement with the color of the square in the original color image.

One of the major advantages of the C-Y color space is that its transformations are much simpler to compute than the HSI color model. Secondly, since the C-Y color space is the standard color model as proposed by the NTSC for use in color television, this makes the real-time hardware implementation of any of the color algorithms developed in the C-Y color space feasible due to the enormous amount of color video hardware that is available.

Another color model that is closely related to the C-Y model and is also used in color television is the *YIQ color model*. The three components for this color model are the luminance as defined by Equation (5.31), the in-phase *I*, and quadrature-phase *Q* component. The *I* and *Q* components define the

(a)

(b)

Figure 5.11 The hue, saturation, and luminance images of Plate 2:
(a) the luminance image and (b) the saturation image.

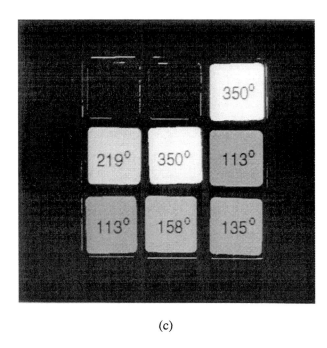

(c)

Figure 5.11 The hue, saturation, and luminance images of Plate 2:
(c) the hue image.

chromaticity of the color and form two orthogonal components.   The
transformation from the RGB color space to the YIQ color space is:

$$\begin{bmatrix} Y \\ I \\ Q \end{bmatrix} = \begin{bmatrix} 0.299 & 0.587 & 0.114 \\ 0.596 & -0.275 & -0.321 \\ 0.212 & -0.523 & 0.311 \end{bmatrix} \begin{bmatrix} R \\ G \\ B \end{bmatrix} . \tag{5.37}$$

Once in the YIQ space, the hue and saturation can be computed in a similar
manner to the C-Y color space:

$$S = \sqrt{I^2 + Q^2} \tag{5.38}$$

and

$$\theta = \begin{cases} \tan^{-1}\left[\dfrac{Q}{I}\right] & \text{for } S \neq 0 \\ \text{undefined} & \text{for } S = 0 \end{cases} , \tag{5.39}$$

where the hue is measured relative to the I axis and varies between 0 and 360°
and is undefined if the saturation is equal to zero implying a grayscale color.

Like the C-Y color space, the zero saturated colors are located at the origin, while the highly saturated colors are located at the outer edges. The YIQ space is the same as the C-Y space except the Q component is compressed in comparison to the B-Y axis and this space is rotated by 33°. The I axis leads the B-Y axis by 33° and the same is true for the Q axis relative to the R-Y axis. The processing of a color image using the YIQ color space is identical to the process used for the C-Y color space, except Equation (5.37) is used to convert from the RGB color space to the YIQ color space. Equations (5.38) and (5.39) are then used to compute the saturation and hue components and the inverse transformation to go from the YIQ components to the RGB components is used:

$$\begin{bmatrix} R \\ G \\ B \end{bmatrix} = \begin{bmatrix} 1.0 & 0.956 & 0.62 \\ 1.0 & -0.272 & -0.647 \\ 1.0 & -1.108 & 1.705 \end{bmatrix} \begin{bmatrix} Y \\ I \\ Q \end{bmatrix} . \tag{5.40}$$

Two color spaces used for printing purposes are the subtractive color spaces *CMY* and *CMYK*. As with the HSI color space, both the CMY and the CMYK color spaces use the normalized RGB color components given in Equations (5.16) through (5.18). The subtractive color space CMY produces the primary colors cyan, magenta, and yellow. The RGB to CMY transformation is defined as

$$\begin{bmatrix} C \\ M \\ Y \end{bmatrix} = \begin{bmatrix} 1 \\ 1 \\ 1 \end{bmatrix} - \begin{bmatrix} r \\ g \\ b \end{bmatrix} . \tag{5.41}$$

The inverse transformation is found by subtracting the C, M, Y components from one. As discussed earlier in this chapter, many printing processes use a fourth color to improve the quality of the black colors printed. The CMYK color space is the model used for the four color printing process and uses the four color components C_k, M_k, Y_k, and K, with the fourth component K representing the additional color black. The RGB to CMYK transformation is first computed using Equation (5.41) to generate the CMY color components. The rest of the transformation into the CMYK space is accomplished using

$$K = \min (C, M, Y) \tag{5.42}$$

and

$$\begin{bmatrix} C_k \\ M_k \\ Y_k \end{bmatrix} = \begin{bmatrix} C \\ M \\ Y \end{bmatrix} - \begin{bmatrix} K \\ K \\ K \end{bmatrix} . \tag{5.43}$$

The final two color models that will be presented in this section are those that have been derived by the CIE to correct for the deficiency that exists in the CIE XYZ coordinate system. They have been summarized here for reference and the interested reader is referred to the text *Fundamentals of Electronic Image Processing* listed in the bibliography for further discussion. Distance changes in either the x or y directions of the CIE chromaticity diagram of Plate 1 do not correspond to the perceived differences in color. In the green region of the CIE chart, larger steps have to be made to obtain the same perceived change in color as compared to the red or blue regions of the chart. Before these other color models are given, the equation that relates the CIE XYZ color space to the NTSC RGB color space will be given. By definition, the transformation from the RGB color space to CIE XYZ color space is

$$\begin{bmatrix} X \\ Y \\ Z \end{bmatrix} = \begin{bmatrix} 0.607 & 0.174 & 0.201 \\ 0.299 & 0.587 & 0.114 \\ 0.000 & 0.066 & 1.117 \end{bmatrix} \begin{bmatrix} R \\ G \\ B \end{bmatrix}, \tag{5.44}$$

and its inverse is

$$\begin{bmatrix} R \\ G \\ B \end{bmatrix} = \begin{bmatrix} 1.91 & -0.534 & -0.289 \\ -0.984 & 1.998 & -0.027 \\ 0.058 & -0.118 & 0.897 \end{bmatrix} \begin{bmatrix} X \\ Y \\ Z \end{bmatrix}. \tag{5.45}$$

The first of the color models that was developed to correct for the perceived color limitation of the original 1931 color space was the *UVW* system. The transformation from the XYZ coordinate system that derives the three components U, V, and W is

$$\begin{bmatrix} U \\ V \\ W \end{bmatrix} = \begin{bmatrix} 0.666 & 0.00 & 0.000 \\ 0.000 & 1.000 & 0.000 \\ -0.500 & 1.500 & 0.500 \end{bmatrix} \begin{bmatrix} X \\ Y \\ Z \end{bmatrix}. \tag{5.46}$$

Its inverse transformation is defined as

$$\begin{bmatrix} X \\ Y \\ Z \end{bmatrix} = \begin{bmatrix} 1.500 & 0.000 & 0.000 \\ 0.000 & 1.000 & 0.000 \\ 1.500 & -3.000 & 2.000 \end{bmatrix} \begin{bmatrix} U \\ V \\ W \end{bmatrix}. \tag{5.47}$$

The three trichromatic coefficients u, v, and w are defined as

$$u = \frac{U}{U + V + W}, \tag{5.48}$$

$$v = \frac{V}{U + V + W} \, , \tag{5.49}$$

and

$$w = \frac{W}{U + V + W} \, . \tag{5.50}$$

Equations (5.48) through (5.50) form the uniform chromaticity scale *USC* system.

A further improvement on the UVW or USC color space is the CIE uniform perceptual color space $U^*V^*W^*$. This color space improves on the UVW color space by shifting the white reference point of the original CIE chart to a better perceived white, as defined by the 1960 CIE standard. The transformation from the XYZ color space to $U^*V^*W^*$ color space is

$$W^* = 25 \cdot (100 \cdot Y)^{1/3} - 17 \qquad \text{for } Y \geq 0.01 \, , \tag{5.51}$$

$$U^* = 13 \cdot W^* \cdot (u - u_o) \, , \tag{5.52}$$

and

$$V^* = 13 \cdot W^* \cdot (v - v_o) \, , \tag{5.53}$$

where u and v are defined by Equation (5.48) and (5.49) and u_o and v_o are the coordinates of the 1960 CIE chromaticity reference white point.

The final color model of interest is the $L^*u^*v^*$. In the $L^*u^*v^*$ color model, the u^* and v^* components completely describe the chromatic part of a color. The $L^*u^*v^*$ components are defined by

$$L^* = \begin{cases} 903.3 \cdot (Y / Y_o) & \text{for } 0 \leq Y < 0.01 \\ 25 \cdot (100 \cdot Y / Y_o)^{1/3} - 16 & \text{for } Y \geq 0.01 \end{cases} \, , \tag{5.54}$$

$$u^* = 13 \cdot L^* \cdot (u - u_o) \, , \tag{5.55}$$

and

$$v^* = 13 \cdot L^* \cdot (v - v_o) \, , \tag{5.56}$$

where u_o, v_o are the coordinates of the reference white point. The advantage of the $L^*u^*v^*$ color system is that it provides a means of modeling the characteristics of different color displays, hard copy, and acquisition devices. Typically, each of these devices produces or uses a different reference white point. The $L^*u^*v^*$ color system allows the selection of the reference white point.

Many of the color models mentioned in this section have a finite limited color space that varies with luminance or intensity. Typically the processing of color images involves transforming from the RGB color space to another color space. Since the RGB color space will be of finite volume, the allowable colors in the new space must be contained within the RGB color cube. The processing of color images in this new space could produce unrealizable colors when transforming back into the RGB color space that was used to represent the original color image. For example, consider the use of the HSI color model, in which the color red has been modified from $L = 85$, $S = 1.0$, and $\theta = 0°$ to $L = 255$, $S = 1.0$, and $\theta = 0°$ by enhancing just the luminance component of the color. The inverse transformation from HSI to RGB to obtain the unnormalized RGB components yields $R = 765$, $G = 0$, and $B = 0$. For a 24-bit color image, in which the range of values for the RGB components is 0 to 255, this color would not be realizable. In fact, for an intensity over 85 in a 24-bit color image, the saturation value cannot be equal to 1.

Unrealizable colors must be brought back into the allowable color space. As previously mentioned, the eye is most sensitive to hue. The approach that is typically taken is to transform these unrealizable colors into new realizable colors in which their hues are left unchanged. This leaves changing either a color's luminance or saturation. Changing the luminance of a color is accomplished first by transforming the color back to the RGB color space. Next, if any one of the three color components exceeds the allowable range, then the value of each color component is decreased equally. For example, for a 24-bit color image each RGB component is multiplied by

$$\frac{255}{\max(R,G,B)} \cdot \qquad (5.57)$$

Equation (5.57) essentially decreases the luminance of the color, while leaving the saturation and hue unchanged. The second approach, in which the saturation is changed, requires knowing the maximum allowable saturation for a given hue and luminance. Once the maximum saturation is known, the unrealizable color's saturation is scaled back to the maximum saturation. Of the two methods, scaling the saturation component is the most difficult in that it requires knowledge of the maximum saturation of a color. If the color space being used does not directly calculate a color's saturation, this adds additional computations to the processing of the color image. The author recommends the use of Equation (5.57) whenever possible.

The color models presented in this section are only a summary of the more popular color models that exist. There are many more color models that are available that have not been mentioned here. The interested reader is referred to the *Fundamentals of Electronic Image Processing* by Weeks for discussion of other existing color models.

5.3 EXAMPLES OF COLOR IMAGE PROCESSING

In this section, several techniques for the processing of 24-bit color images will be presented. Because of the author's familiarity with the NTSC color television standard, the RGB and C-Y color spaces have been chosen in the processing of these 24-bit color images. Any of the other color spaces could have been used to demonstrate the techniques in this section. Color images were acquired either directly from a video camera into a computer using a 24-bit 640 × 480 image acquisition card or via photographic prints that were scanned using a 24-bit 300 × 300 dots per inch color desktop scanner. The color space chosen depended on the type of processing that was desired. If the RGB color model was used, each of the three color components was treated as a separate grayscale image and processed individually. For cases when the hue, saturation, and luminance components were required, the RGB color space was converted to the C-Y color space using Equation (5.32) and the saturation and hue components were then computed using Equations (5.34) and (5.35). After processing of the hue and saturation components, Equation (5.36) was used to transform from the C-Y back to the RGB color space. Finally, the RGB values computed from the inverse transformation were verified as valid colors. If one of the RGB triplet values exceeded the maximum allowable value of 255, Equation (5.57) was used to scale the luminance of the color back into the allowable range of color.

The first two color image processing examples show how the hue and saturation of a color image can be used to enhance the overall appearance of an image. Plate 5 shows an image of a field of tulips taken in central Holland. Except for the row of red tulips in the center of the image, this image is lacking in color saturation overall. The goal of this example is to enhance this image by adding more color saturation, similar to adjusting the color level control on a color television. For every pixel within this image, its luminance, hue, and saturation were computed so that its saturation value could be increased by a factor of 1.8. The enhanced saturation component along with the original hue and luminance components were then used to compute the RGB values of the enhanced image, as shown in Plate 6. A comparison of this image with the original shows the increase in the overall color saturation. In particular, the green colors of the tulip field in the lower part of the image are much more vivid as well as the buildings present in the image.

The enhancement of a color image by increasing the color saturation is one of the techniques that will produce colors outside the allowable range. As the luminance of a color approaches 0 or 255, the maximum saturation approaches zero, as shown in Figure 5.10. Multiplying every pixel's saturation value by a gain coefficient that is independent of luminance can produce hue, saturation, and luminance values that are outside the RGB range of colors when converted back to the RGB space. Equation (5.57) was used to limit the luminance of the color to prevent the generation of any unrealizable colors. This is why there is very little apparent increase in the saturation level of the blue sky in the upper part of Plate 6 as compared to the increase in the saturation of the tulip field. The blue colors of the sky were already at the maximum allowable saturation because of their high luminance values. If Equation (5.57) was not included in the saturation enhancement algorithm, eventually, as the saturation level of the enhanced image was increased, the color of the sky became distorted (red and green colors start to appear) due to the colors being pushed outside the allowable range of RGB colors.

In the observation of an image, the human visual system is most sensitive in detecting the presence of incorrect flesh tones. Slight changes in the hue can produce objectionable reddish or greenish flesh tone colors. Plate 7 shows an image of a boy with a reddish flesh tone. This image can be easily corrected by transforming the RGB components to their hue, saturation, and luminance components. The same process as explained for the saturation enhancement of a color image is used, except a constant value is either subtracted or added to the hue of each pixel. This rotates all the colors about the origin of the C-Y color space, changing the hue of each color within the image, which results in changing the hue of the boy's skin. The C-Y color space given in Plate 4 can be used to determine the angle in which the hue of the image in Plate 7 must be rotated to correct for the reddish flesh tone. Normally, flesh tones would produce colors around 135° relative to the *B-Y* axis. Since the flesh tone of the boy's skin in Plate 7 is on the reddish side, the hue angle of these colors is probably somewhere between 100° and 110°. Adding approximately 30° to the hue of each pixel will shift the overall hue of Plate 7 toward green. Plate 8 shows this result. Notice how correcting the hue of this image has removed the reddish flesh tone, giving the image a more natural appearance.

The two previous examples used the same hue and saturation correction throughout the entire image. The concept behind the next color image processing example is to process only pixels with luminance, saturation, or hue values within a predetermined range. Using this method, one can process only certain colors within a color image, leaving the other colors unchanged. This provides a means of enhancing objects within an image based upon their color. This type of color processing is known as *selective color processing*. Plate 9 shows an image of a farm plow with a blue frame and a red wheel against a stone wall. It is desired to highlight the blue frame of the plow by setting the entire image to a grayscale image except for the blue frame of the plow.

The C-Y color space given in Plate 4 shows that the blue frame of the plow will be in the blue region of the C-Y space and have hue values in the range of 260° to 315°. To convert a color image into a grayscale image is easily accomplished by converting the RGB color components to the C-Y to obtain the saturation and hue components as given by Equations (5.34) and (5.35) and then setting the saturation to zero for all pixels within the image. The goal of this example is to set the saturation of each pixel within the image to 0 except those corresponding to the blue frame of the plow:

$$S_{new} = \begin{cases} S_{old} & \text{for } \theta_1 \leq \theta \leq \theta_2 \\ 0 & \text{otherwise} \end{cases}, \qquad (5.58)$$

where θ_1 and θ_2 specify the range of hues that define the blue color of the plow, S_{new} is the new saturation value, and S_{old} is the saturation value of the original color. After several iterations, the best values for θ_1 and θ_2 to set the image to black and white, except for the frame of the plow, were found to be $\theta_1 = 290°$ and $\theta_2 = 315°$. Plate 10 shows that the entire image is now black and white, except for the blue frame of the plow.

Decreasing the range of hues smaller than the range given begins to set part of the frame to black and white. Determining the range of luminance, saturation, or hue values to implement selective color image processing is easily accomplished when the desired colors are well separated from the other colors within the image. For example, if it was desired to highlight the red wheel instead of the blue frame, the range of hue for Equation (5.58) would be difficult to obtain. A hue range set too small would set the color of the pixels defining the red wheel to black and white, while a hue range too large, would leave the parts of the wall behind the plow in color. This would not accomplish the desired goal of setting the color image to a grayscale image, except for the red wheel of the plow.

The next example shows how to perform linear filtering on color images using the RGB color space. The use of linear spatial filtering methods enables the RGB color image to be treated as three separate grayscale images, representing the red, green, and blue components. The same linear filter is applied to each of the grayscale images, producing three new linearly filtered images that are then recombined to form the new filtered color image. This approach of separating a color image into three separate color images is valid only for linear operations. Linear filters modify the RGB components equally, which maintains the hue. Nonlinear filters could change the percentage of RGB values, which effectively changes the hue of the RGB triplet.

Plate 11 shows a 24-bit color image of a daffodil flower that has been corrupted by Gaussian noise. Not only has the Gaussian noise added a grainy appearance to the image, it has changed the hue of the colors. This is particularly noticeable in the different colors of red and green that appear in the yellow petals of the daffodil. The hue noise that is present in this image shows that the Gaussian noise is uncorrelated between the three RGB components. Additive noise that has a high degree of correlation between each of the RGB components can add only luminance noise, since the noise changes the value of the RGB components equally. Plate 12 shows the 5×5 mean filtered image of Plate 11. Notice how the noise present in the image has been reduced, but at the expense of blurring the image. The same characteristics of the mean filter for grayscale images apply for the filtered image of Plate 12. The improvement of the noise in Plate 12 is not as great as it appears. The use of the mean filter has reduced the noise present in each of the three RGB components. In effect, this has also reduced the hue noise that is present in the image. Since the eye is most sensitive to changes in hue, reducing the noise present within the image reduces the changes in hue, resulting in a large perceived improvement in the filtered color image.

Because the eye is very sensitive to the slightest change in hue, the incorrect tint of low saturated or pastel colors is easily detected. Colors in an image that should be perceived as white are easily detected by the eye and are typically found to be objectionable if these colors have a slight colored tint. An image in which the white colors of the image have obtained a slight colored tint is known as an improper *white balanced image*. An incorrect white balanced image can be attributed to several different factors. The most common is that the images were acquired under improper lighting conditions. Different light sources emit light energy with different spectral responses. For example, an incandescent light appears slightly yellow when compared to a fluorescent light. When the same image is acquired using different light sources, dramatically different results can occur. Color photographs taken with standard indoor/outdoor film under fluorescent lighting conditions appear to have an overall greenish tint, while images acquired using sunlight appear natural. Chromatic filters exist that compensate for different lighting conditions, but in many instances these white balance errors can be removed during the printing process by a laboratory technician manually adjusting the color filters on the printer until the best color photograph is obtained.

An additional problem with photographs is that the overall white balance changes as the photograph ages. Over time, the emulsion of the photographic paper changes color and alters the white balance of the image from a neutral white to a reddish or yellow colored image. Most of these old photographs can be corrected by taking new photographs of them and modifying the white balance. Correct white balance of an image is also important for color video cameras, which must be able to reproduce the colors present in a scene correctly under a variety of lighting conditions. Hence, white balance correction must be performed to compensate for the different lighting conditions to reduce the white balance errors in the images. Typically, many color video cameras offer manual white balance correction via an external user control. This control is varied until the image appears natural for the given lighting condition.

Manual white balance correction typically consists of transforming an RGB color image into a color space, which generates two orthogonal chromatic components and a luminance component. Typically, for NTSC color video either the YIQ or the C-Y color spaces are used. This is easily accomplished using either the *I* and *Q* components in the YIQ color space or the *R-Y* and *B-Y* components in the C-Y color space. Manual white balance correction of a color image ignores the luminance component of the image, while translating the origin of the chromatic space, comprising either the *I* and *Q* or the *B-Y* and *R-Y* components. The translated chromatic components and the unmodified luminance component are then transformed back to the original RGB color space. The resulting RGB image is the white balance adjusted color image. Translation of the chromatic components continues interactively until the white balance corrected image appears natural to an observer.

Plate 13 is an image of a family portrait taken sometime during the early 1960s that has discolored with age. The reddish tint is very evident in the sky and the rocks on the left side of the photograph. This image was transformed from the RGB color space to the C-Y color space using Equation (5.32). Next, the *B-Y* and *R-Y* locations of each pixel were translated by a fixed amount. The unmodified luminance *Y* and the shifted *B-Y* and *R-Y* components were then used in Equation (5.36) to produce the RGB components of the white balanced modified image. After several iterations, the best white balanced image was obtained by shifting the *B-Y* components by -29.5 and the *R-Y* components by 40.5. Plate 14 shows the corrected white balanced image. Note the overall improvement in the colors of this image. The whites of the children's clothes appear to be white and the sky now appears to be blue. In comparing Plate 14 to Plate 13, the overall reddish color of the original image has been reduced. A more advanced method is to divide the image into two or three luminance regions. Next, the *R-Y* and *B-Y* chromatic spaces are computed for each luminance region. White balance correction now involves the translation of each of these *R-Y* and *B-Y* color spaces. For three luminance regions this involves the adjustment of six coefficients. Thus, both the *B-Y* and *R-Y* components are translated for each of the three luminance regions.

5.4 PSEUDOCOLORING AND COLOR DISPLAYS

An important use of color is the enhancement of grayscale images to highlight key features in color. Since the eye can observe color more readily than differences in graylevels, color can be used to emphasize a selected range of graylevels. The process of adding color to a grayscale image is called *pseudocolor* or *falsecolor*. The process of pseudocoloring should not be confused with the process of coloring a grayscale image. The process of coloring a grayscale image is to convert the various objects within it to colors that present what the image should look like in color. For example, the process of colorizing a grayscale image of a red automobile would require that the color image of the automobile contain the correct color red. The process of pseudocoloring, on the other hand, is to highlight a grayscale image based upon the graylevels present in the image, without any regard to the colors of the original objects.

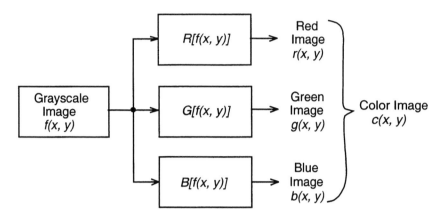

Figure 5.12 A block diagram showing the process of pseudocoloring.

Pseudocoloring of a grayscale image is typically performed using the RGB color model. Figure 5.12 gives a block diagram describing the process of pseudocoloring a grayscale image. The input image $f(x, y)$ is assumed to be an N graylevel image that is mapped into a red $r(x, y)$, a green $g(x, y)$, and a blue $b(x, y)$ image that represents the RGB color components of a color image. Usually, the grayscale image contains 256 graylevels that are mapped to a 24-bit color image using 256 intensity values for each one of the RGB component images. The type of function chosen for $R[f(x, y)]$, $G[f(x, y)]$, and $B[f(x, y)]$ determines the type of pseudocoloring enhancement possible. For example, if

$$c(x,y) = \begin{cases} R[f(x, y)] = f(x, y) \\ G[f(x, y)] = f(x, y) \\ B[f(x, y)] = f(x, y) \end{cases}, \tag{5.59}$$

the output image is simply the original grayscale image.

Individual graylevels in a grayscale image can be highlighted in color by selectively modifying the mapping function to emphasize individual graylevels. For example, consider the mapping of a particular graylevel T in a grayscale image to the color blue. The mapping function for this pseudocoloring operation is

$$c(x,y) = \begin{cases} R[f(x, y)] = 0;\ G[f(x, y)] = 0;\ B[f(x, y)] = b_{max} \text{ for } f(x, y) = T \\ R[f(x, y)] = G[f(x, y)] = B[f(x, y)] = f(x, y) \qquad \text{otherwise} \end{cases}, \tag{5.60}$$

where b_{max} is the maximum graylevel allowed for the blue color image. The mapping function in Equation (5.60) sets the red $r(x, y)$ and green $g(x, y)$ component images to zero when the input grayscale image $f(x, y)$ is at graylevel T, but sets the blue $b(x, y)$ component image to its maximum allowable graylevel. Otherwise this mapping function produces the original grayscale image. For $f(x, y) = T$, the output image displays a bright blue color.

Figure 5.13 is a 256 graylevel image of a Hewlett Packard model 35 calculator. The symbols on the keys in the leftmost column are barely visible in the grayscale image. The goal of pseudocoloring here was to form a 24-bit RGB color image to highlight this column of keys in color to emphasize the symbols located on them. After several iterations, the best pseudocolored image that highlighted these keys was obtained by setting graylevels between 105 and 185 to red and leaving all other graylevels as is:

$$c(x,y) = \begin{cases} R[f(x, y)] = 255; G[f(x, y)] = B[f(x, y)] = 0 & 105 \leq f(x, y) \leq 185 \\ R[f(x, y)] = G[f(x, y)] = B[f(x, y)] = f(x, y) & \text{otherwise} \end{cases}$$

$$(5.61)$$

Plate 15 shows the result of this pseudocoloring operation. Note that the symbols are clearly defined. It becomes immediately obvious by looking at this image which pixels are in the graylevel range of 105 to 185.

Another pseudocoloring mapping function is to map the entire set of graylevels to the rainbow of colors that represent the visible spectrum. This pseudocoloring mapping function is typically used in the display of infrared imagery to highlight the changes in grayscale, that represent changes in temperature. Since the perception of color is that blue is cooler than red, blue colors are mapped to the lower set of graylevels that represent the lower thermal temperatures. The red colors are mapped to the higher graylevels that represent the hot temperatures in a thermal image. The following mapping function will produce a rainbow of colors, ranging from blue to red, for a 256 graylevel image that is mapped to a 24-bit RGB color image:

$$c(x,y) = \begin{cases} R = 0, & G = 254 - 4 \cdot f, & B = 255 & 0 \leq f \leq 63 \\ R = 0, & G = 4 \cdot f - 254, & B = 510 - 4 \cdot f & 64 \leq f \leq 127 \\ R = 4 \cdot f - 510, & G = 255, & B = 0 & 128 \leq f \leq 191 \\ R = 255, & G = 1022 - 4 \cdot f, & B = 0 & 192 \leq f \leq 255 \end{cases}$$

$$(5.62)$$

where $R = R[f(x, y)]$, $G = G[f(x, y)]$, $B = B[f(x, y)]$, and $f = f(x, y)$.

Figure 5.13 A grayscale image of a Hewlett Packard model 35
calculator to be pseudocolored to enhance the keys.

Figure 5.14(a) shows an example of a simple 640 × 480 NTSC compatible
grayscale display system that is capable of displaying 256 graylevels. The 256
graylevel image is stored into 307,200 bytes of memory. At the rate of 30
times a second, each byte of storage is fed to the digital-to-analog converter
and converted from a digital number to an analog voltage. This voltage
modulates the electron beam of the black and white CRT, producing a
grayscale image on the monitor. Usually a graylevel of 0 produces a black
intensity, while a graylevel of 255 produces a white intensity. To display a
pseudocolored image, either a palette display system or a 24-bit color display
system is needed. Figure 5.14(b) shows an example of a palette display system
that is capable of displaying 256 colors out of a palette of 16.7 million colors.
The palette display system works on the same principle as a palletized color
image. The palette display system shown in Figure 5.14(b) can display only a
640 × 480 by 256 color palette image. For pseudocoloring, the original 640 ×
480 grayscale image is stored in the 307,200 bytes of display memory. At the
rate of approximately 30 to 75 times a second (depending on the system) each
byte from this memory is fed to the address lines of three 256 byte lookup
tables. The outputs of each of the RGB lookup tables are then converted from
a digital number to an analog voltage using three digital-to-analog converters.
Each of these RGB analog signals are then used to modulate the red, green, and
blue electron beams in an RGB color monitor, producing the desired color
image.

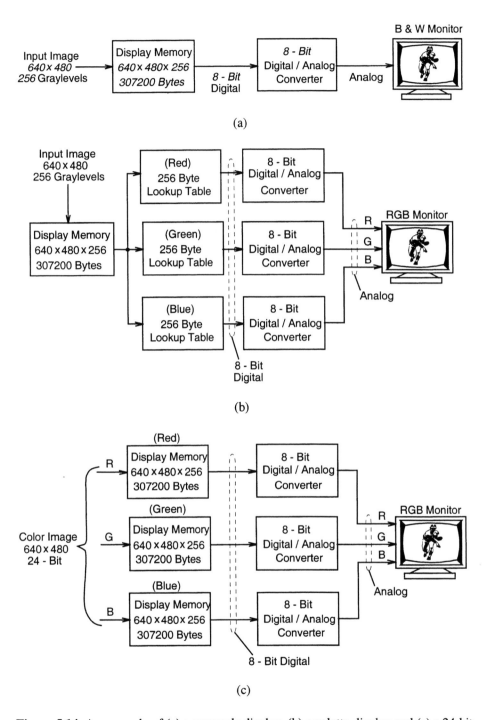

Figure 5.14 An example of (a) a grayscale display, (b) a palette display and (c) a 24-bit color display (images, © New Vision Technologies).

The lookup tables provide the means of implementing the three color mapping functions $R[f(x, y)]$, $G[f(x, y)]$, and $B[f(x, y)]$. The values stored in each of the 256 bytes of memory that compose each lookup table determine the final mapping function. For example, if every location in the three lookup tables is set to zero except the 255*th* location of the red lookup table, which is set to 255, the output image will display, in red, only those pixels within the grayscale image that are at graylevel 255. Another method of displaying a pseudocolored image is to use a 24-bit color display system directly. The grayscale image is converted to a 24-bit color image using the pseudocoloring block diagram given in Figure 5.12 with the desired mapping functions. Then each of the RGB color component images that comprised the 24-bit color image is loaded into its corresponding display memory. At a rate of approximately 30 to 75 times a second, the bytes from the three display memories are fed to the three input digital-to-analog converters. The output voltages from the three digital-to-analog converters are again used to modulate the red, green, and blue electron beams in the RGB color monitor.

In this chapter, the reader was introduced to the concepts of color image processing. Several different color models were given along with several color image processing examples using the RGB and C-Y color models. Pseudocoloring of a graylevel image was given, showing the advantages of highlighting important features within a graylevel image in color. The limited scope of an introductory discussion of color image processing made it impossible to provide a detailed comparison between each of the color models, review the limitations of each model, or discuss the effects of finite numerical precision on each model. Also not discussed is the processing of color images using multichannel techniques. Since most of these topics are relatively new, the interested reader is referred to the open literature or any of the texts mentioned here for more detailed discussion.

REFERENCES

Carterette, Edward C., and Morton P. Friedman, eds., *Handbook of Perception*, Volume 2, Academic Press, New York, 1975.

Dereniak, Eustace L., and Devon G. Crowe, *Optical Radiation Detectors*, John Wiley & Sons, New York, 1994.

Foley, James D., Andries van Dam, Steven K. Feiner, and John F. Hughes, *Computer Graphics: Principles and Practice*, Addison-Wesley, Reading, MA, 1996.

Gonzalez, Rafael C., and Richard E. Woods, *Digital Image Processing*, Addison-Wesley, Reading, MA, 1992.

Hall, Ernest L., *Computer Image Processing and Recognition*, Academic Press, New York, 1979.

Kingston, R.H., *Detection of Optical and Infrared Radiation*, Springer-Verlag, Berlin, 1979.

Martin, A. V. J., *Technical Television*, Prentice-Hall, Englewood Cliffs, NJ, 1962.

Myler, Harley R., and Arthur R. Weeks, *The Pocket Handbook of Image Processing Algorithms in C*, PTR Prentice Hall, Englewood Cliffs, NJ, 1993.

Pitas, Ioannis, *Digital Image Processing Algorithms*, Prentice Hall, New York, 1993.

Russ, John C., *The Image Processing Handbook*, CRC Press, Boca Raton, FL, 1995.

Sid-Ahmed, Maher A., *Image Processing: Theory, Algorithms, and Architectures*, McGraw-Hill, New York, 1995.

Weeks, Arthur R., C. E. Felix, and Harley R. Myler, "Edge Detection of Color Images Using the HSL Color Space," *SPIE Proceedings: Nonlinear Image Processing VI*, Vol. 2424, San Jose, pgs. 291-301, March, 1995.

Weeks, Arthur R. and T. Kasparis, "Fundamentals of Color Image Processing," *DSP Applications*, Vol. 2, No. 10, pgs. 47-60, October, 1993.

Weeks, Arthur R., G. Eric Hague, and Harley R. Myler, "Histogram equalization of 24-bit color images in the color difference (C-Y) color space," *Journal of Electronic Imaging*, Vol. 4, No. 1, pgs. 15-22, January, 1995.

.

6

ENHANCEMENT OF DIGITAL DOCUMENTS

Robert P. Loce
Xerox Corp.
Webster, New York

Edward R. Dougherty
Texas A&M University
College Station, Texas

Digital document resolution conversion and enhancement is a field of image processing with two specific goals that define the field and its problems. One goal is to enable document portability by processing a document bit map in a manner that enables its displayed or printed form to be consistent in appearance when rendered by various devices. A second goal is to arrive at documents with an enhanced appearance over that attainable by directly printing the given bit map. The goals are typically achieved through spatial resolution conversion and quantization range conversion. For the image processing problems encountered, the image class and the processing setting tend to be well defined. This chapter provides a general description of the field and defines the basic terms and methods employed in resolution enhancement technology.

6.1 THE FIELD OF SPATIAL RESOLUTION CONVERSION AND ENHANCEMENT

Technologies that enable portability of document bit maps are of prime importance to digital document processing systems. For example, digital documents are often printed or displayed using different devices, where each device may operate at a different spatial sampling resolution. It is desirable for the image to have an appearance that is consistent between all of these devices, or possibly enhanced on a chosen device. Methods for spatial resampling of binary images are at the root of document portability technology.[1-18]

In a typical office setting, digital documents may be handled by a variety of devices and software packages that interoperate to form a system capable of

---

The views represented herein are those of the authors and are not intended to represent the position of the respective employers.

167

tasks such as scanning, printing, display, transmission, and various forms of processing. Figure 6.1 is a schematic of the devices in a typical networked office environment. The devices that comprise this system may each operate at a different spatial sampling resolution. To enable a comprehensive system, spatial resampling must often be performed to allow porting of the digital image amongst the various devices. To appreciate the role of resampling and the need for specialized methods more fully, a discussion is warranted on the given image class and on typical sampling conditions.

Other fields of image processing, such as remote sensing and medical imaging, often operate in a gray-scale setting, where it is reasonable to model the underlying continuous image as piecewise-linear, -quadratic, -cubic, or other related form. Digitization is usually performed at 8 bits/pixel and the image class is referred to as *continuous tone (contone)*. For purposes of display, printing to film, or compatibility with subsequent image processing algorithms, these imaging communities also need to magnify digitally, or resample, their images. Because of the image class and gray-scale digitization, well-known interpolation methods may be applied to perform resampling. When considering digital documents it must be understood that the underlying image class is essentially piecewise-constant and digitization is usually performed at 1 bit/pixel. Scaling in this setting is best performed by logic-based methods.

Each stage of handling a digital document has associated with it one or more typical sampling resolutions. Digital documents are created by scanning a hardcopy original or by decomposition from a *page-description-language (PDL)* form. Office scanners typically digitize at 8 bits/pixel and either immediately threshold or threshold after some processing and prior to transmission to another device. Note that this other processing may be digital resampling by gray-scale methods. Typical spatial resolutions for office scanners are 300, 400, and 600 spots per inch (spi).

Page Description Language refers to a computer language used in digital document creation. Examples are PostScript[19] and PCL 6. Other document-creation software typically writes the PDL code. Obtaining a digital image from a PDL typically involves several stages starting with generating an ASCII document with word processing software and writing the document to a file in PDL form. Characters and line art are interpreted as a collection of functions that describe curves and shapes in a plane. The file can then be processed by software that decomposes the PDL code to binary bit-map form. This rasterization step is often referred to as *Raster Image Processing* or simply *RIP*. The decomposer may be on a host computer or integrated into a printer. Most decomposers can render a bit map at any of a number of spatial resolutions while others may be limited to decompose at a fixed resolution, say, 240 spi, for a specific printer.

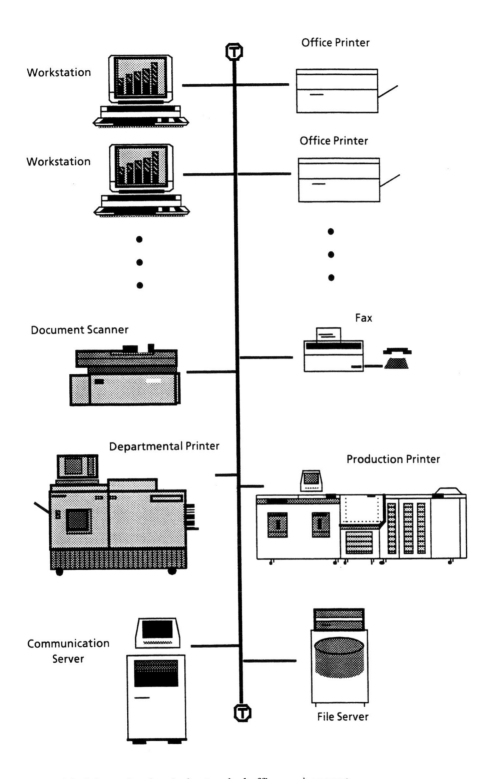

Figure 6.1 Schematic of typical networked office environment.

Because document images are typically *bilevel*, current office printers are generally based on laser/xerographic or ink jet technologies, which are well-suited for making binary marks on paper (two levels: ink or no ink). The sampling resolution employed in an office printer is usually one of the following: 240, 300, 360, 400, or 600 spi. Document bit maps created or decomposed for one printer quite often must be printed on another. This can occur because different manufacturers that design for different resolutions may have produced scanners, decomposers, and printers on a given network.

Another group of sampling resolutions is used in high-quality typesetters and graphic arts printers, some of which print at resolutions up to 3600 spi. Resolution conversion and enhancement must also be performed in this high-resolution environment.

Fax transmission has become the most common method of digital document delivery. It incorporates elements of scanning, image processing, digital communications, and printing. The CCITT Group 3 standard for fax specifies a minimum of 204 spi horizontally and 98 spi vertically for an 8.5" × 11" page (see Ref. 20 for a comprehensive review of fax standards). Alternatively, fax can be represented at 204 spi × 196 spi in the so-called high-resolution mode. Fax documents tend to be ragged and blocky in appearance due to the low sampling resolution. To improve the appearance it may be desirable to capture the image signal and resample it to enable printing on a higher resolution device. Resampling may also need to be performed to provide compatibility of fax documents with image processing software that, for example, may perform OCR on the document.

Several simple techniques are sometimes employed to enable printing with mismatched resolutions. The less complex methods usually do not provide satisfactory results. For example, one could treat the image as if it were sampled at the printer resolution, say, 600 spi, instead of the decomposed resolution, say, 240 spi. In this case the printed image would be minified by 2.5× relative to its intended size, as shown in Fig. 6.2. For noninteger ratio conversions such as this, simple bit duplication cannot achieve the correct size. The magnification problem is sometimes circumvented by alternating between bit replication factors.[15] For 240 spi and 600 spi, bit replication alternates between doubling and tripling along the two axes. This modified nearest-neighbor scaling method only provides acceptable results for large text. Text with fine strokes, periodic patterns, and halftone images are printed with noticeable defects, such as stroke-width variation and beating (moiré). Figure 6.3 shows the line width defect that occurs when this simple alternating replication algorithm is applied to a periodic pattern.

Simplistic scaling methods can also produce image defects for integer ratio conversions. For example, for 300 spi to 600 spi conversion, the correct size can be achieved by simply performing nearest-neighbor scaling. Stroke width would be faithfully transformed, and beating (moiré) artifacts would not appear

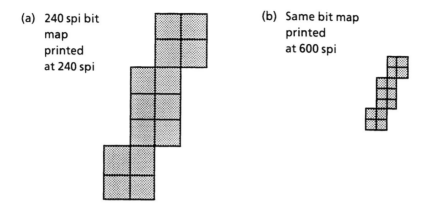

(a) 240 spi bit
 map
 printed
 at 240 spi

(b) Same bit map
 printed
 at 600 spi

Figure 6.2 Magnification problem that occurs when resolution conversion is not performed.

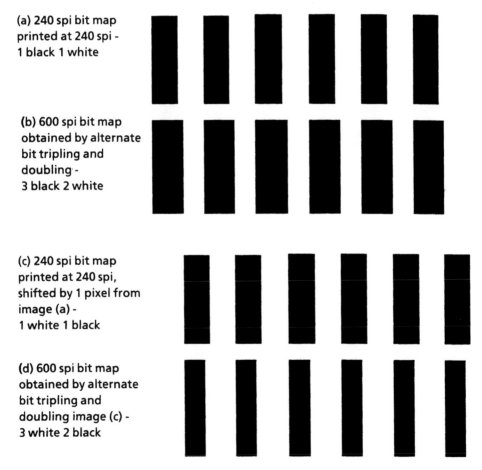

(a) 240 spi bit map
printed at 240 spi -
1 black 1 white

(b) 600 spi bit map
obtained by alternate
bit tripling and
doubling -
3 black 2 white

(c) 240 spi bit map
printed at 240 spi,
shifted by 1 pixel from
image (a) -
1 white 1 black

(d) 600 spi bit map
obtained by alternate
bit tripling and
doubling image (c) -
3 white 2 black

Figure 6.3 Line width problem that can occur with simplistic noninteger bit-map scaling. Lines are converted to different widths depending upon their position on the page. Periodic patterns can transform to possess different densities or beat patterns.

in halftoned images. On the other hand, one would not be fully utilizing the image quality capability of the 600 spi printer. Diagonal and curved strokes would print with the characteristic jaggedness of a 300 spi printer and possibly worse (could be worse because of the sharper writing spot).[5] In this setting, it is not desirable for a resolution conversion algorithm to produce a bit map that is only a coarse bit map that has been scaled to the correct size. The algorithm should utilize the sampling resolution of the printer to best represent the piecewise-constant underlying image. Resolution conversion performed in this manner can result in an enhanced print with less staircasing and blockiness than would have been achieved on the originally intended device. Figure 6.4 demonstrates the limited capability of bit replication compared to a more intelligent resolution conversion algorithm that attempts to enhance the image appearance.

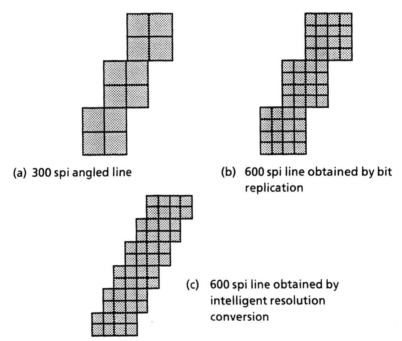

(a) 300 spi angled line

(b) 600 spi line obtained by bit replication

(c) 600 spi line obtained by intelligent resolution conversion

Figure 6.4 Integer resolution conversion by bit replication does not fully utilize the capabilities of a printer. Printing using bit replication can result in more jaggedness at high resolution because high-resolution printers employ sharper writing spots and thus can more clearly show the staircasing of an angled stroke.

Figure 6.5 illustrates the possible pixel-to-pixel transformations for nearest-neighbor scaling and for a more intelligent mapping for 300 spi to 600 spi resolution conversion.[21] In either case a conditional decision is made to generate the output pixel values. For the nearest-neighbor case, conditioning is only upon the value of the pixel of interest and only two possible decisions exist for the output values. The more intelligent mapping makes a conditional

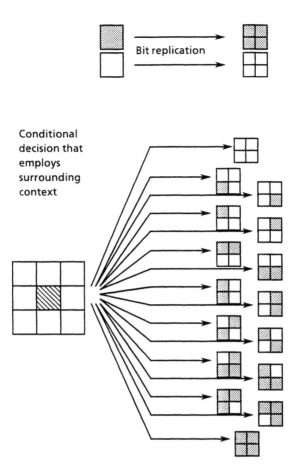

Figure 6.5 Nearest-neighbor scaling employs trivial decision making and thereby replicates the blockiness of the input image. Template matching filters utilize surrounding context to make conditional decisions that select an optimal output set of pixel values.

decision based upon values of neighboring pixels by comparison of those neighbors to a set of stored templates. Any one of 16 possible decisions can be made to arrive at a set of pixel values that yield the best appearance when viewed along with their surrounding context.

6.1.1 Other image quality issues

In addition to the magnification issues mentioned above, digital documents possess other special image quality attributes. A common observable defect is a jagged, pixelated structure that can occur on slanted and curved strokes, referred to as a *jaggie*. This contour-sampling artifact is also referred to as *staircasing* and is considered a form of sampled-signal *aliasing*.

Jaggies can be observable at a remarkably fine scale due to the *hyperacuity* of the human visual system. *Hyperacuity* is a term coined by Westheimer[22] to denote human visual perceptual abilities that went beyond the ability attributed to acuity. There are several forms of hyperacuity; the one most relevant to appearance of digital documents is *vernier acuity*, which describes the eye's ability to perceive the displacement of abutting lines. An example of vernier acuity is shown in Fig. 6.6. While 20/20 vision can resolve 1 cycle/60 arcsec as a periodic pattern, the normal vernier acuity threshold is roughly 4 arcsec and trained observers can detect 1 arcsec of displacement – 60 times greater than the normal resolving power of the eye. At 12" viewing distance, 4 arcsec corresponds to an addressability of 4300 spi. A statement by Curry[23] well describes the need for resolution enhancement methods:

> "What is needed is an imager which can position edges in the process direction with this precision while backing off on the resolution requirements so as to only slightly exceed the required human visual goals. Or, put another way, a laser imaging system [is needed] which naturally has similar characteristics to human vision: its edge placement precision is an order of magnitude beyond its own resolving power."

Figure 6.6 A grating illustrating vernier acuity.

Another consideration is the perceived darkness of text and line art. The perceived darkness of typical character strokes is primarily a function of their width. Darkness is adjusted in appearance matching and appearance tuning operations.

Appearance Matching is an image processing technique applied to a document image prior to printing or display that yields a print having the appearance matched to that observed when printing on a different device. Appearance matching typically involves line growth or shrinkage, often at a subpixel level, and may include some sharpening or shaping effects.   For example, xerographic laser printers that operate with the laser writing the white background of a page (*write-white*, or *Charged-Area Development - CAD*) tend to yield thinner lines for a given digital character stroke than printers that write the black image foreground (*write-black*, or *Discharged-Area Development - DAD*). For matched appearance in this setting, a digital image printed on a write-white device must have thicker digital character strokes than those employed to generate the print using the write-black device. Digital erosion and dilation techniques can be applied to adjust document bit maps to yield prints with matched appearance between devices.[24]

Appearance Tuning is sometimes referred to as *digital darkness control*. In appearance tuning, digital erosion and dilation techniques are applied to adjust a document bit map to yield a print with the desired degree of darkness on a given printing device.[25,26] In application, it is very similar to appearance matching.

6.2 DOCUMENT ENHANCEMENT

Digital document resolution enhancement is an image processing method, usually employing template matching, that transforms a binary bit map into multiple bits per pixel or a higher spatial sampling resolution with the goal being to improve the appearance of the resulting printed image. The image processing setting for resolution enhancement is described below.

6.2.1   Inferential and noninferential image enhancement

Inferential Image Enhancement is a digital image-to-image mapping to higher spatial resolution or finer quantization resolution than the given input bit map, where the goal is to enhance the appearance of the printed image by inferring more information about the underlying source image. The paradigm is to first assume that there is an underlying image that is the source of the input bit map. Then, through a probabilistic understanding of the image class and through observation of the input bit map, more information is inferred about the specific image realization at hand. Typically, a window of pixels is observed in the input bit map and pixels are output at either higher spatial resolution or higher quantization resolution. These pixels are estimates of the underlying source image. If the assumptions about the underlying image are correct, then the higher resolution image prints or displays with an enhanced appearance (e.g., jaggies are smoothed; see description of jaggies above) compared to the given input image.   Inferential methods generally do not possess the information to adjust stroke width and edge position on a local basis. Inferential image enhancement is used for scanned documents that have been thresholded and in cases where the Page Description Language document is not available or not used to generate high-resolution information[27] (see Figs. 6.7 and 6.8).

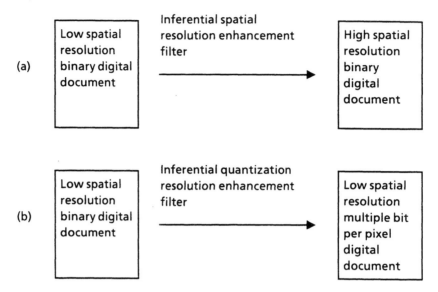

Figure 6.7 Image mappings for inferential filtering: (a) spatial resolution enhancement; (b) quantization range enhancement. The low-spatial-resolution binary digital document may have been obtained from a scanner or from decomposing a Page Description Language form of the document.

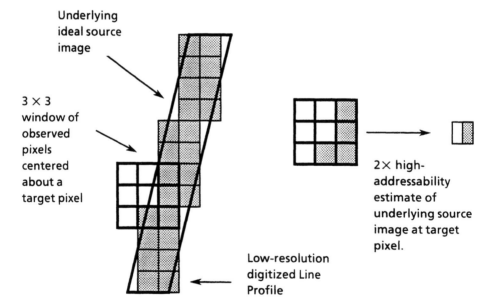

Figure 6.8 Based upon a group of observed pixels, template matching can be used to infer, or estimate, a high-resolution sampling of an underlying source image, that is not available for directly obtaining high-resolution samples.

Noninferential or Alternate Representation Image Enhancement[27-31] is a digital image-to-image mapping to lower spatial sampling resolution. Typically, the image is available in Page Description Language form, which is decomposed to a resolution that is higher than acceptable for the given printer. This high-resolution binary bit map is then filtered to generate a lower resolution digital image with additional quantization levels. The total number of bits may remain roughly the same, but the representation is altered from high-resolution binary to lower resolution gray scale. An advantage of employing this method is that more information is used to generate the multiple-bit/pixel image than is employed in inferential methods; hence, better gray-scale representations may result. More accurate smoothing is attainable and information is present (high-resolution samples) to allow for character stroke width and position to be written in fractional pixel increments. There are three disadvantages, the first being that the user may not possess a PDL or high-resolution version of the document, thus not allowing use of this method. The second problem is that longer decomposition times are required for higher-resolution images, hence resulting in a loss of printer productivity. A third problem involves an increase in size of memory buffers used during decomposition. See Figs. 6.9 and 6.10.

6.2.2 Enhancement by quantization range conversion and redefinition. The contour restoration problem

Consider an electronic reprographic printing process where a hardcopy document is first scanned into electronic form using a digital document scanner. For text documents, prior to storage, transmission, or printing of this image, it is typical to threshold the gray-scale (8 bits/pixel) values to yield a 1-bit/pixel bit-map image. Although hardcopy text images are binary in nature (black characters on white paper), the processes of spatial sampling and thresholding lose contour information; a smooth diagonal line on the original hardcopy is converted to a jagged-edged sampled line. Printing this bit map on (say) a xerographic laser printer operating at less than or equal to 400 spi typically results in a print with text that appears ragged.

Although the gray-scale and thresholded (binary) digital forms of a scanned document are at the same spatial resolution, the gray-scale version contains additional information relative to character contours. This is due to the finite area of the photosensors used in the digitization process. Several scenarios relative to contour digitization can occur. The straightforward case is where an all-black or all-white region of the document is seen by the photosensor resulting in a black or white reading, respectively. A more interesting situation is where a black/white contour falls within the observation region of the photosensor resulting in a "gray" reading whose value depends upon the fraction of black and white areas seen by the sensor. Upon

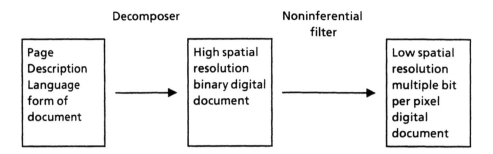

Figure 6.9 Image mappings for noninferential filtering. Decomposers typically cannot decompose saturated (nongray) text to multiple bits per pixels; thus the scheme shown above is often employed to obtain the desired digital image.

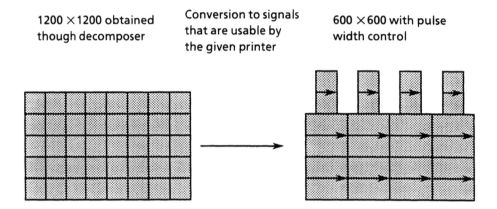

Figure 6.10 Example of noninferential resolution enhancement showing how a 2 ½ pixel line can be produced at 600 spi. A 1200 × 1200 spi bit map is converted to 600 × 600 spi with pulse width control. Arrows indicate the video signals used in writing the pixels.

thresholding, this fractional information is lost. Furthermore, a small amount of noise in the system due to electronics, paper fibers, optical conditions, etc., which could affect the fractional reading in a small way, could have a significant effect upon a contour obtained by thresholding. To insert probable contour information, an optimal filter may be applied to the binary bit map to yield an estimate of a partial gray-scale digital image. The estimation setting is depicted in Fig. 6.11. From a 1 bit/pixel bit map, the filter estimates the digital image that would have resulted from quantizing (possibly with some intelligence) the 8 bits/pixel image to 2 bits/pixel. Note that this estimation paradigm is not limited to 2 bits/pixel, but for common printing resolutions it is

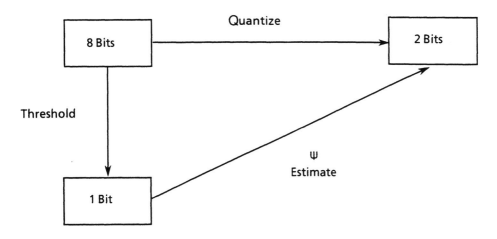

Figure 6.11 Estimation setting for reprographic 1-to-2 bit conversion.

typically limited to 4 bits/pixel or less. The partial gray-scale image, say, 2 bits/pixel (or, *quaternary*), can then be sent to a suitable printer or display device whereupon it is rendered to have smoother contours than the image that results from the binary bit map.

A similar situation occurs for orthographic documents. Suppose a document in a PDL form is decomposed to 300 spi at 1 bit/pixel. Here the loss of contour information is due to the transformation from analytical form to sampled form. As in the reprographic case, an optimal binary-to-gray-scale filter could allow for a more aesthetically pleasing print to be generated from the bit map. The orthographic estimation setting is depicted in Fig. 6.12. From a 1 bit/pixel bit map, the filter estimates the digital image that would have resulted from directly decomposing to 2 bits/pixel.

A related problem occurs with digital display devices. Displayed characters often appear blocky due to typical CRT display resolutions ($\cong$72). Various *antialiasing* schemes have been devised by the computer graphics community for improving the aesthetic quality of displayed text and line art.[32] Because of the early methods used to ameliorate jagged characters, antialiasing sometimes carries the connotation of employing an averaging algorithm to smooth text in a digital display. The look-up table method developed in this chapter does not employ averaging, and therefore typically produces a sharper image than some early antialiasing algorithms. Note that early antialiasing methods tended to average, or low-pass filter, a received bit map, thereby inferring additional quantization levels beyond the input binary condition. More recently, noninferential RIP-based antialiasing has been developed for printing text and line art on hardcopy devices. The decomposer employs several bits of quantization to better represent contours that partially cover pixels.

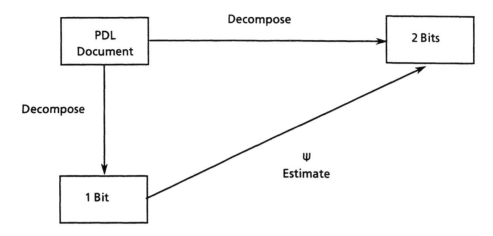

Figure 6.12 Estimation setting for orthographic 1-to-2 bit conversion.

6.2.3 Methods of writing enhanced pixels

The concept of a gray, or intermediate level, pixel must be viewed in the context of the particular writing process at hand. Certain processes are capable of displaying true gray scale and others operate primarily in a binary mode. Consider three display and printing processes: CRT, ink jet, and laser xerography. The physical differences of these marking systems necessitate an individualized understanding of the gray-pixel concept itself and of the method for binary-to-gray filtering and gray-pixel writing. In a true gray writing device, a gray pixel is written at an intermediate luminance level across the area of the pixel. In a binary writing device, partial area coverage is usually employed.

As an example of a gray-scale writing system, consider a CRT display device, which is relatively linear and stationary. At points where raster lines of electron exposure overlap, the resulting perceived brightness is equal to the brightness generated by the sum of the electron exposures. A digital image possessing multiple bits per pixel for gray-level quantization may be displayed by driving the electron beam at intermediate power levels, where the power level used for a given pixel corresponds to the pixel value. An example of a gray-scale printing device is the dye-sublimation printer. In these gray-scale-capable systems, image enhancement filters that increase the quantization range generate multiple-bit-per-pixel values that correspond to image light intensities. Antialiasing operators are an example of such filters.

An intermediate-level pixel in a binary marking device possesses a different meaning and function than its counterpart in a gray-scale-capable system. Consider two examples of binary-marking devices: ink jet printers and xerographic laser printers. In an ink jet printer, an area on a paper print is either covered with a drop of ink or it is not covered. In an image printed by xerographic means, it is highly desirable to print with a saturated amount of toner or no toner at all. Figure 6.13 illustrates the noise problem that can occur when attempting to print a xerographic image that possesses intermediate gray levels of exposure. Due to the physics of these marking systems, intermediate-level pixels are typically written in a partial-area-coverage mode, where the toner, for example, partly covers a pixel area. The covered area must be properly positioned within the pixel area to achieve the desired visual effect and be compatible with the system physics.

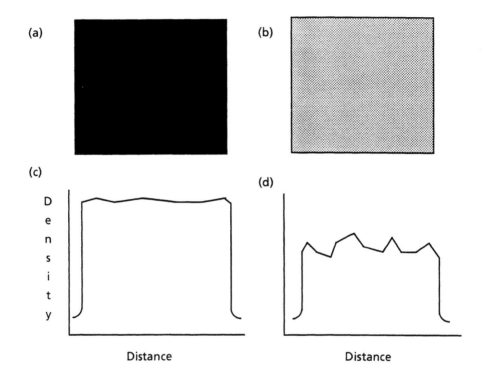

Figure 6.13   (a) Uniform saturated image; (b) desired uniform intermediate-level image; (c) density slice through developed saturated image; (d) density slice through developed intermediate-level image. Minor xerographic process fluctuations can result in significant noise when attempting to develop intermediate levels or small isolated patches of toner.

There are two key physics-based positioning issues. One is that a small patch of toner or ink tends to print with a more consistent size and mass when it is adjoining another toner patch because development and transfer of an isolated small patch of toner can be sensitive to minor system fluctuations. Another issue is that the writing member (e.g., laser) passes, say, from left to right, through the center of a pixel area. The allowable positions for partial toner patches are thus limited. It is relatively easy to generate partial coverage on the left or right side of a pixel and more complicated to generate partial coverage on the top, bottom, or corners of a pixel area. In binary-writing systems, image enhancement filters that increase the quantization range, output multiple-bit-per-pixel values that correspond to waveforms that drive the writing member. The types of waveforms employed in image enhancement can be quite varied from device to device. Laser printers are one of the most common binary or near-binary marking devices, so subsequently we focus on image enhancement in that setting.

There have been various approaches to resolution enhancement in xerographic laser printers. Image enhancement in a laser writing system typically employs a form of *pulse width* and *position modulation* (PWPM)[25,33–42] or *high addressability*.[23,43–46] These two laser control methods allow for proper positioning of partial pixels within a nominal pixel area. PWPM utilizes several bits per pixel to access a video waveform described by a PWPM look-up table. Simple PWPM has three events within a pixel time: an initial "off" time, the laser "on" time, and a trailing "off" time. The video waveforms are generated by analog electronics clocked out at the nominal pixel clock rate. Figure 6.14 shows a simple 4 bits/pixel PWPM scheme. Depending upon the surrounding image structure, PWPM pulses are typically shifted to the left, or right, or placed in the center. Figures 6.15 and 6.16 show how PWPM may be used to perform jaggie reduction in near-vertical and near-horizontal lines. Note that it is relatively easy to download various PWPM tables into a given printer, thus enabling the use of partial pixels in lighter or darker modes.

Image enhancement employing high addressability divides a nominal output pixel into several equal divisions and each division possesses a binary value (0,1) indicating how the laser should be activated over that fractional area. In this setting, a high-speed digital pixel clock is employed to clock out pulses at the high-addressability rate. In a raster printer or display device, a high-speed digital clock is used to address the writing member, such as a laser beam, at a finer spatial resolution in the raster direction than in the direction perpendicular to the rasters. High addressability also typically implies that the writing member (laser) is addressed at spatial increments finer than the size of the writing spot.[47] For example, a printing device that operates at a nominal 400×400 spi resolution may have an enhanced mode that operates at 1600×400 spi. The high-addressability factor employed in laser printers tends to be at some fixed multiple of a nominal symmetric addressability (e.g., 2×, 4×, 8×) but in general is not limited to this configuration (see Fig. 6.29). The design and implementation of template matching filters for generating high-addressable bit maps is very similar to integer resolution conversion (see Section 6.5.1). Figure 6.17 shows pixel waveforms for 4× high addressability and Figs. 6.18 and 6.19 show enhancement of near-vertical and near-horizontal lines, respectively, using high addressability.

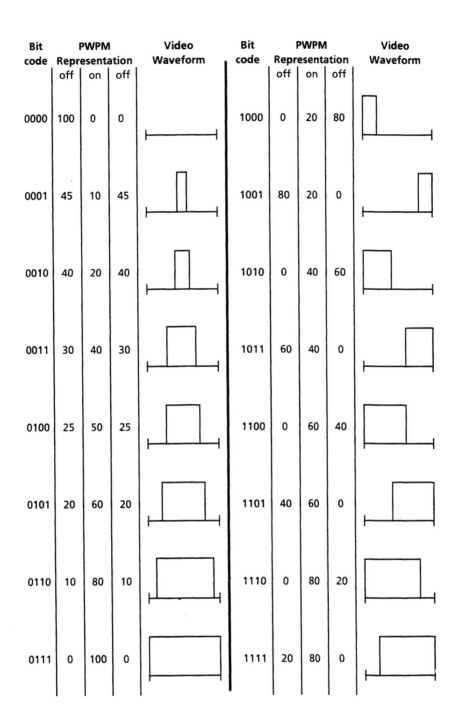

Figure 6.14 A 4-bit PWPM table. On and off times are given as percentages of the nominal pixel time.

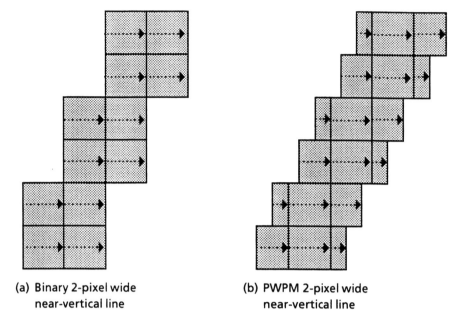

(a) Binary 2-pixel wide
near-vertical line

(b) PWPM 2-pixel wide
near-vertical line

Figure 6.15 Quantization range enhancement of a near-verticle line using the PWPM table of Fig. 6.14. A smoother realization of the line is achieved by employing pulses 1001 and 1011 on the left and 1000 and 1010 on the right. Arrows represent the waveform driving the laser.

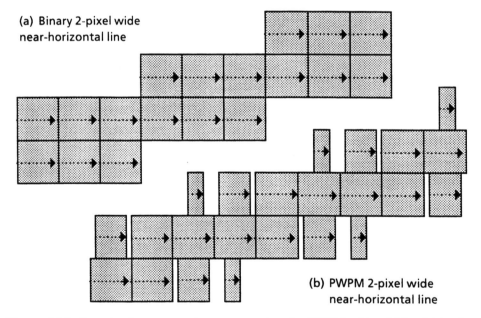

(a) Binary 2-pixel wide
near-horizontal line

(b) PWPM 2-pixel wide
near-horizontal line

Figure 6.16 Quantization range enhancement using the PWPM table of Fig. 6.14. A smoother realization of the line is achieved by employing pulses corresponding to 0010 and 0110. In this case, enhancement is dependent upon blurring in the subsequent printing processes.

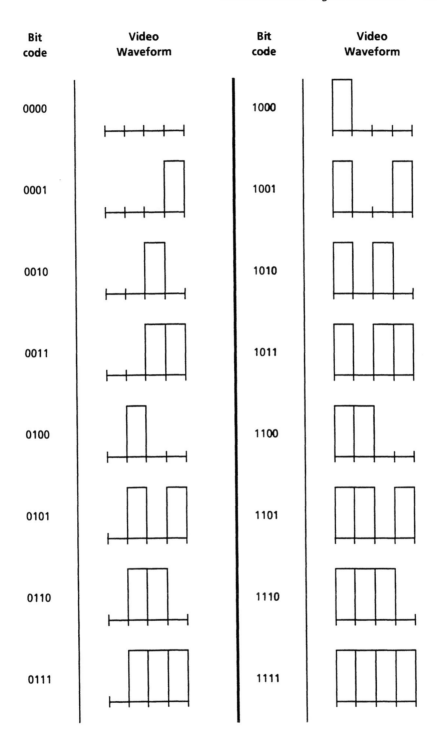

Figure 6.17 Pixel waveforms for 4× high addressability.

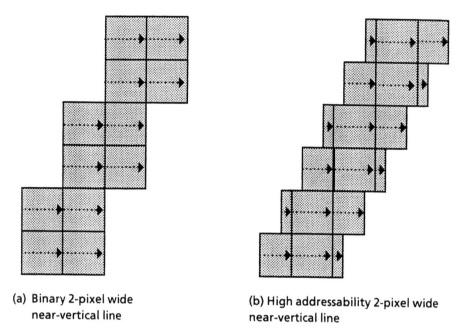

(a) Binary 2-pixel wide
    near-vertical line

(b) High addressability 2-pixel wide
    near-vertical line

Figure 6.18 Enhancement of a near-vertical line using signals of Fig. 6.17. A smoother realization of the line is achieved by employing pulses 0001 and 0111 on the left and 1000 and 1110 on the right. Arrows represent the waveform driving the laser.

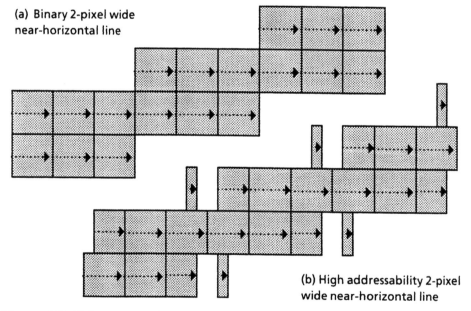

(a) Binary 2-pixel wide
near-horizontal line

(b) High addressability 2-pixel
wide near-horizontal line

Figure 6.19 Enhancement of a near-horizontal line using the signals of Fig. 6.17. A smoother realization of the line is achieved by employing pulses corresponding to 0010, 0100, 0111, and 1110. In this case enhancement is dependent upon blurring in the subsequent printing processes.

A method for writing enhanced image structures using binary pixels also exists. *Halfbitting* refers to a method employing alternating whole binary pixels to encode stroke widths and edge positions at a resolution finer than the nominal pixel resolution. For example, a stroke with a desired width of 2 1/2 pixels can be printed at the correct width by employing a bit map containing a line that is 2 columns of pixels wide with the third row containing pixels that alternate in a 1-on-1-off pattern. After passing through certain printing processes and the human visual system, the alternating pattern can blur to form a line that is 1/2 pixel wide and is adjoining the 2-pixel-wide line.[48] See Fig. 6.20.

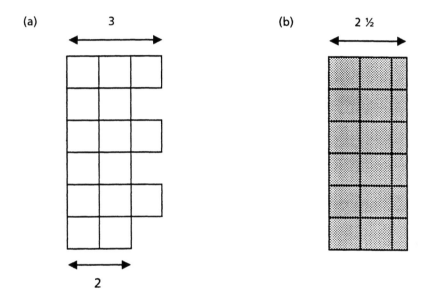

Figure 6.20 (a) Bit map with halfbits; (b) developed line.

6.3 BINARY IMAGE FILTERS

For the most part, the subject of this book employs binary image processing using translation-invariant windowed filters. Reference 49 discusses these filters in detail from a careful mathematical standpoint in the context of Boolean logic; for the present, we informally introduce basic filtering concepts, relate them to both logical processing and pixel geometry, and use these concepts to introduce several fundamental document-processing applications.

A binary digital image consists of pixels that are either 0-valued or 1-valued. Because our aim is to enhance and restore documents, we consider black the foreground and white the background. Hence, white and black pixels are valued 0 and 1, respectively. A filter is an operator that transforms one image into another image. A windowed binary filter computes an output value (0 or 1) at a given pixel location by using windowed pixel values as input. The windowed pixels are centered about the given target pixel whose filtered value is to be computed. This form of operation requires that there be some function whose arguments are the 0's and 1's in the observation window and whose functional value gives the filter output at the given target pixel. If the same function is employed for all window locations to transform all pixels in the image, then the filter is said to be *translation invariant (spatially invariant).* For the purpose of this introductory material, we employ an observation window that is a 5-pixel cross of the form

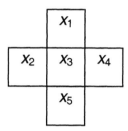

where x_1, x_2, x_3, x_4, and x_5 are pixel values, each value being 0 or 1, and x_3 is the input value of the pixel whose output is being computed by the filter. Since five pixels are observed in the window and each pixel has two possible values, there are 32 vectors of possibly observed pixel values.

To specify a function of the windowed pixel values requires specification of the output value y in terms of the input values x_1, x_2, x_3, x_4, and x_5. One of the simplest functions that we may employ is the logical single-product function, which is of the form

$$y = h(x_1, x_2, x_3, x_4, x_5) = x_1^{p(1)} x_2^{p(2)} x_3^{p(3)} x_4^{p(4)} x_5^{p(5)}, \qquad (6.1)$$

where the "power" $p(j)$ is interpreted for binary to be either "not present" or be complementation c. When the power is not present, the pixel value x_j itself is in the product; when complementation is specified, the invert, or complement, of x_j is in the product.

As an example of a single-product function, consider

$$y = h(x_1, x_2, x_3, x_4, x_5) = x_1 x_2 x_3^c x_4 x_5. \qquad (6.2)$$

Only the variable x_3 corresponding to the observed value at the center pixel (the one whose new value is being computed) is complemented; the others are in the product directly. For this product function, $y = 1$ if and only if $x_1 = x_2 = x_4 = x_5 = 1$ and $x_3 = 0$. All other 31 possible observation patterns yield 0. In general, when a product function is formed from all possible variables (either complemented or uncomplemented), the output value is 1 for only one observed pattern and zero for all others. Because each input value is binary, every product function defines a logical computational architecture. The architecture corresponding to the function h of Eq. (6.2) is shown in Fig. 6.21.

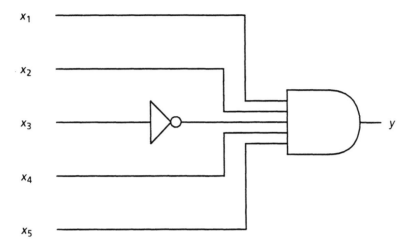

Figure 6.21 Computational architecture for the single-product function h of Eq. (6.2).

Geometrically, the function h defines a *template match*: if the 0-1 template

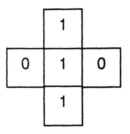

is placed at the pixel, the filter value is 1 if and only if the values in the template exactly match the windowed pixel values of the input image. If filter Ψ is defined by this template, then the Boolean function h is applied at all translations of the window within the input image S. The filter $\Psi(S) = 1$ for white-valued (0) input pixels whose neighbors directly above, below, to the left, and to the right are black-valued (1) — and only such pixels yield a 1; all other pixel patterns result in $\Psi = 0$. Figure 6.22 shows an image S processed by Ψ.

For a second illustration, consider the single-product function

$$y = h(x_1, x_2, x_3, x_4, x_5) = x_1 x_2^c x_3 x_4^c x_5. \tag{6.3}$$

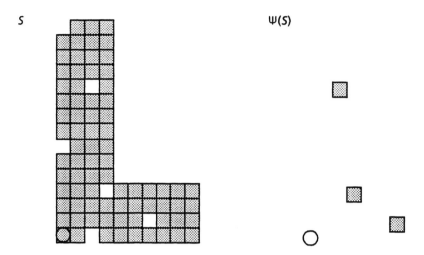

Figure 6.22 Input image S and filtered image $\Psi(S)$ for Ψ defined by the single-product function h of Eq. (6.2). The circle denotes the origin pixel in each image.

For this product, $y = 1$ if and only if $x_1 = x_3 = x_5 = 1$ and $x_2 = x_4 = 0$. Geometrically, this function corresponds to the template

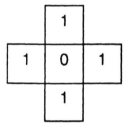

If the filter Ψ is defined by this single template, then $\Psi(S) = 1$ for black input pixels whose neighbors directly above and below are black and those directly to the left and right are white — and only such pixel patterns; all other patterns yield $\Psi(S) = 0$ (white). The corresponding computational architecture is shown in Fig. 6.23 and an image S processed by the filter Ψ defined by h is shown in Fig. 6.24.

Single-product logical functions can be used as building blocks to form more complex functions. Consider the function

$$
\begin{aligned}
y &= h(x_1, x_2, x_3, x_4, x_5) \\
&= x_1 x_2^c x_3 x_4^c x_5 + x_1 x_2 x_3 x_4^c x_5 + x_1 x_2^c x_3 x_4 x_5 + x_1 x_2 x_3 x_4 x_5 .
\end{aligned} \tag{6.4}
$$

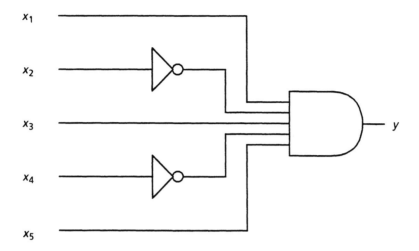

Figure 6.23 Computational architecture for the single-product function h of Eq. (6.3).

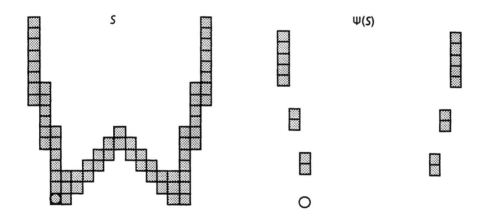

Figure 6.24 Input image S and filtered image Ψ for Ψ defined by the single-product function h of Eq. (6.3).

Each of the four products composing the sum has value 1 for a unique windowed observation pattern. Corresponding to the order of summation, these observations are the following: (1) the center pixel and its neighbors directly above and below are black, and its neighbors directly to the left and right are white; (2) the center pixel, its neighbors directly above and below, and its neighbor directly to the left are black, and its neighbor directly to the right is white; (3) the center pixel, its neighbors directly above and below, and its neighbor directly to the right are black, and its neighbor directly to the left is white; and (4) all pixels in the window are black. The filter output is 1 if and only if one of these conditions holds (and no more than one of these conditions can hold because they are mutually exclusive). From the perspective of template matching, the filter outputs 1 at the center pixel location if and only if one of the four templates shown below matches the input image when placed at the pixel:

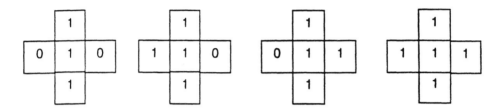

Geometrically, one can try to match each of the preceding four templates and then paint a given pixel in the output image black if and only if there is a match with one of these templates at that pixel.

For software implementation of a filter, it is common to use its *look-up table* (truth table) form. The look-up table corresponding to a logical function is simply a list of templates to be compared to observed values, each template in the list being associated with an output value y. The look-up table corresponding to the function of Eq. (6.4) is given in Table 6.1.

Architecturally, the function defined by these four templates is defined by the gate structure of Fig. 6.25.

Examination of the template matching corresponding to the logical function of Eq. (6.4) shows that the output value is 1 if and only if the center pixel and its neighbors directly above and below are 1-valued. Geometrically, this means the output pixel is 1-valued if and only if there is a vertical string of black pixels in the input image of at least length 3 such that the pixel at which the window is centered is the center black pixel of that string. While one can reach this conclusion from the template-matching geometry, it can also be reached by applying Boolean logic minimization to the function of Eq. (6.4) to see that it is equivalent to the function

$$y = x_1 x_3 x_5.$$ (6.5)

Table 6.1 Look-up table for the logical function in Eq. (6.4)

x_1	x_2	x_3	x_4	x_5	y
0	0	0	0	0	0
0	0	0	0	1	0
0	0	0	1	0	0
0	0	0	1	1	0
0	0	1	0	0	0
0	0	1	0	1	0
0	0	1	1	1	0
0	1	0	0	0	0
0	1	0	0	1	0
0	1	0	1	0	0
0	1	0	1	1	0
0	1	1	0	0	0
0	1	1	0	1	0
0	1	1	1	0	0
0	1	1	1	1	0
1	0	0	0	0	0
1	0	0	0	1	0
1	0	0	1	0	0
1	0	0	1	1	0
1	0	1	0	0	0
1	0	1	0	1	1
1	0	1	1	0	0
1	0	1	1	1	1
1	1	0	0	0	0
1	1	0	0	1	0
1	1	0	1	0	0
1	1	0	1	0	0
1	1	0	1	1	0
1	1	1	0	0	0
1	1	1	0	1	1
1	1	1	1	0	0
1	1	1	1	1	1

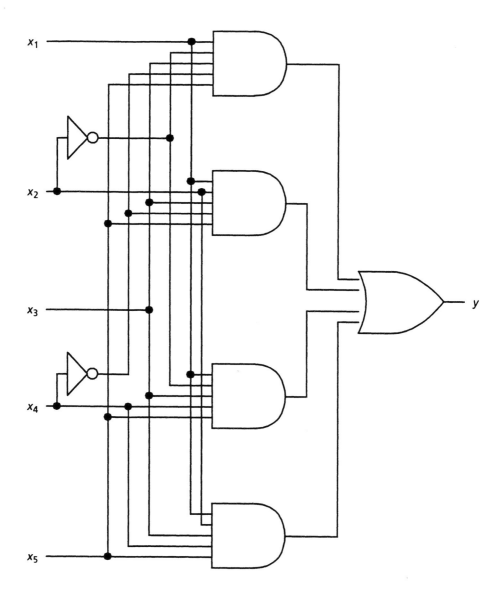

Figure 6.25 Gate structure for the four-product function of Eq. (6.4).

This equivalent form of the function constitutes a reduction from the original. Such reduction is discussed in detail in Ref. 49. For the present, let us simply note that there are fewer variables in the reduced expression and that it does not correspond to pattern matching that utilizes the entire window. For instance, in the present case one need only check to see if the template

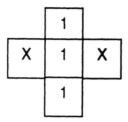

matches, where an "X" is used to represent that we *don't care* whether or not a pixel matches. Equivalently, to determine which output pixels are 1-valued, one need only check if the 3-point template

fits within (is a subset of) the input image when placed at the pixel of interest. Figure 6.26 shows an image S processed by the corresponding filter Ψ.

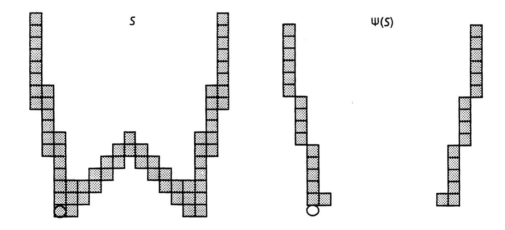

Figure 6.26 Input image S and filter output image $\Psi(S)$ for Ψ defined by the four-product function h of Eq. (6.4) or, equivalently, the single-product function of Eq.(6.5).

While we have deduced the functional representation of Eq. (6.5) from Eq. (6.4) by Boolean algebra, very often logical functions defining binary-image filters are derived directly from geometric considerations, not by logical reduction from a collection of exact, or full-window, template matches. For instance, if we wish to return a 1 at a pixel whenever that pixel is interior to a black column or interior to a black row, the appropriate filter is defined by the logical function

$$y = x_1 x_3 x_5 + x_2 x_3 x_4 \qquad\qquad (6.6)$$

and implemented in hardware by the gate structure of Fig. 6.27. If desired, one can work backwards to find the set of full templates that produces the desired output via matching. Figure 6.28 shows an image S processed by the corresponding filter Ψ.

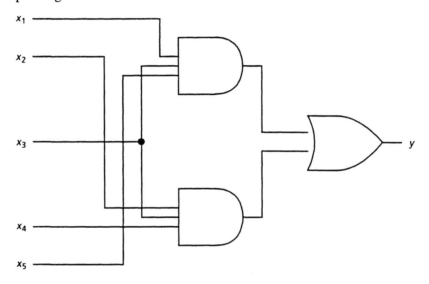

Figure 6.27 Gate structure for the two-product function of Eq. (6.6).

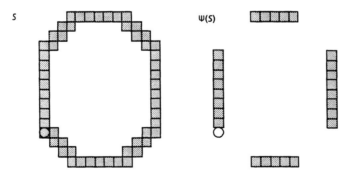

Figure 6.28 Input image S and filter output image $\Psi(S)$ for Ψ defined by the two-product function of Eq. (6.6).

For a second example in which a filter is designed in a reduced form, consider finding pixels that are holes in vertical or horizontal lines, where a white pixel is a vertical hole if it lies vertically between two black pixels and it is a horizontal hole if it lies horizontally between two black pixels. The desired filter is given by the logical function

$$y = x_1 x_3^c x_5 + x_2 x_3^c x_4 \qquad (6.7)$$

and implemented in hardware by the gate structure of Fig. 6.29. Geometrically, it corresponds to a template match by either of the following templates involving *don't care* pixels:

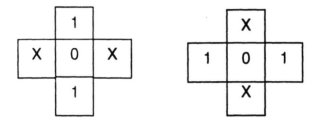

Figure 6.30 shows an image S processed by the corresponding filter Ψ.

A salient logical difference between the representations of Eqs. (6.6) and (6.7) is that there are no complements in Eq. (6.6) and there are complements in Eq. (6.7). While it is possible to provide an equivalent representation of the function defined by Eq. (6.6) that includes complemented variables, it is not possible to give an equivalent representation of the function defined by Eq. (6.7) that has no complemented variables. There are key structural and logical differences between filters defined by logical functions that can and cannot be represented without complemented variables, and design techniques for these two types of filters are often distinguished. If a filter possesses a complement-free representation, then it is called an *increasing filter*[49]; otherwise it is called a *nonincreasing filter*. In terms of templates, an increasing filter can be constructed using templates having no 0s (possibly with *don't cares*), whereas a nonincreasing filter cannot be constructed with 0-free templates.

Template-matching filters fall into the domain of *morphological image processing*. In general, morphological image processing is shape-based and the geometric content of the image is probed. Because a template can be considered to be a pattern (shape), *ipso facto*, the processing is morphological. For a more detailed discussion on morphological operations see Ref 49.

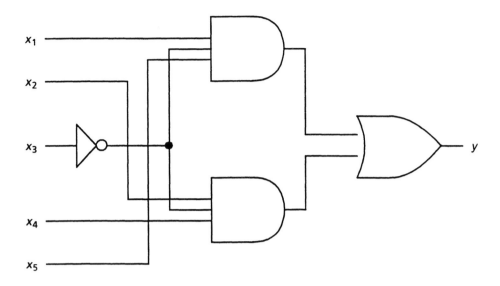

Figure 6.29 Gate structure for the two-product function of Eq. (6.7).

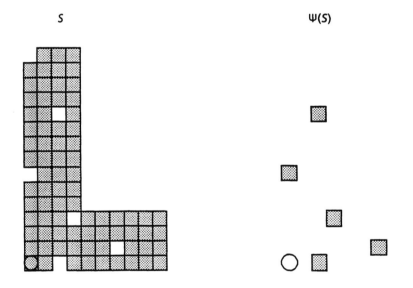

Figure 6.30 Input image S and filtered image $\Psi(S)$ for the two-product function h of Eq. (6.7).

6.4 BASIC DOCUMENT PROCESSING OPERATIONS

In general, the operators utilized in restoration, enhancement, and appearance tuning of digital documents can possess rather involved design procedures and implementations. However, a common collection of document problems may be addressed with relatively simple operators. Several of these document processing tasks are described below.

6.4.1   Thickening character strokes that are too thin

Consider the need to thicken the width of character strokes.[24,26,37,50] One may be in possession of a digital image where the character strokes are too thin for any of a variety of reasons. In the orthographic setting, this problem occurs when a PDL image has been decomposed for a printer with a large writing spot and the actual target printer has a small writing spot. Stroke adjustment can be performed in the bit map to yield a print that appears as if it was produced with the originally intended printer. In the reprographic or scanned document setting, excessively thin image features may have been introduced by any of a number of physical mechanisms. A common office document problem concerns digitization of multi-generation photocopies where some density has been lost on each generation. When dealing with archived or ancient material, a hardcopy original may have been subject to aging or weathering that lightened the density of the text. Digitization and thresholding results in thinned broken text.

Simple thickening can be achieved by ORing the values of selected neighboring pixels throughout an image. Consider the observation window

x_1	x_2	x_3
x_4	x_5	x_6
x_7	x_8	x_9

with origin at the pixel possessing value x_5. This window may be used to define the logical function

$$y = h(x_1, x_2, ..., x_9) = x_1 + x_2 + ... + x_9. \tag{6.8}$$

The computed value y is 1 if any pixels in the observed window possess value 1. When the logical sum h is employed to define a translation-invariant filter Ψ, the filter has the effect of dilating image structures. In fact, *morphological dilation* is defined as a translation-invariant filter that may be represented as an

OR of neighboring pixels. Figures 6.31(a) and 6.31(b) show the effect of dilation on an excessively thin character. The gate structure for Eq. (6.8) is shown in part 6.31(c).

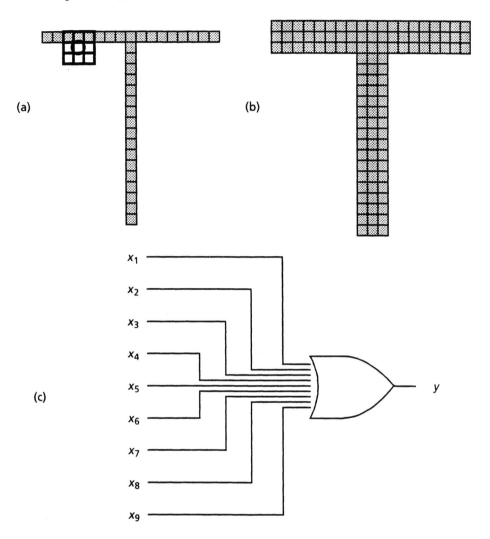

Figure 6.31 (a) Excessively thin input image. The window is shown at a location that yields an activated pixel In the dilation; (b) result of dilating by 3×3, Ψ defined by Eq. (6.8); (c) gate structure for the 3×3 dilation of Eq. (6.8).

Often, dilating an image structure by an integer number of pixels does not allow for sufficiently fine adjustment of stroke width in a printing system. In this case, one often proceeds by identifying the *outside border* of a character and applying some special treatment, such as partial exposure, to those pixels. The outside border can be found by taking the difference between the input image and the dilated image. This operation may be implemented by defining a translation-invariant filter Ψ via a Boolean function that ANDs the complement of the origin pixel with the logical sum. For the 3×3 window, Ψ is defined by

$$y = h(x_1, x_2, ..., x_9) = (x_1 + x_2 + ... x_9) - x_5 = (x_1 + x_5 + ... x_9) \, x_5^c. \qquad (6.9)$$

Figure 6.32(a) shows the result of applying an outside border operator to the image of Fig. 6.31(a). The gate structure for Eq. (6.9) is shown in Fig. 6.32(b).

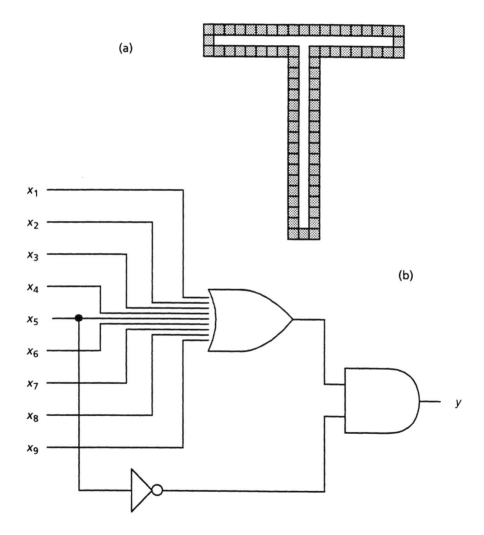

Figure 6.32 (a) Result of outside border operator defined by Eq. (6.9). The outside border pixels may be identified and encoded in some special way to allow for partial pixel growth around a character. (b) Gate structure for Eq. (6.9).

6.4.2 Thinning character strokes that are too thick

Characters and other image structures often require thinning for the opposite reasons cited above for thickening. Simple thinning may be achieved by a single uncomplemented Boolean product, which is known as *morphological erosion*. For the 3×3 window, the defining Boolean function may be written

$$y = h(x_1, x_2, ..., x_9) = x_1 \ x_2 \ ... \ x_9. \tag{6.10}$$

An example of thinning a character by erosion is shown in Fig. 6.33.

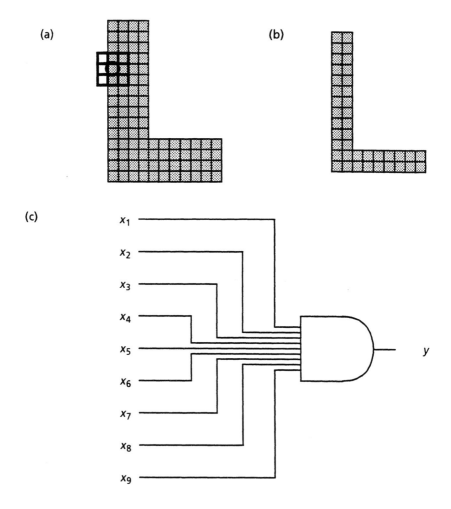

Figure 6.33 (a) Excessively thick input image. The window is shown at a location that yields an unactivated pixel In the erosion; (b) result of eroding by 3×3, Ψ defined by Eq. (6.10); (c) gate structure for the 3×3 erosion of Eq. (6.10).

Analogous to Eq. (6.9), an inside border may be obtained by defining a translation-invariant filter by a Boolean function that ANDs the origin pixel with the complement of the logical product,

$$y = h(x_1, x_2, ..., x_9) = x_5 - (x_1 \ x_2 ... x_9) = x_5(x_1 x_2 ... x_9)^c. \tag{6.11}$$

The result of the inside border operator is shown in Fig. 6.34.

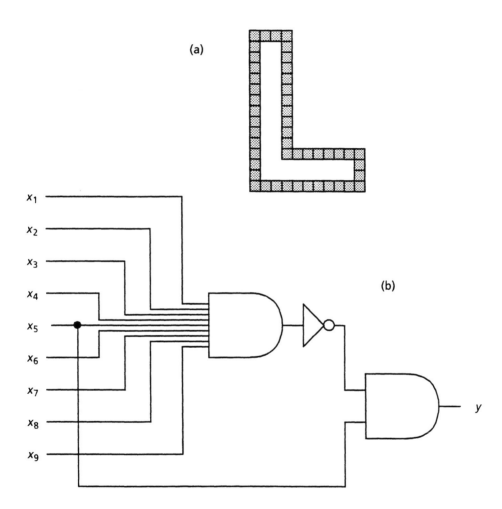

Figure 6.34 (a) Result of inside border operator defined by Eq. (6.11). Inside border pixels may be identified and encoded in some special way to allow for partial pixel shrinkage around a character. (b) Gate structure for Eq. (6.11).

6.4.3 Characters with holes or breaks

Character strokes can become thin and broken for reasons cited above, and it was shown how dilation can improve the appearance of these strokes. In some cases, characters can break or possess holes while the stroke width may be acceptable.[51] Many print papers are textured and can result in density dropouts while stroke width is maintained. Also, skeleton-based recognition algorithms often require connection of breaks in the skeleton.[52] Connecting broken characters is key to enabling accurate machine character recognition (OCR).

There are a variety of filters that can be applied to fill holes. The filter that you choose to employ depends upon the nature of the image and how the holes and breaks lie within the image. To understand the basic operation of hole fillers, let us consider a simple set of templates that can treat many cases:

The filter defined by these templates possesses the Boolean function representation

$$y = h(x_1, x_2, \ldots, x_9) = x_5 + x_4 x_6 + x_1 x_9 + x_2 x_8 + x_3 x_7. \tag{6.12}$$

When employed as a translation-invariant filter, the singleton template x_5 behaves as an identity operator: whatever pixels are valued 1 and 0 in the input image are valued 1 and 0 in the output image, respectively. ORed onto that identity image is the result of each 2-pixel product. The templates corresponding to those products possess 1s that straddle the origin pixel. In this configuration a template can "fit," or yield a 1, when positioned about a hole or break in a character stroke. Figure 6.35 shows the gate structure for Eq. (6.12) and an example of operating on an image. Note that all of the breaks have been repaired, thereby enabling more accurate character recognition, but the restoration is not perfect. A more complicated (more and different products) filter could achieve better restoration.

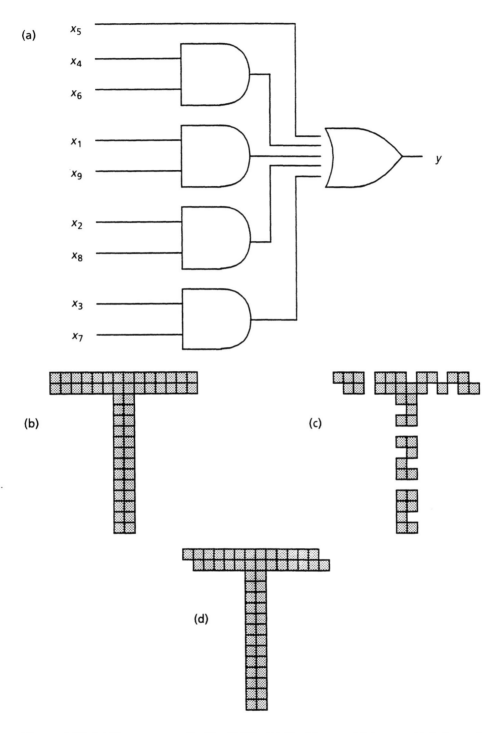

Figure 6.35 (a) Gate structure for Eq. (6.12); (b) ideal image; (c) image with holes and breaks; (d) image resulting from application of filter defined by Eq. (6.12).

6.4.4 Black point noise removal

Black point noise is a common degradation that can result from dirt or marks on the platen of a document scanner or on the original hardcopy of the document. Removal of these isolated extraneous black points is a common scanned-document cleaning operation.[51] Isolated black pixels can be located with the template

0	0	0
0	1	0
0	0	0

which can be represented by the Boolean function

$$f = h(x_1, x_2, ..., x_9) = (x_1^c x_2^c x_3^c x_4^c x_6^c x_7^c x_8^c x_9^c) \, x_5 \, . \tag{6.13}$$

Once located, the isolated black pixels are deleted from the image (see Ref. 49 for discussion on differencing filters). The value output by the overall filter is defined by the Boolean function

$$y = x_5 - f = x_5 f^c$$

$$= x_5 [(x_1^c x_2^c x_3^c x_4^c x_6^c x_7^c x_8^c x_9^c)^c + x_5^c]$$

$$= x_5 (x_1 + x_2 + x_3 + x_4 + x_6 + x_7 + x_8 + x_9) \, . \tag{6.14}$$

6.5 SPATIAL RESOLUTION CONVERSION AND ENHANCEMENT

Technologies that enable portability of document bit maps are of prime importance to digital document processing systems. For example, digital documents are often printed or displayed using different devices, where each device may operate at a different spatial sample resolution. It is desirable for the image to have an appearance that is consistent between all of these devices, or possibly enhanced on a chosen device. Methods for spatial resampling of binary images are at the root of document portability technology. Other fields of image processing often operate in the gray-scale setting, and resampling may be adequately performed by some spline-fitting technique. For binary digital documents these methods are typically not suitable. In this chapter we examine two methods for performing spatial resampling using nonincreasing operators that are more appropriate for the binary image setting.

Spatial resolution conversion is the mapping from one spatial sampling resolution to another, e.g., 300 spi to 600 spi. Strictly speaking, spatial resolution conversion can be performed without attempting to enhance the appearance of the image. But, most often, enhancement is attempted during the conversion step, and the more accurate terminology is *spatial resolution conversion with enhancement.* Throughout this chapter, we assume an enhancement is being performed in the spatial resolution conversion operation.

In a typical office setting, digital documents may be handled by a variety of devices and software packages that interoperate to form a system capable of tasks such as scanning, printing, display, transmission, and various forms of processing. The devices that comprise this system may each operate at a different spatial sampling resolution. To enable a comprehensive system, spatial resampling must often be performed to allow porting of the digital image amongst the various devices. Most often the digital documents are in binary form and therefore resampling is best performed by logic-based techniques.

Simplistic scaling methods are sometimes applied to the resampling problem but usually don't yield satisfactory results. As described above, simplistic scaling is usually based on bit replication or direct printing at an incorrect resolution. The dominant image defects that result are excessive blockiness, beating in periodic image structures, or magnification (minification) of the image. Below, we examine two general logic-based methods that employ nonincreasing (template matching) filters. One method passes the input image through several filters in parallel, and the resulting image planes are interleaved to form a resampling of the underlying image. The other method first performs a nearest neighbor resampling, then filters this image to estimate samples of the underlying image.

6.5.1   Categories of resolution conversion

A resolution conversion and enhancement algorithm may be categorized by the ratio of the input and output sampling resolutions. Let us define resolution parameters; let N_1 and N_2 be the input sampling resolutions in spi in the horizontal and vertical direction on a page, respectively, and let M_1 and M_2 be the respective output sampling resolutions. The input image and ideal output image may be denoted as A_{N_1, N_2} and A_{M_1, M_2}, respectively. For resampling at a higher resolution ($M_1/N_1>1$, $M_2/N_2>1$) we typically desire an algorithm that performs *conversion with enhancement.* The resampled bit map should utilize its higher resolution to better represent fine features in the underlying image and minimize jaggedness in curved and angled strokes.

In laser printers, control of a high-speed diode laser allows for much finer addressing in the scan line direction compared to the direction perpendicular to the scan lines, which is controlled by the motion of the image substrate.

Printing with an image bar, such as a linear LED array, presents the opposite situation where there is a relative ease to print at high resolution in the direction of photoreceptor motion. Due to device physics, it is common to perform image enhancement by resampling from an isomorphic resolution (N_1=N_2) to a resolution that is high in only one direction (M_1>M_2, M_2=N_2, or M_2>M_1, M_1=N_1).[43–46] High addressability is typically performed as an *integer* resolution conversion.

Integer conversion to a higher resolution involves resampling such that M_1/N_1 and (or) M_2/N_2 are integers; the inverse ratios are integers for decimation. Integer resampling to a higher resolution in only one direction is the addressability enhancement case described above. Integer resampling tends to require simpler implementation architecture and result in fewer conversion artifacts than *noninteger* resampling. Noninteger resampling is often performed via an *area mapping*. Area-mapping methods map an input block of whole pixels to an output block of whole pixels. For example, when converting a document sampled at 240 to 600 spi, an input pixel maps to a 2½ ×2½ block of output pixels. An area-mapping algorithm would map a 2×2-pixel block in the input bit map to a 5×5-pixel block in the output image.[1,7,15,17] See Fig. 6.36.

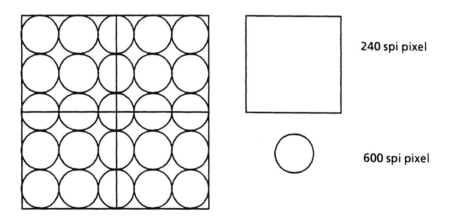

240 spi pixel

600 spi pixel

Figure 6.36 Area mapping a 240 spi image to 600 spi. Note that pixels at either resolution tile the plane in the same manner. The circle and square representation is used here only to show the relative dimensions of the grids.

For resampling at a lower resolution (M_1/N_1<1, M_2/N_2<1), the algorithm should perform *decimation with minimized coarse-sample degradation*. Decimation is often performed on electronic signature capture and verification devices. Digitizing devices often capture the stylus information at approximately 500 spi, while the LCD display of that data may be at 75 spi.[14]

Simple subsampling could result in image dropouts at regions of fine structure. When converting to a coarser resolution these structures must be made thicker or intelligently sampled[14] to allow their representation in the image (see Fig. 6.37). The need to thicken or thin image features for modified sampling resolution is related to the approach taken with PDL decomposers when rendering a bit map at different resolutions. The decomposer must render all image structures to sampled form without dropouts. In the function-based description of PDL form, the underlying image structures are piecewise constant, but any given piece of a character may be adjusted in size for the given resolution in a manner that assures fine features are represented in bit-map form.

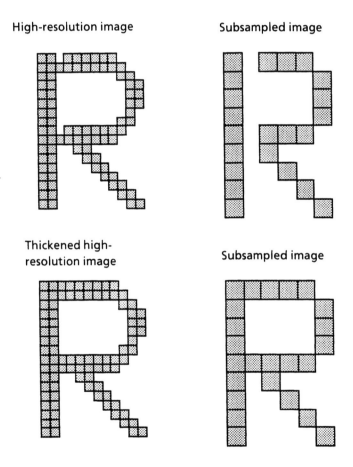

Figure 6.37 Illustration of the need to thicken selected image features to enable their representation in a coarsely sampled form.

Another sample conversion setting receives a binary image at a spatial resolution that is inappropriate for the given printer and the printer can support several levels of gray-scale printing. For noninteger conversion and for general magnification operations, it is difficult to maintain local density and binary quantization. Hence, quantization range expansion is employed to preserve local density in the resampling operation. Often, averaging or error diffusion is employed in these conversions.[2,3,5-7,53]

6.5.2 Resolution conversion by multiple parallel filters

6.5.2.1 Integer conversion

Integer resolution conversion[4,8,12,15,18,21] is accomplished by mapping each pixel in the low-resolution input bit map to a corresponding block of pixels in the high-resolution output bit map. For input sampling resolutions N_1 and N_2 and output sampling resolutions M_1 and M_2 with integer ratios $M_1/N_1>1$ and $M_2/N_2>1$ let A_{N_1,N_2} and A_{M_1,M_2} be the low-resolution input and high-resolution output images, respectively, which we shall write in compact form A_N, A_M for the analysis below. If each pixel i in the low-resolution input is mapped into L pixels, then the underlying pixel transformation is of the form

$$
i \rightarrow \begin{pmatrix} (i,1) \\ (i,2) \\ \cdot \\ \cdot \\ \cdot \\ (i,L) \end{pmatrix}.
\tag{6.15}
$$

Each pixel value $A_N(i)$ of the low-resolution bit map is a single sample of a region of some underlying image and the corresponding high-resolution bit map has multiple samples for the same region. We treat these multiple samples as components of a vector defining the ith block of the output image A_M, namely,

$$
\mathbf{A}_M[I] = \begin{pmatrix} A_M(i,1) \\ A_M(i,2) \\ \cdot \\ \cdot \\ \cdot \\ A_M(i,L) \end{pmatrix}.
\tag{6.16}
$$

Assuming there are R pixels forming the low-resolution image, for $i = 1, 2, ...,$ R, each component $A_M(i, j)$ of $\mathbf{A}_M[i]$ gives the output high-resolution image value at pixel (i, j), $j = 1, 2, ..., L$. In vector form, the input low-resolution image is written as

$$A_N = \begin{pmatrix} A_N(1) \\ A_N(2) \\ \cdot \\ \cdot \\ \cdot \\ A_N(R) \end{pmatrix}, \tag{6.17}$$

and, in block form, the full high-resolution image is written as

$$A_M = \begin{pmatrix} \mathbf{A}_M[1] \\ \mathbf{A}_M[2] \\ \cdot \\ \cdot \\ \cdot \\ \mathbf{A}_M[R] \end{pmatrix}. \tag{6.18}$$

For each $j = 1, 2, ..., L$, the collection (over all i) of pixels $A_M(i, j)$ forms a digital image that is the j subsampling phase of A_M, which in vector form can be written as

$$A_M(\cdot, j) = \begin{pmatrix} A_M(1, j) \\ A_M(2, j) \\ \cdot \\ \cdot \\ \cdot \\ A_M(R, j) \end{pmatrix}. \tag{6.19}$$

Figure 6.38 illustrates blocking and subsampling phases.

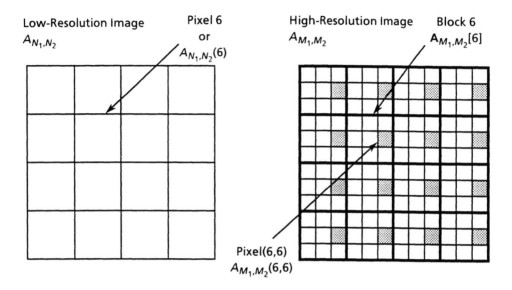

Figure 6.38 Blocking and subsampling phases. Gray pixels denote the $j = 6$ subsampling phase of A_M.

A resolution-conversion filter Ψ is defined by observing pixel values in a window about pixel i in the low-resolution image and using these values to estimate the L values forming the corresponding high-resolution block $\mathbf{A}_{M_1,M_2}[i]$. The filter is of the form

$$\Psi \equiv \begin{pmatrix} \Psi_1 \\ \Psi_2 \\ \cdot \\ \cdot \\ \cdot \\ \Psi_L \end{pmatrix}, \tag{6.20}$$

where the component filters are used to obtain estimates $\hat{A}_M(i, j)$ of the components $A_M(i, j)$ of $\mathbf{A}_M[i]$. The filter output $\Psi_j(A_N)(i)$ serves as the inferential estimate of $A_M(i, j)$:

$$
\begin{pmatrix} \hat{A}_M(i,1) \\ \hat{A}_M(i,2) \\ \cdot \\ \cdot \\ \cdot \\ \hat{A}_M(i,L) \end{pmatrix} = \begin{pmatrix} \Psi_1(A_N)(i) \\ \Psi_2(A_N)(i) \\ \cdot \\ \cdot \\ \cdot \\ \Psi_L(A_N)(i) \end{pmatrix}. \tag{6.21}
$$

In vector form,

$$
\hat{\mathbf{A}}_M[i] = \Psi(A_N)(i). \tag{6.22}
$$

Blockwise, $\Psi(A_N)(i)$ estimates the image samples defining the high-resolution block $\mathbf{A}_M[i]$.

Each component vector Ψ_j forming Ψ has a Boolean function f_j associated with it and f_j is defined over an n-pixel window W. Define the vector-valued Boolean function $\mathbf{f}$ on n variables $x_1, x_2, ..., x_n$ by

$$
\mathbf{f}(x_1, x_2, ..., x_n) = \begin{pmatrix} f_1(x_1, x_2,..., x_n) \\ f_2(x_1, x_2,..., x_n) \\ \cdot \\ \cdot \\ \cdot \\ f_L(x_1, x_2,..., x_n) \end{pmatrix}. \tag{6.23}
$$

The function $\mathbf{f}$ is an L-valued function on n variables and it is the window function for Ψ because

$$
\Psi(A_N)(i) = \mathbf{f}(A_N(i+w_1), A_N(i+w_2), ..., A_N(i+w_n)). \tag{6.24}
$$

Imagewise,

$$
\Psi(A_N) = \Psi \begin{pmatrix} A_N(1) \\ A_N(2) \\ \cdot \\ \cdot \\ \cdot \\ A_N(R) \end{pmatrix}
$$

$$
= \begin{pmatrix}
\mathbf{f}(A_N(1+w_1), A_N(1+w_2),..., A_N(1+w_n)) \\
\mathbf{f}(A_N(2+w_1), A_N(2+w_2),..., A_N(2+w_n)) \\
\cdot \\
\cdot \\
\cdot \\
\mathbf{f}(A_N(R+w_1)A_N(R+w_2),..., A_N(R+w_n))
\end{pmatrix}, \tag{6.25}
$$

which is an RL-vector estimating the high-resolution image

$$
A_M = \begin{pmatrix}
A_M(1,1) \\
A_M(1,2) \\
\cdot \\
\cdot \\
\cdot \\
A_M(1,L) \\
A_M(2,1) \\
A_M(2,2) \\
\cdot \\
\cdot \\
\cdot \\
A_M(2,L) \\
A_M(3,1) \\
\cdot \\
\cdot \\
\cdot \\
A_M(R,L)
\end{pmatrix} \tag{6.26}
$$

Although we have described the vector-valued window function $\mathbf{f}$ for Ψ in the context of resolution conversion, the concept is general and applies to any circumstance in which a vector of observation values is being used to determine a vector of output values. Two different terminologies can be applied in the present context. Since each pixel in the input image corresponds to L pixels in the output, the mapping Ψ performs $1 \rightarrow L$ pixel conversion. Since the window function is an L-valued function of n variables (in the window), the window function is an $n \rightarrow L$ transformation.

Figure 6.39 illustrates a 3 × 3 window at 300 spi being used in a conversion to 600 spi (the large circle denoting the window origin). For 300 to 600 conversions, Ψ estimates four pixels in the 600 spi document at each location in the 300 spi document.

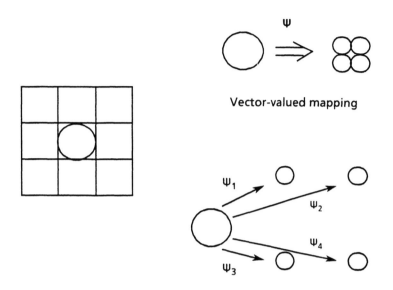

Figure 6.39 A 3×3 window at 300 spi used in mapping to 600 spi. The large circle is the origin of the window. For this resolution case, the vector-valued mapping Ψ generated four pixels at each location of the window origin.

If for this example we label the values in the 3×3 window according to Fig. 6.40, then there are four Boolean functions f_1, f_2, f_3, and f_4 corresponding to the component filters $\Psi_1, \Psi_2, \Psi_3, \Psi_4$, and, for $j = 1, 2, 3, 4$,

$$\Psi_j(A_N)(i) = f_j(x_1, x_2, ..., x_9),\tag{6.27}$$

where $x_1, x_2, ..., x_9$ are the values in the translated window W_z.

Example: Orthographic 300 spi conversion to 600 spi

Consider the problem of orthographic 300 spi to 600 spi conversion with enhancement[21,24] or, equivalently, 240 spi to 480 spi.[8] An *orthographic bit map* is a bit map that is computer-generated from a Page Description Language. Such a conversion problem is encountered when a document bit map has been created at 300 spi by decomposition from PDL form and the current printing device is capable of 600 spi resolution.

Although filters constrained by a small observation window (3 × 3) may be well designed manually by human experts, larger windows that enable greater enhancement are best designed by analytical optimization methods.[7] For this example we employ the optimal design methods of Ref. 49 to obtain the filter templates for a 5 × 5 window. We chose a 5 × 5 window because it is large enough to treat most image structures that could benefit from this resolution enhancement. Figure 6.41 shows an example of increased enhancement enabled by a larger window. Ideally, a larger window could perform a better enhancement, but at the cost of a greater number of templates and an increased number of line buffers in the printing device (see Refs. 43 and 54 for hierarchical template matching methods that mitigate some of the problems associated with large windows).

Photomicrographs of actual prints show the effect and quality of the enhancement that is achievable with a statistically optimized 5 × 5 filter. Figure 6.42(a) is a print of italic text using the ideal 600 spi bit map, which was decomposed directly from PDL form. Figures 6.42(b) and 6.42(c) show the results of converting a 300 spi bit map to 600 spi by (b) simple nearest-neighbor interpolation (bit double, line repeat) and (c) the 4-phase conversion filter. Notice that the conversion filter generates a significantly smoother print than simple replication. Also note that the ideal 600 spi image is slightly smoother than the filter-converted image. To obtain increased performance a larger window could be used or a second iteration could be employed.

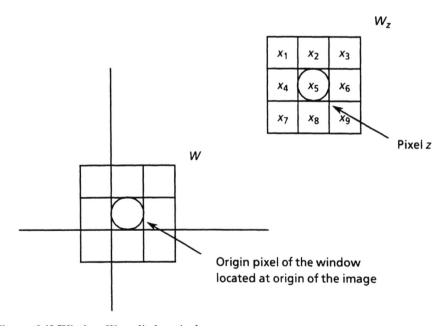

Figure 6.40 Window W applied at pixel z.

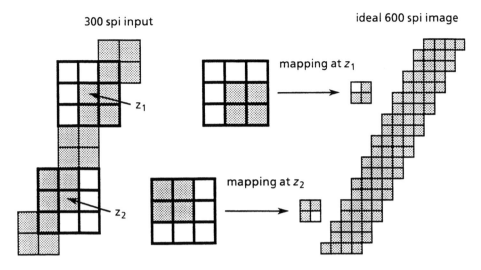

3 × 3 window can achieve ideal mapping for 2-up-1-over jaggie

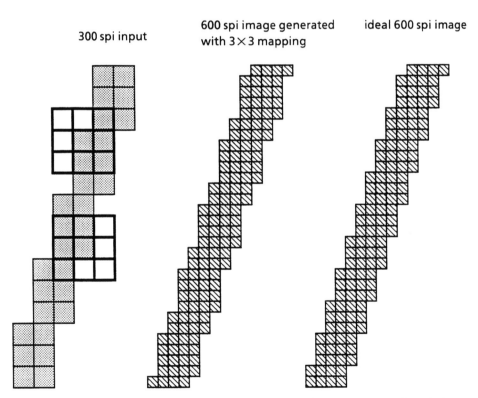

3 × 3 window cannot recognize a 3-up-1-over slope. The mapping employed above generates a nonideal 4-up-1-over-2-up-1-over line.

Figure 6.41 A 3×3 window cannot distinguish slopes beyond 2-up-1-over and therefore performs a mapping that is less than ideal. Greater slopes could be treated ideally with a larger window.

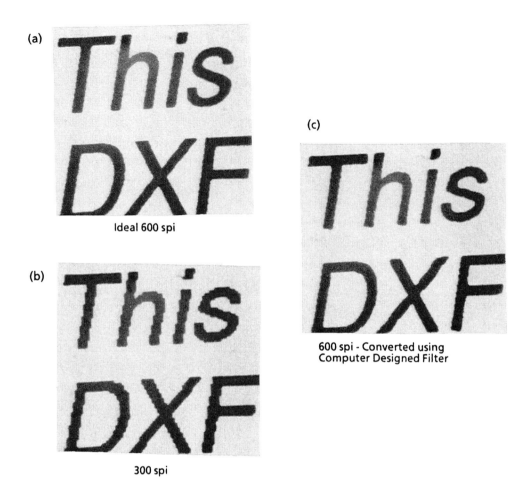

(a)

Ideal 600 spi

(b)

300 spi

(c)

600 spi - Converted using
Computer Designed Filter

Figure 6.42 Photomicrograph of italic text: (a) ideal 600 spi image; (b) nearest-neighbor interpolation; (c) 4-phase filter conversion.

6.5.2.2 Noninteger conversion

Noninteger resolution conversion can be performed using a scheme related to the integer resolution conversion method.[10,15] Consider the problem of converting an image sampled at 400 spi to one sampled at 600 spi. The observation window, when placed in the 400 spi image, can fall at a finite number of grid phase relationships in the 600 spi image. Because of the 2/3 × 2/3 sampling ratio ($N_1/M_1 \times N_2/M_2$) there are four grid phases that generate a total of nine output pixels. In general,

$$P = \frac{N_1}{GCD(N_1, M_1)} \times \frac{N_2}{GCD(N_2, M_2)} , \qquad (6.28)$$

$$L = \frac{M_1}{GCD(N_1, M_1)} \times \frac{M_2}{GCD(N_2, M_2)} , \qquad (6.29)$$

where P is the number of grid phases, L is the total number of pixels generated for all the grid phases, and $GCD(a, b)$ is the *greatest common divisor* of a and b.

For integer conversion, the image operator Ψ can be defined by Eq. (6.20) in which each Ψ_j forming the vector provides a sample phase of the output image. Each pixel i is mapped into L pixels, the underlying pixel transformation given by Eq. (6.15), where (i, j) denotes the pixel in the jth phase of the high-resolution image corresponding to pixel i in the low-resolution image. This $1 \rightarrow L$ pixel correspondence is manifested in Eqs. (6.23) and (6.25), where the vector-valued Boolean function $\mathbf{f}$ has L components and depends on values in the window centered at pixel i.

For noninteger conversion, the matter is more complicated. For each block of P pixels in the low-resolution input image, there corresponds a block of L pixels in the output high-resolution image. The pixel transformation of Eq. (6.15) now takes the form

$$\begin{pmatrix} (k,1) \\ (k,2) \\ \cdot \\ \cdot \\ \cdot \\ (k,P) \end{pmatrix}_{N_1, N_2} \rightarrow \begin{pmatrix} (k,1) \\ (k,2) \\ \cdot \\ \cdot \\ \cdot \\ (k,L) \end{pmatrix}_{M_1, M_2} \qquad (6.30)$$

where k now refers to the kth block in the input image A_N and output image A_M. In electronic printing, pixel block transformation is often referred to as area mapping.[1,7,15,17]

6.5.3 Resolution conversion by filtering in the resampled space

For digital printer applications, spatial resolution conversion must often be performed in real time. Hardware implementations are usually required to process several pages per minute. A given filter architecture translates into a hardware architecture. For reasons of available memory partitions, for template storage, and compatibility with other image processing hardware, it may be desirable to filter using an architecture other than the one described above. One straightforward alternative is first to perform a nearest-neighbor resampling of the image and filter for enhancement or artifact correction in the resampled space.[5,55] One advantage of this method over the previously described method is that the Boolean functions defining the conversion filter Ψ can be represented as single N-bit-in-to-1-bit-out look-up tables. A disadvantage occurs when resampling to a higher resolution, because a large number of pixels must be used in the window to cover the same physical portion of the document image. When considering cost associated with a given window, two key factors are (1) the logic gate count for all the pixels employed, (2) the number of scan lines spanned by the window, which typically determines how many lines must be buffered during the filtering operation.

REFERENCES

1. Coward, R., "Area Mapping Table Look-up Scheme," US Cl 355/287, *Xerox Disclosure Journal*, Vol. 18, No. 2, pp. 217-222, Mar./Apr., 1993.
2. Coward, R., and J. Parker, "Unquantized Resolution Conversion of Bitmap Images Using Error Diffusion," US Patent 5,363,213, Nov. 8, 1994.
3. Eschbach, R., "Method for Making Image Conversions with Error Diffusion," US Patent 5,226,094, Jul. 6, 1993.
4. Handley, J., and E. R. Dougherty, "Model-Based Optimal Restoration of Fax Images in the Context of Mathematical Morphology, *Journal of Electronic Imaging*, Vol. 3, No. 2, April 1994.
5. Kang, H. R., "Resolution Conversion of Bitmap Images," US Patent 5,270,836, Dec. 14, 1993.
6. Kang, H. R., "Resolution Conversion with Simulated Multi-Bit Gray," US Patent 5,301,037, Apr. 5, 1994.
7. Kang, H., and R. Coward, "Area Mapping Approach for Resolution Conversion," *Xerox Disclosure Journal*, Vol. 19, No. 2, Mar./Apr., 1994.
8. Kantor, S., "Pel Resolution Addressing Conversion," US Patent 4,975,785, Dec. 4, 1990.
9. Kato, Y., "Image Communication System," US Patent 4,814,890, Mar. 21, 1989.
10. Loce, R., M. S. Cianciosi, and R. V. Klassen, "Non-Integer Image Resolution Conversion Using Statistically Generated Look-Up Tables," US Patent 5,387,985, Feb. 7, 1995.
11. Loce, R., E. R. Dougherty, R. E. Jodoin, and M. S. Cianciosi, "Logically Efficient Spatial Resolution Conversion Using Paired Increasing Operators," *Journal of Real-Time Imaging*, 3(1) 7–16, Feb. 1997.
12. Mailloux, L. D., and R. E. Coward, "Bit-Map Image Resolution Converter," US Patent 5,282,057, Jan. 25, 1994.
13. Marko, K. R., and C. A. Storlie, "Continuously Variable Resolution Laser

Printer," US Patent 5,239,313, Aug. 24, 1993.

14. Memarzadeh, K., "Method for Converting High Resolution Data into Lower Resolution Data," US Patent 5,283,557, Feb. 1, 1994.

15. Miller, S., "Method and Apparatus for Mapping Printer Resolution Using Look-up Tables," US Patent 5,265,176, Nov. 23, 1993.

16. Nakajima, S., and K. Izawa, "Image-Resolution Conversion Apparatus for Converting a Pixel Density of Image Data," US Patent 4,841,375, Jun. 20, 1989.

17. Papaconstantinou, P., "Hardware Architecture for Resolution Conversion Using Area Mapping," US Cl 382/041, *Xerox Disclosure Journal*, Vol. 18, No. 5, pp. 553-562, Sep./Oct., 1993.

18. Walsh, B. F., and D. E. Halpert, "Low Resolution Raster Images," US Patent 4,437,122, Mar. 13, 1984.

19. Adobe Systems Inc., *PostScript® Language Reference Manual, Second Edition*, Addison-Wesley, Reading, 1995.

20. Urban, S. J., "Review of Standard for Electronic Imaging for Facsimile Systems," *J. Electronic Imaging*, Vol. 1, No. 1, pp. 5-21, Jan. 1992.

21. Morimoto, K., S. Murakami, M. Aiba, and Y. Kamei, "An Image Recording Apparatus for Providing High Quality Image," US Patent 5,289,564, Feb. 22, 1994.

22. Westheimer, G., *Visual Hyperacuity*, Progress in Sensory Physiology 1, Springer-Verlag, New York, 1981.

23. Curry, D. N., "Hyperacuity Laser Imager," *Journal of Electronic Imaging*, Vol. 2, No. 2, pp 138-146, Apr. 1993.

24. Mailloux, L., and T. Robson, "Dilation of Images without Resolution Conversion for Printer Characteristics," US Patent 5,483,351, Jan. 9, 1996.

25. Bassetti, L. W., "Fine Line Enhancement," US Patent 4,544,264, Oct. 1, 1985.

26. Crawford, J., and J. Cunningham, "Boldness Control in an Electrophotographic Machine," US Patent 5,128,698, Jul. 7, 1992.

27. Carely, A. L., "Resolution Enhancement in Laser Printers," Copyright 1993, XLI Corp., Woburn, MA.

28. Frazier, A. L., and J. S. Pierson, "Resolution Transforming Raster-Based Imaging System, " US Patent 5,134,495, Jul. 28, 1992.

29. Frazier, A. L., and J. S. Pierson, "Interleaving Vertical Pixels in Raster-Based Laser Printers," US Patent 5,193,008, Mar. 9, 1993.

30. Gilbert, J., L. Lukis, and L. Steidel, "Non-Gray Scale Anti-Aliasing Method for Laser Printers," US Patent 5,041,848, Aug. 20, 1991.

31. Steidel, L., "Technology Overview: Resolution Enhancement Technologies for Laser Printers," Publication by LaserMaster Corp., Sep. 1991.

32. Crow, F. C., "The Use of Gray-scale for Improved Raster Display of Vectors and Characters," *Computer Graphics*, Vol. 12, Aug. 1978.

33. Bassetti, L. W., "Interacting Print Enhancement Techniques," US Patent 4,625,222, Nov. 25, 1986.

34. Bunce, R., "Pixel Image Enhancement Employing a Reduced Template Memory Store," US Patent 5,237,646, Aug. 17, 1993.

35. Cianciosi, M., "Digital Video Pulse Width and Position Modulator," US Patent 5,184,226, Feb. 2, 1993.

36. Cianciosi, M., R. Loce, M. Banton, and R. Jodoin, "Method and Apparatus for Enhancing Discharged Area Development Regions in a Tri-Level Printing System," US Patent 5,479,175, Dec. 26, 1995.

37. Crawford, J. L., and C. D. Elzinga, "Improved Output Quality by Modulating Recording Power," Conference Record, Electronic Imaging 88—Institute for Graphic Communication, October 3–6, 1988, pp. 516–522.

38. Lung, C. Y., "Edge Enhancement Method and Apparatus for Dot Matrix Devices," US Patent 5,029,108, Jul. 2, 1991.

39. Sanders, J. R., W. Hanson, M. Burke, R. S. Foresman, J. P. Fleming, "Behind Hewlett-Packard's Patent on Resolution Enhancement® Technology," B. Colgan, ed., Prepared by Torrey Pines Research Carlsbad CA, Distributed by BIS CAP International, Newtonville MA, 1990.

40. Tung, C., "Piece-Wise Print Image Enhancement for Dot Matrix Printers," US Patent 4,847,641, Jul. 11, 1989; US Patent 5,005,139, Apr. 2, 1991.

41. Tung, C., "Resolution Enhancement in Laser Printers," in Proc. SPIE 1912, *Color Hard Copy and Graphics Arts II*, Jan. 31, 1993, San Jose, CA.

42. Yoknis, M. E., "Method and Apparatus for Print Image Enhancement," US Patent 4,933,689, Jun. 12, 1990.

43. Curry, D. N., R. St. John, and S. Filshtinsky, "Enhanced Fidelity Reproduction of Images by Hierarchical Template Matching," US Patent 5,329,599, Jul. 12, 1994.

44. Murata, K., "Image Processing Method and Apparatus," US Patent 5,450,208, Sep. 12, 1995.

45. Nagata, K., T. Kojima, and Y. Tadama, "Method for Smoothing Image," US Patent 5,404,233, Apr. 4, 1995.

46. Watanabe, T., T. Uenishi, H. Sahara, and K. Yamsoka, "Smoothing Circuit for Display Apparatus," US Patent 4,544,922, Oct. 1, 1985.

47. Auyeung, V., "Raster Output Scanner with Subpixel Addressability," US Patent 5,325,216, June 28, 1994.

48. Zack, G., and W. Nelson, "The Font Solution - Optimized Scaling and Production of Raster Fonts from Contour Masters," US Patent 5,459,828, Oct., 17, 1995.

49. Loce, R., and E. Dougherty, *Enhancement and Restoration of Digital Documents*, SPIE Press, Bellingham WA, 1997.

50. Barski, L., and R. Gaborski, "Image Character Enhancement Using a Stroke Strengthening Kernel," US Patent 4,791,679, Dec. 13, 1988.

51. Petrick, B., and P. Wingfield, "Method and Apparatus for Input Picture Enhancement by Removal of Undesired Dots and Voids." US Patent 4,646,355, Feb. 24, 1987.

52. Lougheed, R., and J. Beyer, "Method for Repairing Images for Optical Character Recognition Performing Different Repair Operations Based on Measured Image Characteristics," US Patent 5,142,589, Aug. 25, 1992.

53. Eschbach, R., "Pixel Quantization with Adaptive Error Diffusion," US Patent 5,208,871, May 4, 1993.

54. Denber, M., "Image Quality Improvement by Hierarchical Pattern Matching with Variable Size Templates," US Patent 5,365,251, Nov. 15, 1994.

55. Kolb, J., and K. Woodruff, "Method and Apparatus for Smoothing an Expanded Bitmap for Two-State Image Data," US Patent 5,526,468, Jun. 11, 1996.

7

DIGITAL HALFTONING FOR PRINTING AND DISPLAY OF ELECTRONIC IMAGES

Robert P. Loce
Paul G. Roetling
Ying-wei Lin
Xerox Corporation
Webster, New York

7.1 INTRODUCTION

This chapter presents encoding methods, referred to as halftoning, that are used to reduce the number of quantization levels per pixel in a digital image, while maintaining the gray appearance of the image at normal viewing distance. Halftoning is widely employed in the printing and display of digital images. The need for halftoning encoding arises either because the physical processes involved are binary in nature or the processes have been restricted to binary operation for reasons of cost, speed, memory or stability in the presence of process fluctuations. Examples of such processes are most printing presses, ink jet printers, binary cathode ray tube (CRT) displays, and laser xerography. In most printing and display applications, the halftoned image is composed ideally of two gray levels, black and white. Spatial integration, plus higher level processing performed by the human visual system, and local area coverage of black and white pixels, provide the appearance of a gray level, or "continuous tone," image. Halftone techniques are readily extended to color and to quantization using more than two levels, but within this chapter there is space to cover these topics only briefly .

Within this chapter, a brief history of the field is presented, starting with analog methods of halftone image rendering and proceeding to the digital techniques of template dot halftones, noise encoding, ordered dither, and error diffusion. Key advances and lessons in the development of the technology are summarized. Following the discussion on historical development, there is a discussion of visual perception, touching only upon the concepts pertinent to an understanding of halftoning.   Then, several current methods, such as ordered dither and error diffusion, are described in detail. For each technique there is a discussion of the methodology as well as issues such as tone

---

The views represented herein are those of the authors and are not intended to represent the position of the employer.

reproduction, screen visibility, image artifacts and robustness. Current directions of research are mentioned. See Eschbach[1] for an additional summary of recent research and Jones[2] for a historical view and trends in halftoning found in the patent literature.

A significant portion of the present chapter has been adapted from Ref. 3.

7.2 HISTORICAL PERSPECTIVE AND OVERVIEW

7.2.1 Analog screening

The technologies for mass reproduction of pictorial imagery include the methods of relief printing, gravure and lithography. A key step in the evolution of this technology was the invention of the photo engraving halftone process by Talbot.[4] In Talbot's method a screen of black gauze is placed between a photosensitive plate and the object whose image is to be reproduced. The resulting sandwich is then exposed to light where the transmittance of the screen and object are multiplied, producing a screen-encoded latent image on the photosensitive plate (Fig. 7.1). Chemical development yields a final image that is binary and screen-encoded such that spatial integration of the human visual system causes the image to have the gray appearance of the original.

The primitive screen of Talbot has evolved through several significant stages of design and analysis. For many years, screens were typically ruled glass, employed out-of-contact, and could be rotated to be used in conjunction with other screens in the reproduction of color prints. Streifer et al.[5] describe the theory of operation of out-of-contact ruled screens. A key step in the evolution of halftoning was the invention of the contact screen.[6] This invention simplified the process by obviating the need for careful spacings and simultaneously yielded improved image quality (diffraction was no longer a degrading factor). Most relevant to the topic of the present chapter is the evolution of the contact screen to digital clustered dot screens (Section 7.7.2.4). The close relationship allows much of the wisdom of the graphic arts printing industry concerning tone reproduction, screen visibility, moiré, etc., to be applied directly to printing using digital halftoning methods.

In summary, almost one and a half centuries of practical experience with analog halftone methods provide a rich source of information, which has not always been fully utilized in optimizing digital methods.

7.2.2 Template dots

The advent of computer driven line printers inspired the desire to print images by various *template-dot* methods. Little or no use of graphic arts knowledge was employed in these early digital prints. In template-dot, each image pixel is represented by a character, or template cell, with its area coverage of ink

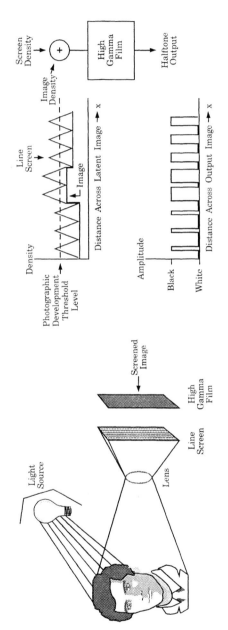

Figure 7.1 Schematic of analog screening.

corresponding to the desired density of that image pixel. To obtain gray-level appearance using these two-level printing apparatus, Perry and Medelsohn[7] employed alphanumeric characters of different sizes and shapes, combined with one level of overstriking. Figure 7.2 shows an example of this early template-dot technique. Macleod[8] extended this method, obtaining more gray levels thereby suppressing false gray-level contours by allowing a greater number of overstrikes and adding pseudo-random noise to the signal prior to conversion to a basic cell. Schroeder[9] described a technique for producing pictures on a microfilm recorder by adjusting the f-stop of the camera and exposing each point on the film several times. Schroeder[10] and Knowlton and Harmon[11] describe other related methods, such as varying dot size, combining spots within a unit cell, and using several output pixels within a cell to represent the brightness of a single input pixel. When printers became more versatile, Arnemann and Tasto[12] proposed a run-length-coding method and Hamill[13] proposed nonalphanumeric cells that can be grown in fine increments with no overstriking. Other names given to this method are *orthographic gray scale*, *surface area modulation - SAM* and *font dot*. Little use or research has been performed recently with template dots because the resolution of the image is limited to the resolution of a halftone cell, whereas other methods can yield resolution as fine as a pixel in certain image situations.

In reviewing the history of the template-dot halftone, we recognize that a significant contribution to halftone image quality often comes from spatial detail too fine to be represented by dots that vary only in size. Template-dot methods are no longer in use for most standard imaging applications because other methods use dot shape (often called *partial dots*) to carry finer detail and thus offer better image quality. To see the relative loss of resolution of the template-dot method, compare the sharpness of Fig. 7.3(a) and 7.3(b), which are halftoned using template dots and ordered dither, respectively (Ordered dither is discussed in Sections 7.2.4 and 7.4.2).

7.2.3 Noise encoding

In spite of the similarity to graphic arts analog screening, digital screening developed more directly from digital communications and display methods. The link to analog methods was made later. Therefore, we discuss the early digital communication developments before proceeding into the halftone imaging of today. *Noise Encoding* is a method of bit reduction that was employed early in the fields of digital image display and communications. Goodall,[14] working with pulse-code modulated (PCM) TV signals, observed that adding a small amount of noise to an input digital image almost eliminated the appearance of gray-level false contouring in the displayed image, but at the expense of a small increase in granularity. Roberts[15] found that under certain noise and quantization conditions, the bit density for observable contours can

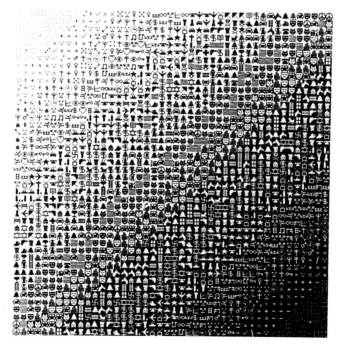

Figure 7.2 An example of template-dot halftoning using a set of symbols. Reproduced with permission from Knowlton and Harmon.[11]

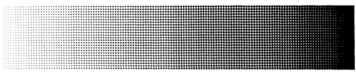

(a)

Figure 7.3 Comparison of (a) template-dot and (b) ordered-dither methods. In these examples, both use a 0° 65 level halftone screen at 300 pixels per inch. Note the raggedness of the tree trunk and roof in (a) as well as a loss in sharpness on edges of the building.

(b)

Figure 7.3 Comparison of (a) template-dot and (b) ordered-dither methods. In these examples, both use a 0° 65 level halftone screen at 300 pixels per inch. Note the raggedness of the tree trunk and roof in (a) as well as a loss in sharpness on edges of the building.

be reduced from 6 or 7 bits to 3 or 4 bits/sample. Roberts added pseudo-random noise of amplitude equal to one quantization level to the video signal before quantizing, then subtracted the same noise at the receiver. Combining this approach with compression and expansion of the intensity scale to match the human visual system, Roberts concluded

> Transmitted data may be reduced to 3 bits per sample for most TV requirements and 4 bits for more demanding applications, and that because the eye tends to average out noise in local areas, distributed quantizing noise is considerably more pleasing to the eye than quantizing contours.

The noise-encoding process is shown in Fig. 7.4(a), and 7.4(b) with an example of noise encoding a one-dimensional signal using a uniformly distributed random variable. A noise-encoded image is shown in Fig. 7.5.

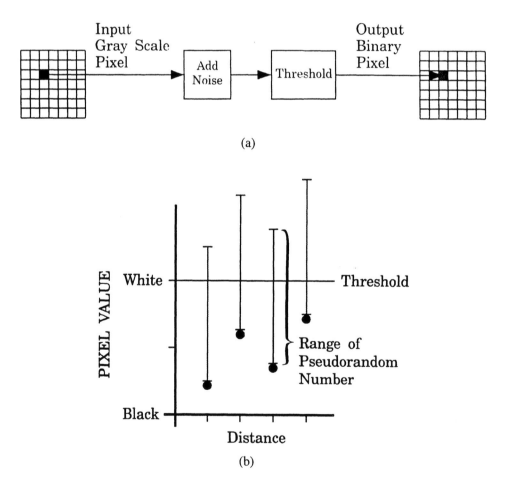

(a)

(b)

Figure 7.4 (a) Schematic of noise-encoding process and (b) one-dimensional noise-encoding example.

Figure 7.5 Example of noise-encoded image. Pixels and noise sampling at 75 spots/inch to be similar to typical CRT display resolution.

Researchers began analyzing the noise, or *dither signal*, in an attempt to optimize its statistical properties. Schuchman[16] has shown that the quantizing noise at the receiver is least dependent on the signal when the dither signal has a uniform amplitude distribution over a quantization level. From an information theory viewpoint, this would be expected to yield the most favorable results, because Widrow[17] has shown that this corresponds to the minimum loss of the statistical data of the picture. It must be noted that the optimal noise amplitude is greater than one quantization level when considering the desire to achieve spatially uniform granularity of a halftoned image. Thompson and Sparkes[18] further optimized and extended this method of quantizing by considering multiple frames in television imaging. Limb[19] considered the visibility of the granulation in the quantizer output. Using a simple model of the visual process, Limb determined the dependence of the visibility of granulation resulting from independent random samples having a uniform probability density function. Limb also showed that visibility can be reduced by introducing negative correlation between samples. Limb[20] then applied his analysis to differential quantization, *deterministic dither* (ordered), and three-dimensional picturephone applications (2-D plane and time). From Limb's analysis of deterministic dither, noise encoding has evolved into various *ordered-dither* methods.

There has been a renewed interest in noise encoding in the way of dithering with *blue noise*, which is related to Limb's negative-correlation dither signal. Steinberg, Easton and Rolleston[21] examined the second-order statistics of an image encoded with blue noise. To enable use of efficient implementation hardware, Mitsa and Parker[22] developed a blue-noise mask and a comparator method of noise encoding. A dithering matrix of finite dimensions is used in a way similar to conventional dithering matrices, except that the matrix is designed such that the halftoned image, at all gray levels, has the frequency spectrum of blue noise. Other methods for designing blue noise dithering matrices have been proposed.[23,24] The blue noise dithering matrices are also known as stochastic screens or FM (Frequency Modulated) screens in the printing industry. The similarities and differences between stochastic screens and conventional clustered dot screens have been described[25] and design practices for several types of stochastic screens reviewed.[26]

Noise encoding with random patterns gave significant new understanding of the binarization process, namely that on a pixel-by-pixel basis, the gray level of each pixel is converted to a probability of that pixel being made white or black by noise addition followed by thresholding. This can readily be seen by noting that for uniformly distributed random numbers, the probability of a pixel value being above or below the threshold is proportional to the length of the interval above or below the threshold, as shown in Fig. 7.4(b). A result of this probabilistic process is that halftone spatial detail may be as fine as one output pixel, not just one halftone cycle. Moreover, the work just described that tried to improve the process by using non-random patterns has lead to the major advances in methods called *ordered dither*.

7.2.4 Ordered dither

There are advantages to adding patterns other than random noise before quantizing or thresholding. The approach of adding a fixed pattern of numbers prior to thresholding to binary has been given the name *ordered dither*. Ordered-dither methods may be divided into two categories: *dispersed dot* and *clustered dot*. The dispersed dot was designed for use in binary CRT displays, where isolated pixels are reproduced reliably and the overall response of the system is linear. Although CRTs are capable of gray-scale display, binarization may be necessary for compatibility with electronics designed to display saturated text. Clustered-dot patterns were designed for printing devices, which typically have a nonlinear response and have difficulty producing isolated single pixels. Clustering works in conjunction with the nonlinearities to provide fine steps in tone reproduction and also to provide stability in the presence of marking process fluctuations. The factor of four or more in spatial resolution of typical printers of today, compared to CRTs, also changes the optimization of parameters, allowing clustered patterns to yield acceptable levels of visual artifacts. An implementation schematic of ordered-dither halftoning is similar to the noise-encoding schematic [Fig. 7.4(a)] where the "add" and "threshold" are replaced with a step of comparing to a periodically varying threshold [Fig. 7.6]. Examples of dispersed-dot and clustered-dot ordered dither are given in Fig. 7.7 (The methods are described in detail in Section 7.4.2).

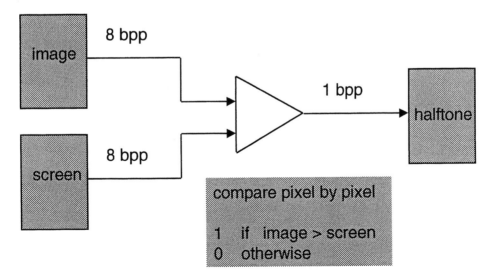

Figure 7.6 Schematic of comparator implementation of ordered dither.

(a)

Figure 7.7 Examples of (a) dispersed-dot and (b) clustered-dot halftone patterns, where (a) is shown at 75 spots/inch, typical of CRT display operation and (b) is shown with a 71 dots/inch halftone screen and pixels at 300 per inch, typical of current laser printers.

(b)

Figure 7.7 Examples of (a) dispersed-dot and (b) clustered-dot halftone patterns, where (a) is shown at 75 spots/inch, typical of CRT display operation and (b) is shown with a 71 dots/inch halftone screen and pixels at 300 per inch, typical of current laser printers.

A key source of the ordered dither methodology was the analysis of optimal noise statistics of random noise encoding. When Limb[19,20] applied visual modeling to minimizing perceived granularity, he examined deterministic noise encoding, which is ordered dither. From Limb's analysis of deterministic dither, noise encoding has evolved into various ordered-dither methods. Other researchers were interested in improving template-dot methods. Klensch and coworkers[27] proposed dispersed-dot and clustered-dot ordered-dither halftone cells as a method of gaining resolution over template methods. They also proposed angled screens for color applications and made the observation that clustered-dot ordered dither resembled conventional halftone photography. Lippel and coworkers[28,29] verified Limb's prediction of two-dimensional ordered dither producing better visual quality than noise encoding methods. Bayer[30] derived an optimality rule for minimum visibility of halftone patterns in terms of minimizing the presence of low frequency components. Using Fourier analysis, he showed that the dispersed cells of Limb[19,20] and Lippel and Kurland[29] followed this optimality rule. Judice, Jarvis and Ninke[31] further extended the optimally dispersed-dot method by developing a recursion method for designing larger cells.

Widespread use and favorable results of ordered dither led to a great deal of analysis of the method. Firm links were made to the analog methods by several authors and the field began to unify. It was noted that ordered dither is, effectively, a sampled version of the analog contact screen process. Researchers[32–34] developed analytic expressions for the Fourier spectrum of a halftone image as a function of the original continuous tone image and of the halftone process. They showed how *aliasing*, or *moiré*, depends jointly upon object contrast, object frequency, and the halftone process. These spectral analyses can be linked to earlier work by scientists,[35] who used coherent optical spatial filtering to examine the spectra of analog halftones. More intuitive explanations of some of the phenomena were provided by Roetling.[37] Spatial resolution and tone reproduction are described by Roetling[36–38] and Bryndgahl.[39] Methods for characterizing noise and image fidelity are described by Refs. 40 and 41, respectively. The effect of "oversized" pixels on tone reproduction was examined by Allebach,[42] where oversized refers to the fact that printed pixels tend toward circular shape and spread beyond idealized rectangular lattice sites. Roetling and Holladay[43] employed a model of the printing process to examine tone reproduction, stability of the printing process and design of clustered dot screens in raster printers. Rotated halftone screens are used to make the color printing processes less sensitive to multipass registration errors. An algorithm for efficiently generating and storing screens at various angles was published by Holladay.[44] Vision models were employed in determining image coding limits[45] and developing halftone algorithms.[46] Randomly nucleated screens, which are used to minimize moiré, have been described by Allebach and Liu[33] and Allebach.[34] Attempts were made to obtain the advantages of the ordered-dither and error-diffusion methods by combining the processes. An early synthesis of the methods was described by Billotet-Hoffman and Bryngdahl.[47]

More recent research on ordered dither has taken several directions. In an attempt to increase effective halftone frequency, *multiple-nuclei screens* were developed and have been disclosed by Riseman et al.[48] The *dual-dot* and *quad-dot* screens that they described are variants of the traditional clustered dot algorithm, which uses a single nucleus per cell. The need to design screens with accurate angles and frequencies for image setters led to the development of *supercells*, which contain a large number of dot centers or nuclei. Supercells use the multiple nuclei within a cell to accurately achieve an average angle and frequency of the dots. In digital printers, gear-tooth noise, solenoid noise, etc., can cause uneven motion and positioning errors as the image signal is being written onto a photosensitive medium. Reflectance banding and other image defects caused by pixel placement errors are analyzed in Refs. 49 through 52 and for color images in Ref. 53. Versatility of modern electronics and improved stability of printing processes has prompted research on multiple bit-per-pixel halftoning algorithms. Image quality of ink jet systems employing trinary pixels was examined by Naing et al.[54] Tone reproduction capabilities of halftones utilizing trinary and quaternary pixels have been described by Lama et al.[55] With the introduction of high addressability[56] another multilevel method came into use on laser printers. High addressability utilizes control of the width of laser pulses in the laser fastscan direction. Using this mechanism, the size and edge positions of the clustered dot can be changed at subpixel increments and the number of gray levels for a halftone at a given frequency can be increased drastically. The increased use of office scanners has prompted analysis of moiré caused by digitizing halftoned images.[57,58] Although not all inclusive, these examples should provide a sense of direction of current research.

In several of its embodiments, clustered-dot ordered dither is essentially the sampled equivalent of the analog contact screen halftone process. As such, it is the most popular current technique for digital printing of pictures. Because of the similarity, ordered-dither halftone patterns are often referred to as a "screen." It is examined in detail in Section 7.4.2.

7.2.5 Error diffusion

Although the development of digital halftoning techniques did not always follow from the analog methods, the methods described above use processes similar to analog contact screens. We next consider a method that uses an inherently different approach to binarization.

A binary encoding scheme where feedback of quantization error was employed for one-dimensional signals was described by Inose and coworkers.[59,60] Quantization error of the signal at a given pixel, caused by thresholding, was added onto the signal at the next pixel prior to thresholding that pixel. The Δ-Σ modulation of Inose, et al. now often referred to as one-dimensional *error diffusion* (ED), propagated accumulated threshold error. (The error that is propagated from a pixel is the difference between its value after thresholding and its modified value, which includes quantization error that was passed from other pixels previously operated upon.). A schematic of the process is shown in Fig. 7.8. The result is that the encoded image has an average value equal to the average value of the image prior to thresholding.

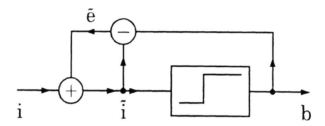

Figure 7.8 Schematic of error diffusion process for one-dimensional signals.

An early application of quantization error feedback employed on images has been described by Schroeder.[9] Schroeder's method, known as minimum-average-error (MAE) quantization, propagated the difference between the value of a pixel before and after quantization. Not included in the error was the quantization error propagated to that pixel. Floyd and Steinberg[61] are generally credited with the first application of propagating accumulated threshold error in binarizing images, where the accumulated error is calculated from the modified brightness of a pixel. The error was distributed in a weighted manner to four neighboring pixels. An example of an image halftoned by their process is shown in Fig. 7.9. The number of neighbors and the weights were arrived at empirically. Hale[62] applied the one-dimensional error-diffusion methodology to images in what he referred to as *dot-spacing modulation* (DSM). Billotet-Hoffman and Bryngdahl[47] examined the resolution of the error-diffusion method as well as the use of a periodic threshold and a larger diffusion matrix (the one employed in MAE studies). To minimize worm-like artifacts, Woo[63] used seven different diffusion masks in dispersing the quantization error. With a similar goal, Dietz[64] described a method to distribute the error randomly to neighboring pixels with certain probability weightings. An analytic description of one-dimensional error diffusion was published by Eschbach and Hauck.[65] It is based on the introduction of a carrier function followed by a signal dependent threshold. A related quantization technique of *pulse density modulation* (PDM), where signal is accumulated until a threshold is met, was applied to images by Eschbach and Hauck.[66] They

Figure 7.9 Example of error-diffused image. Pixels at 75/inch to simulate typical CRT display application.

presented an iterative method for achieving pulse density modulation, then more fully described pulse density modulation in two dimensions.[67]

Error diffusion is currently a very active research topic. More recent investigations have studied pulse density modulation on rasterized media,[68] and have employed visual models in error diffusion design.[69] Optimum intermediate levels in a multilevel pixel scheme were examined from a minimum granularity perspective.[70] Most notable is that this highly nonlinear process is now being understood in terms of linear processes. The lack of low-frequency screen artifacts was described by Ulichney.[71] Using linear methods, Eschbach and Knox[72] have examined edge enhancement of the error diffusion process, and Weissbach et al.[73] and Knox[74] describe the error diffusion image as a sum of the original and high-pass filtered images. For color images, error diffusion can be used to process the individual color planes, but image quality can be improved by using vector error diffusion methods,[75] where the color of each pixel is quantized and the color error in 3-D color space is calculated and distributed among neighboring pixels.

Error diffusion is the most pervasive of binarization methods except for ordered dither, which is examined in detail later in the chapter. Error diffusion offers an attractive alternative way of thinking about binarization. As noted above, ordered dither (and noise encoding) converts pixel gray level to a probability of being white, whereas error diffusion directly forces total gray content to be fixed, and attempts to localize the distribution of gray content.

7.3 VISUAL PERCEPTION

For the applications of printing and display of digital images, the goal of halftoning is to preserve the visual impression of gray tones in spite of the fact that pixel-by-pixel the image is ideally black or white. Before discussing various halftone methods in detail, we consider several aspects of how the human visual system interprets halftone images. It would be inappropriate to attempt a thorough explanation of the visual system here. Owing to space limitations; we briefly outline relevant aspects of the problem. Interested readers are directed to texts, such as Cornsweet's,[76] and the extensive literature on the subject.

To scope the problem, we need to review a few image parameters and their typical values. At normal reading distance, persons with correct (or corrected) vision can resolve roughly 8 to 10 cycles/mm. The spatial frequency of ordered-dither printed halftones ranges from the low quality of newspapers and many digital printers, using about 65 dots/in (2.5 dots/mm), to high-performance systems, using 150 to 200 dots/in (6 to 8 dots/mm). Thus, the halftone pattern is usually visually resolvable even for high-quality systems. For displays, Cohen and Gorog[77] have described that, given free choice, people view TV screens at a distance where scan lines are just resolved. If the same holds for computer displays, then the halftone patterns are certainly also resolved. From these data, it is clear that for both printing and display, the impression of gray does not come simply from the visual system blurring the screen and thereby smoothing the image to a gray appearance. Moreover, dot shape and location, not just dot size, preserve image detail beyond the halftone screen frequency, which is necessary for a sharp appearance of the image.

Fortunately, halftone screens do not have to reproduce every gray level at all spatial frequencies. Roetling[78] has provided a rough analysis of how many gray levels can be perceived as a function of spatial frequency. His result, shown in Fig. 7.10, is an overestimate of performance, because a number of simplifying assumptions were biased toward showing an upper limit to viewer capability. In particular, resolution of the human visual system is roughly a factor of two lower for frequencies oriented at 45° to the horizontal as compared to resolution at 0° or 90°; a fact usually used in practice by placing the screen (or the black screen, in color images) at 45°. Allebach[46] has used this as a guide in designing some ordered-dither patterns as well. Roetling[78] also showed a theoretical limit of how many gray levels could be represented by an ideal binary digital image as a function of spatial frequency, Fig. 7.11. For a periodic digital screen, only the area inside the rectangle (terminated at one-half of the screen frequency) is reproduced by simple dot size differences (template dots). Higher frequency detail requires "partial dots," that is, the pattern must change within a halftone period (for ordered dither). More gray levels are achieved by employing adjacent halftone cycles, where the cycles are not identical, i.e., by introducing lower spatial frequencies into the screen pattern. Both of these constraints are discussed in the algorithm explanations later in this chapter.

The above discussion treated perceived gray level as though the visual system averaged the amount of area covered, whereas we already noted that the screen is usually visible to some degree. One effect of being able to resolve the screen is a tendency to treat it like a window screen, ignoring it to the extent it is observed. This creates the perception of a lighter image for the same dot area coverage in coarser screens, which is one of several reasons why the Tone Reproductin Curve (TRC) needs to be adjusted as a function of screen frequency.

A related but separate problem from false gray-level contouring is that of texture contouring. Especially in many ordered-dither techniques, the fundamental frequency of the screen pattern can rotate 45° between adjacent gray levels. A texture change can show as a contour even if the gray levels themselves would not cause a perceived contour. Allebach[46] has included a condition in some screen designs to minimize such artifacts. Experimentally, one can distinguish between texture and gray level contours by backing away from the image. Gray level contours occur at very low spatial frequencies and

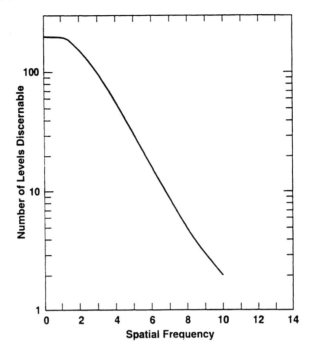

Figure 7.10 Visual performance limits.

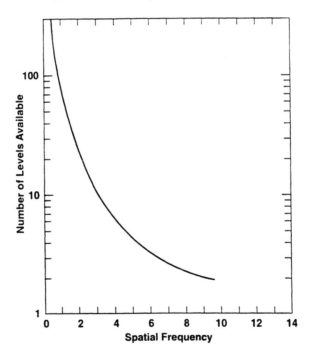

Figure 7.11 Theoretical limit for number of gray levels that can be represented by a binary image sampled at 20/mm, an idealized "adaptive" halftone cell. See Ref. 78 for details.

are visible at a considerable distance. Texture contours, on the other hand, disappear when one cannot resolve the textures and therefore are not visible at larger viewing distances.

Noise and artifact visibility are typically low spatial frequency effects. In noise encoding or random screening, the low spatial frequencies of the screen create a perceptible mottle in the image. In attempts to minimize screen visibility of ordered dither, Bayer[30] and Allebach[46] have tried to maximize the frequency of the lowest frequency harmonic or the strongest harmonics in the screens that they have designed (see ordered-dither descriptions). Note that built-in shape recognition of the visual process can upset expected results. A low spatial frequency possessing low amplitude can often yield disproportionate visibility when patterns are recognized (for example, in a checkerboard of pixels, changing one pixel creates very obvious plus signs that stand out). For the error-diffusion process Ulichney[71] has examined the spatial frequency spectra of binarized images in terms of the amount of power occurring at low spatial frequencies. Several workers (see sections on history and methods of error diffusion) have examined methods of reducing worm-like patterns, which are relatively low spatial frequency structures.

Unfortunately, no one has yet determined a single or small set of optimization criteria for binary image perception. We know, for example, that well-known optimization criteria, such as minimizing mean-square error (MSE), are incorrect. Note that at each pixel, simple midrange thresholding minimizes the expected error between the gray pixel value and the binary result, thus minimizing the total MSE over the image. However, this result does not generally maintain the impression of gray tones. First applying a spatial filter that models the visual system to both the continuous-tone image and halftone image before determining MSE improves the estimate of quality, but this clearly fails to take into account the observer's ability to ignore some visual artifacts and not others. It also ignores effects like human perception not matching physical averaging, such as described above. A very promising approach has been proposed by Daly[79] where the difference between a processed image and an ideal reference image is calculated and a human visual model is used to predict the probability of perception of the difference. A similar method has been proposed for color images.[80]

The noise encoding and ordered-dither tenet of converting gray to a probability of being white offers a means to give the "impression of gray," but does not provide a good means to determine the relative merits of systems that converge to the correct gray over different sized areas, although convergence over smaller areas is clearly better. Error diffusion achieves the total gray content, but like the above, does not give a quantitative measure depending on the distance over which the gray is spread. None of the methods give clear criteria in terms of the visibility of undesirable image noise patterns, nor do they give methods to trade-off between image content and appearance of noise.

Despite the state-of-the-art of understanding of visual interpretation, much of halftone design is still empirical. The fact that a halftone image is usually affected by the physical printing process in ways that are also not well modeled simply increases the uncertainty of mathematical prediction of output image quality.

7.4 METHODS OF HALFTONING

In the present section we describe details of several key halftoning methods. We discuss the algorithms as well as describe where the algorithm is applicable, effects on lines and edges, spatial spectral effects, tone reproduction and other notable effects. Algorithm schematics and examples of the methods are given. The algorithms discussed are variations of ordered-dither and error-diffusion techniques. Before ordered dither, we discuss noise encoding for its simple embodiment of the dither method as well as for historical significance. In describing resolution (frequency response) capabilities of ordered dither we also discuss template-dot methods for comparison.

7.4.1 Noise encoding

Recall that the problem at hand is to reduce the number of quantization levels in a digital image while maintaining the appearance of gray tones. Consider simple midrange thresholding of an 8 bit/pixel image that contains a uniform gray region producing a binary [0, 1] representation of that image. Although, as noted above, this thresholding method results in minimum mean-square-quantization error, the uniform region suffers a severe shift in gray appearance, becoming either completely black or white. To obtain a truer appearing image we may employ the probabilistic method of thresholding known as *noise encoding*. Noise encoding is a point-process halftoning method intended to maintain gray appearance and edge sharpness.

To understand the gray appearance effect via probabilistic generation of activated pixels, consider a uniform gray digital image $f(i, j)$ where a pseudo-random number in the range $[-a, a]$ is added to each pixel to produce the intermediate image $f'(i, j)$,

$$f'(i, j) = f'(i, j) + \text{ran}[-a, a] \ . \tag{7.1}$$

Thresholding each pixel at level t is then applied to produce

$$b(i,j) = \begin{cases} 0, & f'(i, j) < t \\ 1, & f'(i, j) \geq t \end{cases} \tag{7.2}$$

where b is the binary halftoned image.

In the final halftoned image the value of a pixel at (i, j) is dependent upon the original value of the pixel $f(i, j)$, the threshold level t and the range and particular value of the pseudo-random number a. For the simple case where the random number is uniformly distributed, t is midway in the gray scale range $[0, M]$, and $a = M / 2$, we can write probability expressions for the graylevel of a binarized pixel:

$$P[b(i, j) = 1] = f(i, j) / M ,\qquad (7.3)$$

$$P[b(i, j) = 0] = 1 - f(i, j) / M .\qquad (7.4)$$

Assuming a uniform continuous-tone image region and stationarity of the random signal process, the fractional area coverage of 1's and 0's in the halftone image region is equal to the probabilities given in Eqs. (7.3) and (7.4). When viewed on a print or display, spatial integration performed by the human visual system produces a gray appearance corresponding to fractional area coverage, which Eqs. (7.3) and (7.4) show to be the original gray level of the continuous-tone image.

The edge preservation capabilities of noise encoding can also be seen in Eqs. (7.3) and (7.4). At high-contrast edges, one side of the edge possesses high-brightness values that tend to be encoded to 1, and conversely for the low-brightness edge. Although all edge transitions tend to occur at the correct pixel locations, higher contrast edges are better rendered because the greater difference in the probabilities results in the halftoned edge transition being more defined about the true edge location.

Although noise encoding has the virtues described above, it is not ideal in the sense that granularity is introduced into the image. In early applications of noise encoding to bit reduction, Goodall[14] and Roberts[15] noted the trade-off between false contouring and graininess. The statistical method described above is a binary implementation of their method, where they employed a pseudo-random dither signal with a white (uniform) power spectrum and flat histogram. This method applied to an 8-bit image was shown above (Fig. 7.5). Although grayness and edge definition are preserved, low-frequency components of the noise signal are not filtered out by the human visual system and a very grainy image results. A goal of early noise encoding research was the design of optimal dither signals. Limb,[19] employing a human visual model, designed minimum visibility dither signals that evolved into ordered dither.

There has been a renewed interest in noise encoding in the way of screening with a blue noise mask.[22–24,26] *Blue noise* is a term derived from optics denoting that only high spatial frequency components are present. Thus, in a blue noise encoded image, low frequency granularity is not present and the underlying structure is less distracting to the viewer.[71] Because it is the halftone patterns that need to have certain spatial frequency characteristics, current methods for generating the blue noise mask focus on the halftone patterns, then derive the mask from them.

In the method used by Mitsa and Parker,[22] the halftone patterns are designed gray level by gray level. For each gray level, the pattern must satisfy two constraints: Its frequency spectrum must possess no low-frequency components, and the pattern must be binary. For the first gray level, usually at midtone, the method starts with an initial guess of the pattern, transforms the pattern to the frequency domain and filters it to shape the frequency spectrum, then transforms back to the spatial domain. At this point the pattern has the desired frequency spectrum but is no longer binary; it is then thresholded to form a binary pattern, the pattern is adjusted so that the number of dots is correct for the gray level and the above process is repeated until convergence occurs. The patterns for other levels are built from this first level in sequence by removing or adding dots using the same set of constraints. From the complete set of patterns, a single mask is derived, which is a matrix of threshold values, and can be implemented with computational efficiency. Note that simply shaping the frequency spectrum of the mask is not enough, because the thresholding step is nonlinear, which generates new frequency components, including undesirable low-frequency components. This problem is well known and has been studied in the context of blue noise mask generation by Steinberg, Rolleston and Easton.[81]

Sullivan, Ray and Miller[23] use a visual cost function in the frequency domain to optimize the halftone patterns. The visual cost function is an estimate of the visibility of the halftone pattern and is given by

$$\text{Cost} = \iint \left| P\left(f_x, f_y\right) V\left(f_x, f_y\right) \right|^2 df_x df_y, \qquad (7.5)$$

where $P(f_x, f_y)$ is the frequency spectrum of the halftone pattern and $V(f_x, f_y)$ is the frequency response of the human visual system. Ideally, the cost of all possible halftone patterns with the correct number of dots should be calculated and the one with the lowest cost chosen as optimal. But the number of possible patterns is too large for such exhaustive search, so a simulated annealing method is used to find the pattern possessing the lowest cost. The same process is used for each gray level independently, resulting in halftone patterns that are not correlated from level to level. Such halftones are useful for large uniform areas in computer graphics, but not suitable for processing pictorials. Sullivan and Ray[82] extended this method to include gray level to gray level correlation. An example of Sullivan and Ray's method versus white noise encoding is shown in Fig. 7.12.

(a)

(b)

Figure 7.12 Examples of (a) blue noise mask of Spaulding, Miller and Schildkraut[26] and (b) white noise mask. Reprinted with permission from Eastman Kodak Company. ©Eastman Kodak Company.

Ulichney[24] uses a simple technique for generating blue noise patterns called "void-and–cluster." It is based on the principle that in a blue noise pattern, large clusters of dots or voids should not be present. Optimization of the patterns consists of two steps. First, a pattern is derived for an initial gray level. For this step, an initial guess of the pattern is made, the pattern is blurred by convolving it with a lowpass filter and then a dot is removed from the place with the tightest cluster of dots and moved to the place with the largest void. This process is repeated until convergence results. Once the pattern for the first gray level is determined, the patterns for other levels are derived in sequence. Dots are added to form the patterns for darker levels or removed to form the patterns for lighter levels, following the same principle. Lin[83,84] described several variations and refinements of the void-and-cluster method.

7.4.2 Ordered dither

As the statistical understanding of noise encoding evolved, deterministic dither signals were developed. In currently employed techniques, a two-dimensional *dither mask* contains threshold values that are used in a comparison algorithm to determine the value of an output halftone pixel. The dither mask, *threshold array*, or *halftone cell*, is replicated to tile the image plane, thereby producing periodically varying thresholds throughout the image. Key factors in designing an ordered-dither halftone are the arrangement of thresholds within the cell, cell size and shape and offset of successive rows of cells. We first describe the basic algorithm of implementation. Then we discuss two threshold arrangement methodologies. One, the dispersed-dot method, minimizes objectionable low-frequency components induced by the halftone screen. It is ideally suited for display devices where isolated pixels tend to be reproduced faithfully and the overall response of the system is linear. The *clustered-dot* method groups pixel types within a cell to form, typically, a single black dot per cell. In an attempt to suppress low- frequency components induced by clustering, some variations of the method form several small clustered dots within the cell (multinuclei cell). Clustered dots are employed in printing devices, that tend to be nonlinear in response and are subject to process fluctuations. This method of halftoning yields repeatable low-noise images with acceptable tone reproduction characteristics. The clustered-dot method, in its basic form, is essentially a sampled version of analog halftone techniques. Note that low-resolution printers (<300 pixels per inch) sometimes use dispered-dot dither because the printing processes tend to be more stable relative to the larger spots, thereby eliminating the necessity for clustering. For the clustered-dot method, we include a discussion on angled screens for both monochrome and four-color images. We end the ordered-dither section with discussions on topics common to either threshold arrangement: The trade-off between screen visibility and the number of achievable gray levels and spatial resolution effects such as frequency response and moiré.

7.4.2.1 Implementation

When halftoning a digital image, the threshold array is replicated to tile the image plane (Fig. 7.13) and then a comparison is performed (Fig. 7.14). The comparison is performed on a pixel-by-pixel basis to retain spatial resolution in the output image. For a threshold array T_{mn} having dimensions m by n, the image $f(i, j)$ is halftoned via the operation

$$b(i,j) = \begin{cases} 0, & f(i, j) < T_{mn}(k,l) \\ 1, & f(i, j) \geq T_{mn}(k,l) \end{cases}, \qquad (7.6)$$

where $k = \text{mod}(i, m)$ and $l = \text{mod}(j, n)$ for screens at $0°$ [The notation $\text{mod}(i, m)$ denotes i modulo m]. When screens are desired at some other rational tangent angle, successive cells of the tiling are offset by an amount s (see discussion on Holladay's algorithm[44] in Section 7.4.2.3.2), filling the page much like bricks in a wall. Using this local comparison method results in *partial dots*, where the internal shape of a printed cell varies depending on the image detail within the cell boundary. Sharp edges and discontinuities, to which the human visual system is very sensitive, are preserved in a manner similar to the noise encoding method. This high resolution (or high-frequency response) is in contrast to older *template-dot* methods, which used many output pixels in fixed patterns to represent a single input pixel. An example comparing the perceived sharpness of the two methods was given in Fig. 7.3. The relationship to noise encoding may be seen by allowing T to be the size of the image and employing uniformly randomly distributed numbers for the threshold values.

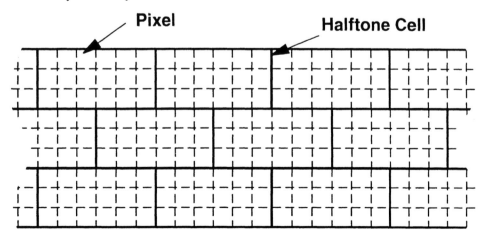

Figure 7.13 Schematic of tiling a halftone cell on a page.

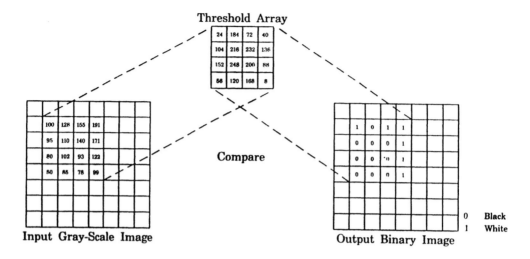

Figure 7.14 Schematic of implementation of ordered dither using sample-by-sample comparison.

7.4.2.2 Dispersed-dot ordered dither

Dispersed-dot ordered dither is used primarily in display devices where isolated pixels can be reproduced faithfully or the system response is approximately linear. The threshold arrays are designed to minimize low-frequency texture induced by the screen. Threshold arrays are typically designed to meet optimality criteria defined by Bayer.[30] We first state Bayer's criteria and then show how cells may be designed using the recursion relationship of Judice et al.[31] As an index of merit for texture induced by a halftone cell, Bayer employed the longest finite wavelength of the non-zero amplitude sinusoidal components of a halftoned uniform area. This quantity, denoted by Λ, can be written

$$\Lambda = \max[\ \lambda_{uv} : A_{uv} \neq 0, \lambda_{uv} < \infty]\ , \qquad (7.7)$$

where λ_{uv} and A_{uv} are the wavelength and amplitude of the u, v frequency component, respectively. The preferred halftone cell is defined to be one such that Λ is as small as possible for all possible gray levels. This choice of definition is based on the premise that if we view an entire sequence of gray levels at one time, we base our first impression of overall texture on those patterns having the largest value of Λ.

To express the optimality conditions mathematically, let us consider a uniform gray area that is represented by repeating cells of black and white pixels. Let k equal the number of black dots in a printed cell halftoned by threshold array T. The array must be square with a side length equal to an integer power of 2 (dimension $2^m \times 2^m$). Therefore, halftoned image b obeys

$$b(i, j) = b(i + s2^m, j + t2^m), \qquad (7.8)$$

where s and t are integers.

For each k, let 2^n be the largest power of 2 that divides k. That is, $k = p2^n$ where p is an odd integer. With $b(i, j)$ denoting the halftoned image, the optimality conditions can now be stated in two parts:

(1) For even n, b must obey

$$b(i + 2^{m-n/2}, j) = b(i, j+2^{m-n/2}) = b(i, j). \qquad (7.9)$$

(2) For odd n, b must obey

$$b(i + 2^{m-(n+1)/2}, j+2^{m-(n+1)/2}) = b(i + 2^{m-(n+1)/2}, j-2^{m-(n+1)/2}) = b(i, j). \qquad (7.10)$$

The halftone cell yields the maximum index of merit if and only if these conditions are satisfied for all k. Furthermore, under such conditions, the maximum wavelength is given by

$$\Lambda = 2^{m-n/2}, \qquad (7.11)$$

for each k, $0 < k < 2^{2m}$, $k = p2^n$, p odd.

Table 7.1 provides an example of the parameters described above for a 4×4 dither matrix.

Dither arrays developed by Limb[20] and Lippel and Kurland[29] were found to satisfy Bayer's condition. Judice et al.[31] developed a recursion method of designing larger dither arrays. Given Limb's 2×2 array,

$$T^2 = \begin{bmatrix} 0 & 2 \\ 3 & 1 \end{bmatrix}, \qquad (7.12)$$

(normalized to four levels) and defining U^n to be dimension $n \times n$,

Table 7.1 Dither Matrix Parameters for 4×4 Cell, $m = 2$.

Number of black pixels				Maximum wavelength
k	p	2^n	n	Λ
0	–	–	–	–
1	1	1	0	4
2	1	2	1	$2\sqrt{2}$
3	3	1	0	4
4	1	4	2	2
5	5	1	0	4
6	3	2	1	$2\sqrt{2}$
7	7	1	0	4
8	1	8	3	$\sqrt{2}$
9	9	1	0	4
10	5	2	1	$2\sqrt{2}$
11	11	1	0	4
12	3	4	2	2
13	13	1	0	4
14	7	2	1	$2\sqrt{2}$
15	15	1	0	4
16	1	16	4	1

$$
U^n = \begin{bmatrix}
1 & 1 & \cdots & 1 \\
\cdot & \cdot & & \cdot \\
\cdot & \cdot & & \cdot \\
\cdot & \cdot & & \cdot \\
1 & 1 & \cdots & 1 \\
\cdot & \cdot & & \cdot \\
\cdot & \cdot & & \cdot \\
\cdot & \cdot & & \cdot \\
1 & 1 & \cdots & 1
\end{bmatrix}, \tag{7.13}
$$

allows us to state the recursion relationship,

$$T^n = \begin{bmatrix} 4T^{n/2} + T^2[0,0]U^{n/2} & 4T^{n/2} + T^2[0,1]U^{n/2} \\ 4T^{n/2} + T^2[1,0]U^{n/2} & 4T^{n/2} + T^2[1,1]U^{n/2} \end{bmatrix}, \qquad (7.14)$$

where $T^2[i, j]$ denotes the i, jth element of Limb's 2×2 array. Using the recursion relationship, the 4×4 dither array is found to be

$$T^4 = \begin{bmatrix} 0 & 8 & 2 & 10 \\ 12 & 4 & 14 & 6 \\ 3 & 11 & 1 & 9 \\ 15 & 7 & 13 & 5 \end{bmatrix}. \qquad (7.15)$$

An example of dispersed dot dither was shown in Fig. 7.7(a). We see that spatial resolution is high and gray tones are well rendered. An artifact of the process can be seen in the gray wedge, where texture contouring can be observed between the gray levels. The possible cells generated by uniform input images are shown in Fig. 7.15.

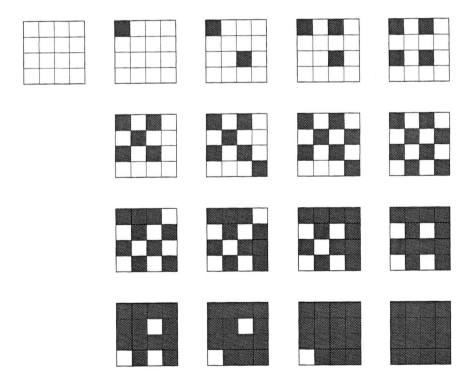

Figure 7.15 Dispersed-dot cells generated by uniform input images.

7.4.2.3 Clustered-dot ordered dither

Clustered-dot ordered dither is the most widely employed method of halftoning in the digital printing industry. In one of its basic forms, it is a sampled version of analog contact screen printing. In many digital printing devices, the reflectance response is nonlinear: Due to the size, shape and density-saturated nature of printed pixels, the reflectance is not equal to the ratio of number of white pixels to total number of pixels in a cell. In such a device, tone reproduction is optimized by clustering pixels, thereby, forcing maximum overlap. Other considerations are pixel position accuracy, single pixel reproduction fidelity and dot overlap, which are engineering issues. Most often, due to cost or physical limitations, a more stable image, free of noise artifacts and nonuniformities is obtained by clustering black pixels within a cell. The cost of correct tone reproduction and robustness in the presence of process fluctuations is a screen with lower frequency content, and hence is more visible than a dispersed-dot halftone. We first discuss the 45° dot screen, along with fundamental tone reproduction and spatial resolution issues. Then Holladay's method[44] of describing angled screens for color applications is presented. Finally, there is a discussion on various multidot clustering schemes.

Before we begin detailed discussion on clustered-dot ordered dither, we note that clustered-dot halftones are often confused with template-dot screens. For a given density the template-dot method always prints the same internal halftone cell pattern. The technique arose out of the early days of computer halftones where one image pixel generated one cell. Although the constant dot shape allows for well-controlled tone reproduction, it causes a loss in spatial resolution in comparison to clustered-dot techniques. The partial dots of the modern techniques produce images of the same high- contrast resolution as the gray-scale image. For comparison, template-dot and clustered-dot halftoned images were shown in Fig. 7.3.

7.4.2.3.1 Dot growth pattern and tone reproduction, and stability

To understand the effect of the dot growth pattern on print reproducibility, consider the 3 × 6 clustered-dot threshold array example of Eq. 7.16, which simulates an analog screen and grows approximately in a spiral pattern.

$$T = \begin{bmatrix} 70 & 30 & 80 & 110 & 160 & 120 \\ 20 & 10 & 40 & 150 & 180 & 170 \\ 60 & 50 & 90 & 100 & 140 & 130 \end{bmatrix}. \tag{7.16}$$

As image darkness varies from white to black, the corresponding cell turns black one pixel at a time in a spiral fill pattern. There is an attempt to minimize the migration of the dot center of gravity from gray level to gray level. At low density, a small cluster of black pixels on the left side is grown to a larger cluster by spiraling outward. At 50% area coverage, a checkerboard pattern is formed. At progressively higher densities, black pixels are grown inward on the right side to fill the remaining white space. For the given threshold array, successive rows of cells are offset by three pixels so that the lowest nonzero screen frequency component is at 45° and, thus, less visible to the viewer. A halftone dot grown in this manner tends to have a minimal perimeter. Since most deviations from ideal printing occur around the perimeter of the dot, this growth scheme stabilizes the gray levels produced. The growth sequence for this cell applied to a uniform image is shown in Fig. 7.16.

When designing a threshold array, a key concern is tone reproduction. Tone reproduction is affected by the threshold-value sequence (fill order) as well as the threshold values themselves. The goal is to produce a halftone image free of gray-level false contours with the same apparent tone rendition as the gray-scale image. Roetling and Holladay[43] and others have discussed these issues in detail. First, let us discuss contouring.

The human visual system is more sensitive to gray steps in reflectance in shadow regions of an image than in the midtones or highlights. For that reason we attempt to utilize more gray levels per unit of reflectance in the shadow region of the tone reproduction curve. In general, to avoid false contours, a minimum of ~100 gray levels is required. This number depends on the noise level of the printing system; a system with a higher level of image noise can tolerate fewer gray levels. To analyze tone reproduction, printer models have been employed. For example, for an electrographic printer, Roetling and Holladay[43] assumed black pixels are circular, infinitely dense and centered on addressable grid locations. For such printers, spot diameter is generally greater than pixel spacing to allow full coverage for the all black case and so diagonal lines are printed without discontinuities (See Fig. 7.17). Since neighboring spots overlap, area coverage (and optical density) increases less rapidly in the highlights when the spots are clustered as in the growth pattern described above. This printing system is in contrast to a linear device, where, with each additional pixel turned black, reflectance would decrement by a fixed amount. Accurate tone reproduction also depends upon gray-level spacing between threshold levels. The printer model described above indicates that spot overlap causes cell reflectance to vary nonlinearly with increasing fill number. Thus, the threshold levels should not be equispaced if we desire a tone response that is linear in reflectance. A common method of selecting threshold levels is to first measure the set of possible reflectances for the chosen fill order $\{R_i : i \in [0, p]\}$, where p is the number of pixels in the cell and R_p reflectance, decreases with increasing i as the cell is filled. Using this notation, in an

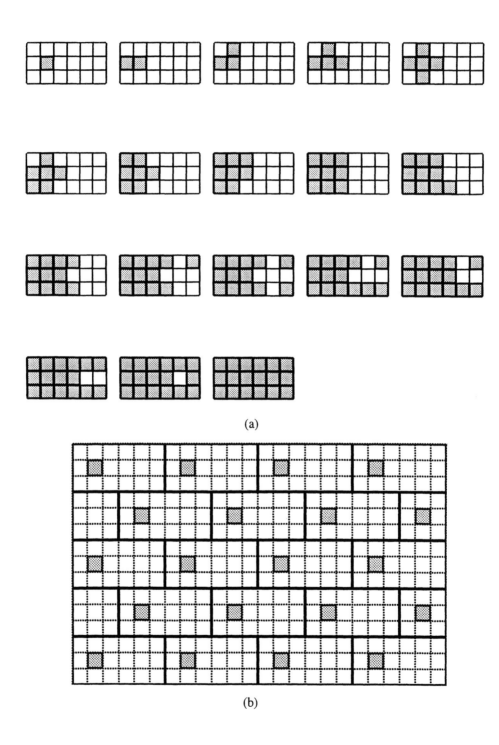

(a)

(b)

Figure 7.16 Example using threshold array of Eq. (7.16), (a) growth sequence. Image processed with this array for a uniform image of value (b) 15.

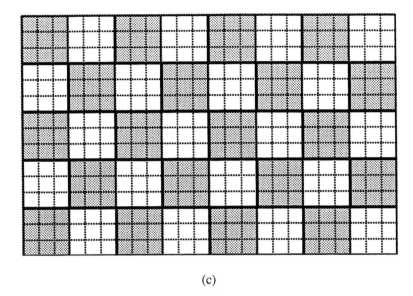

(c)

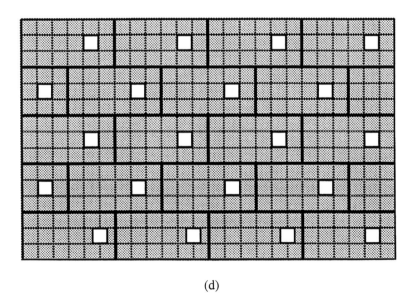

(d)

Figure 7.16 Example using threshold array of Eq. (7.16), image processed with this array for a uniform image of value (c) 95 and (d) 175.

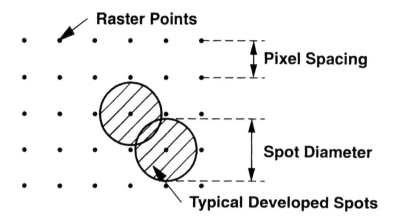

Figure 7.17 Illustration of pixel spacing and pixel diameter showing how individual spots are turned black and overlap.

idealized system, $R_0=1$ and $R_p=0$. To make the reflectance a linear function of the input pixel value f, we can pick the threshold levels t_n according to the following rule:

$$t_n = \frac{M(R_{n-1}+R_n)}{2},\qquad (7.17)$$

where n denotes fill order, $n=1, ..., p$ and M is the largest digital value of image reflectance (e.g., 255 for 8 bits/pixel). Using this rule, when the input image pixel value f is MR_n, then $t_n<f<t_n+1$. When there is a large uniform area of pixels possessing value f, then there will be n darkened pixels in a halftone cell, which gives a reflectance of MR_n. Therefore, the output average reflectance is a linear function of the input pixel values for uniform areas.

We end this section by mentioning two variations of ordered dither. The variants, *multidot cells* and *line screens*, involve different cell growth patterns than described above. One aspect of a multidot cell is that it is an attempt to increase the apparent frequency of a clustered-dot screen for a given cell size. Spiral type growth is employed, but alternating from, typically, two or four nuclei in the cell. The virtues are less low-frequency content and almost equal stability compared to a single clustered-dot cell. The primary drawback is slight texture contouring which results from the growth alternating from nucleus to nucleus. In addition, highlight tone reproduction may be somewhat more sensitive to process fluctuations. Examples of *dual-dot* T_2 and *quad-dot* T_4 threshold arrays are shown below:

$$T_2=\begin{bmatrix} 70 & 30 & 80 & 110 & 160 & 120 \\ 20 & 10 & 40 & 150 & 180 & 170 \\ 60 & 50 & 90 & 100 & 140 & 130 \\ 115 & 165 & 125 & 75 & 35 & 85 \\ 155 & 185 & 175 & 25 & 15 & 45 \\ 105 & 145 & 135 & 65 & 55 & 95 \end{bmatrix}, \qquad (7.18)$$

$$T_4=\begin{bmatrix} 70 & 30 & 80 & 110 & 160 & 120 & 72 & 32 & 82 & 112 & 162 & 122 \\ 20 & 10 & 40 & 150 & 180 & 170 & 22 & 12 & 42 & 152 & 182 & 172 \\ 60 & 50 & 90 & 100 & 140 & 130 & 62 & 52 & 92 & 102 & 142 & 132 \\ 117 & 167 & 127 & 77 & 37 & 87 & 115 & 165 & 125 & 75 & 35 & 85 \\ 157 & 187 & 177 & 27 & 17 & 47 & 155 & 185 & 175 & 25 & 15 & 44 \\ 107 & 147 & 137 & 67 & 57 & 97 & 105 & 145 & 135 & 65 & 55 & 95 \end{bmatrix}. \qquad (7.19)$$

The multicenter dots are generated by starting with a subcell possessing the desired dot frequency, appending flipped version of the subcell to the subcell then adjusting the thresholds of the flipped subcells. The overall cell will have a strong component at the subcell frequency and will have weak components at lower frequencies for some gray levels. See Fig. 7.18 for an example of a growth sequence of dual-dot and quad-dot cells. Additional aspects of multidot cells are described in Section 7.4.2.3.2.

Line screen digital halftones are sometimes employed in reducing motion quality and registration requirement in printers. Compared to dot screens, line screens that run in the process direction (e.g., direction of photoreceptor motion in a laser printer) are less susceptible to reflectance banding due to vibratory motion of the image. In addition, when misregistration of color separations occurs in the process direction, line screens yield less color error than would occur in dot screens. Line screens are very susceptible to color shifts for cross-process misregistration, but less position error usually occurs in this direction in real systems. Although line screens may be more visible than dot screens, the trade-off may be worthwhile when considering the robustness against printer defects and cost savings in motion-quality engineering. Line screens are typically generated with an analog screen generator, where the digital pixel value is converted to an analog video signal, and a saw tooth waveform is used as a reference (Fig. 7.19). The result of comparison of the video signal with the reference is a series of rectangular pulses, and the width of each pulse is proportional to the corresponding digital pixel value. This is a form of Pulse Width Modulation (PWM). The frequency of the reference waveform can be changed. A lower frequency can be used to yield better

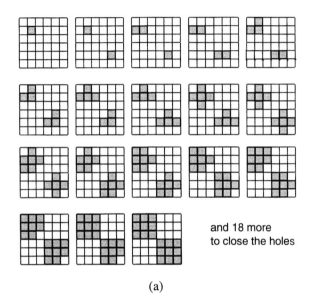

and 18 more
to close the holes

(a)

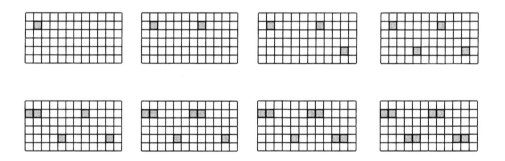

and 64 more levels

(b)

Figure 7.18 Example of growth sequence for (a) dual-dot and (b) quad-dot.

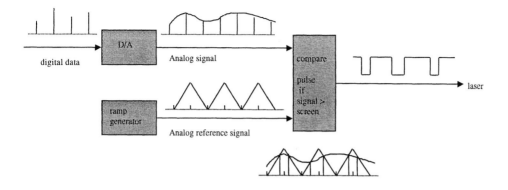

Figure 7.19 Schematic of analog screen generator for line screen halftoning.

stability and uniformity, suitable for pictorials; a higher frequency can be used to produce better detail rendition, suitable for text and line art. The phase of the reference signal can also be changed from scanline to scanline, thus the line screen can be oriented at an angle, which is useful when lateral registration of different colors cannot be maintained accurately.

7.4.2.3.2 Screen angle and color separations

To minimize visibility, single color images (e.g., black and white) use screens that are typically designed to have the fundamental frequency at 45° from horizontal. Color prints require an overlaying of three- or four- color separation images, each halftoned. Dot screens typically possess a fundamental frequency in each quadrant while line screens possess fundamental frequencies only in each of two half planes. Typical colorants have overlapping spectral absorptions, thus interfering with each other. The interference between screens gives rise to the need to employ screens for each separation at specific and exact angles. While analog methods allow for arbitrary angles and frequencies, digital screening can only achieve certain specific angles and frequencies. This limitation leads to several engineering concerns and trade-offs that are briefly discussed below.

When combining periodic signals, such as two colored halftone screens of vector frequency f_1 and f_2, interference produces a beat at the vector difference frequency, $f_b = f_1 - f_2$. This beat frequency is typically referred to as moiré. It can take on one or more of several possible characteristics. If the individual color screens were made at the same angle and frequency, any slight frequency modulation due to vibration in the printing process forms a low-frequency, visually objectionable, beat known as color banding. If misregistration error between color separations occurred ($f_1 = f_2$), but the relative phase of the frequencies deviates from design conditions), a color error results across the entire page ($f_b = 0$). To avoid these types of color errors, the screens are typically oriented at different angles, usually about 30° apart. At 30° separation, the moiré is at about half the screen frequency, thereby producing a high-frequency "rosette" shaped beat pattern. When utilizing rotated screens, misregistration errors cause color errors only at high frequency and therefore are not typically visible. Thus, the visual appearance of rotated screens is much less sensitive to misregistration and vibration effects but at a cost of moiré at half the screen frequency. To achieve the highest frequency rosette pattern, it is typical to orient cyan and magenta at ±15°, yellow at 0° and black at 45°. Because yellow and black have the least and most impact on visual density, respectively, they are oriented at angles where the eye is most and least sensitive, respectively, to observing the screen pattern. Note that although yellow is at 15° to the other screens, its intercolor moiré is not as objectionable because of low contrast. Deviating from these ideal angles can introduce undesirable beats at frequencies lower than the rosette frequency.

Note that although a relatively high-frequency rosette is desirable, the rosette itself has characteristics that must be considered. Researchers[85] have shown that the appearance of the rosette is related to the relative phases of the halftone screens. Particular relative positions yield *clear-centered* or *dot-centered* rosettes. The dot-centered rosette is considered to be more visually objectionable in highlights while clear-centered screen rosettes tend to be more visible in shadows. The dot-centered rosette is obtained when the screens are aligned such that the center of the inked dots lie on the origin. Clear-centered screens occur when the center of the white portions of cells lie on the origin.

Another approach to color screening has also been taken. Because the rosettes are roughly half the screen frequency, it is tempting to eliminate these lower frequencies by using aligned screens (also called dot-on-dot screens) for lower resolution digital printers. If registration is good enough, this is a viable trade-off (and necessary at low addressability), but at higher resolutions, printers usually use rotated screens.

The required accuracy of the screen angle and screen frequency is best analyzed in the two-dimensional spatial frequency domain.[25,86] The moiré to be minimized is the interference among the screens of the three strong colors: cyan (C), magenta (M) and black (K). A general rule for minimizing the moiré is that the vector sum and difference of all the screen vectors do not generate a strong low-frequency component. For common cases, the following simple analysis is adequate. Assume cyan is at about 15°, magenta at around 75°, and black at 45°. We first form the frequency vectors for the three screens, as shown in Fig. 7.20. Then we form the difference vector M−C, which gives the frequency of the beat between the magenta and cyan screens. This beat has a high frequency and is not a problem. Note, however, that M−C is in the same direction, and has about the same magnitude as K. The beat between M−C and K therefore can generate a low-frequency component, which can be objectionable. To avoid this low-frequency beat, we require that the vector

(M–C)–K equal zero. It is easy to see that if the magnitudes of the three frequency vectors C, M, and K are the same, and the angles are exactly 15°, 75° and 45°, respectively, then (M–C)–K is zero. Practical designs aim at generating a beat pattern with a period comparable to the size of a page. Sophisticated design tools, usually proprietary, are used to design screens meeting such requirements.

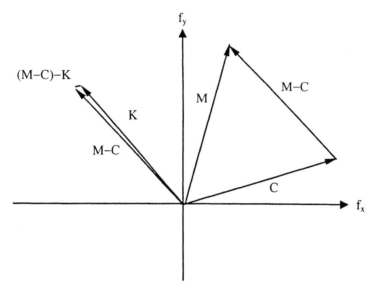

Figure 7.20 Frequency vectors for three screens.

The particular choice of screen angle and the desired angular accuracy have algorithmic implications. Angled screens are defined as *rational* or *irrational*, where rational indicates that the tangent angle of a given screen is a ratio of integers. In this case, halftone dot centers fall neatly on the pixel grid. While rational screens are unable to achieve ±15° separations accurately and therefore offer poorer moiré control, they offer algorithmic advantages over irrational screens. Recent developments have focused on large multidot rational tangent screens, where the dot centers are arrayed such that their average angle approximates a desired angle. This approach is termed supercell halftoning. Conventional multidot cells (dual dots, quad dots,...) are composed of subcells of the same shape and they are internally arrayed at exactly the same angle as the overall cell is aligned with its neighboring cells. In a supercell the internal subcells vary in shape and position in a manner such that their average angle approximates the angle of the overall cell-to-cell alignment. Typically, a large number of dots are used within a supercell (>100) so that the screen angle may be chosen with high precision. Fig. 7.21 illustrates the improved angular accuracy when the screen size (and number of centers) is increased. Supercells are usually used on output devices of high resolution, such as image setters. At low resolution, the position and size difference between the many centers can be quite objectionable, hence supercells are seldom used. Even at high resolution, the design of artifact-free supercells is not trivial.[87] Irrational screens may be produced by randomly dithering between rational screens or by other methods.[88] When employing multiple rational screens, screen storage problems may be encountered.

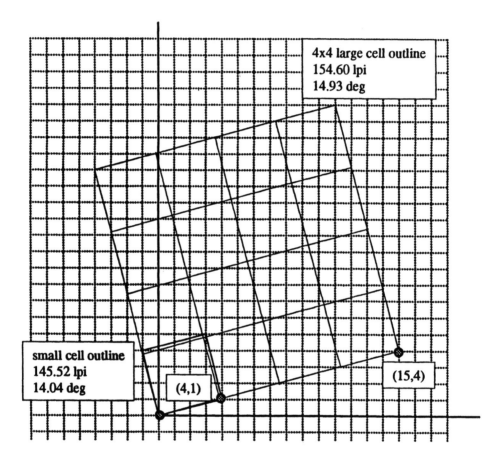

Figure 7.21 Improved angular accuracy with increasing screen size.

Holladay[44] described an efficient method of specifying a rationally angled screen where only one rectangular cell of area equal to one halftone period is stored and used in processing. Not considering fill order, only three parameters are needed to describe the screen: height, width and shift of the rectangle. A general angled parallelogram shaped cell may be described by two vectors corresponding to its side lengths and directions: $\mathbf{Z} = z_1 \mathbf{x} + z_2 \mathbf{y}$, $\mathbf{W} = w_1 \mathbf{x} + w_2 \mathbf{y}$. Given $\mathbf{Z}$ and $\mathbf{W}$, the height p, width l and offset s may be calculated from

$$p = \mathrm{GCD}(z_2, w_2), \tag{7.20}$$

$$l = \frac{A}{p}, \tag{7.21}$$

$$s = 1 - \frac{tA - pw_1}{w_2}, \tag{7.22}$$

respectively, where A is the area of the parallelogram, $A = z_1 w_2 - z_2 w_1$, GCD refers to the greatest common divisor and t is an integer chosen so that $0 < (tA - pw_1) / w_2 = 1$. An example of an equivalent rectangular cell tiling the image plane is shown in Fig. 7.22.

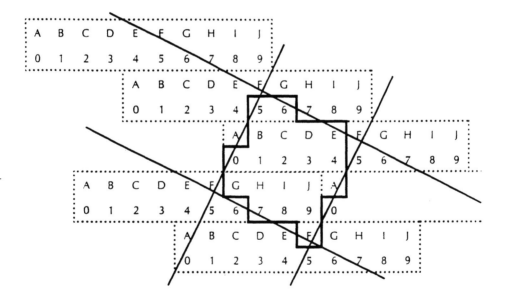

Figure 7.22 Example showing how an angled halftone screen may be represented by rectangular cells aligned with a raster. Method described by Holladay.[44] Reprinted with permission from Knox, Proceedings of the International Conference on Lasers '90.

In the graphic arts industry it is common to use what is known as an *orthogonal screen*. This widely used special case of a rotated screen consists of square cells oriented at an arbitrary angle. The orthogonal screens may also be described by a minimal rectangular cell and a shift. The screens are typically specified by the components x and y of the vector describing the base of the square cell. Rational screens are obtained for integer x and y.

For rational cells with a single dot center we may list the relevant geometric attributes. Table 7.2 gives the number of pixels in the cell and the screen angle for x and y in the range 0 to 7, where N/θ signifies N gray levels and a rotation of $\theta°$ rounded to the nearest integer. Screens are given for the angle range of $0°$ to $45°$, and other angles may be obtained by interchanging x and y. These same screens are described by Holladay's parameters[44] in Table 7.3.

Table 7.2 Number of Pixels in the Cell and Screen Angles for the Orthogonal Screens.

			N/θ			
y / x	2	3	4	5	6	7
0	4/0°	9/0°	16/0°	25/0°	36/0°	49/0°
1	5/27°	10/18°	17/14°	26/11°	37/9°	50/8°
2	8/45°	13/34°`	20/27°	29/22°	40/18°	53/16°
3		18/45°	25/37°	34/31°	45/27°	58/23°
4			32/45°	41/39°	52/34°	65/30°
5				50/45°	61/40°	74/36°
6					72/45°	85/41°
7						98/45°

Table 7.3 Minimal Rectangle Parameters for the Orthogonal Screens.

			$p/l/s$			
y / x	2	3	4	5	6	7
0	2/2/0	3/3/0	4/4/0	5/5/0	6/6/0	7/7/0
1	1/5/2	1/10/3	1/17/4	1/26/5	1/37/6	1/50/7
2	2/4/2	1/13/8	2/10/4	1/29/17	2/20/6	1/53/30
3		3/6/3	1/25/18	1/34/17	3/15/6	1/58/41
4			4/8/4	1/41/32	2/26/16	1/65/18
5				5/10/5	1/61/50	1/74/31
6					6/12/6	1/85/72
7						7/14/7

The entry $p/l/s$ gives the height, width and offset of the cell. The parameters are calculated from Eqs. (7.20) through (7.22), where $\mathbf{Z} = x\mathbf{x} - y\mathbf{y}$, $\mathbf{W} = y\mathbf{x} + x\mathbf{y}$.

7.4.2.4 Screen frequency vs. number of gray levels

Two conflicting design requirements exist for ordered-dither halftones. It is desirable to possess the capability to print many gray levels, e.g., roughly 100 to 200 levels for high- quality pictorials. It is also desirable to maximize the screen frequency, thus minimizing its visibility. High-quality printers may employ screens at 100 to 200 cells per inch. As the number of pixels in a cell increases, a halftone dot can represent more gray levels, thereby decreasing the likelihood of false contours; but, as the cell size increases for single clustered dots, the screen becomes coarser and more visible. The coarser screen is also less able to represent low-contrast fine detail (for low-contrast periodic patterns the spatial resolution limit is half the screen frequency, as stated in Section 7.4.2.5). Conversely, as the number of pixels decreases, the screen frequency becomes higher and less visible and finer detail may be represented. This gain is at the cost of a potential for false contours. Examples of low- and high-frequency screens are shown in Fig. 7.23. The trade-off between screen frequency and number of gray levels is ever present.

A direction of recent research in ordered-dither halftoning concerns alleviating the compromise between screen frequency and number of gray levels. Utilization of multilevel (>2) pixels has been examined[55] and it has been shown that where printers can support limited gray-level printing, the number of cell gray levels produced by a halftoning system may be significantly increased through the use of gray pixels. Hybrid halftone systems with trinary and quaternary pixels (having 1 and 2 intermediate gray levels in addition to black and white) produce many more unique reflectance cells than a binary system with the same cell size. For a cell with p number of pixels and r reflectance levels per pixel, the number of cell gray levels is given by

$$N = \prod_{k=1}^{r-1} \frac{p+k}{k} = \frac{(p+r-1)!}{p!(r-1)!} \ . \tag{7.23}$$

For binary, trinary and quaternary pixels ($r=2$, 3 and 4) the number of cell gray levels becomes

$$\text{Binary: } N_B = \sum_{k=1}^{p+1} 1 = \frac{(p+1)!}{p!} = p+1 \ , \tag{7.24}$$

$$\text{Trinary: } N_T = \sum_{k=1}^{p+1} k = \frac{(p+2)!}{2p!} = \frac{(p+2)(p+1)}{2} \ , \tag{7.25}$$

(a)

Figure 7.23 Examples of screen frequency and gray level trade-off. (a) lower screen frequency with minimal false contours.

(b)

Figure 7.23 Examples of screen frequency and gray level trade-off. Lower screen frequency with (b) higher screen frequency, but very visible false contours.

$$\text{Quarternary: } N_Q = \sum_{k=1}^{p+1} (p+2-k)k$$

$$= \frac{(p+3)!}{6p!} = \frac{(p+3)(p+2)(p+1)}{6}. \qquad (7.26)$$

Increasing the number of pixel levels greatly increases the number of cell reflectance levels. At the present time, algorithms that can utilize many of these hybrid cells while retaining the image resolution of partial dot methods have not been developed. One limitation of the use of gray pixels to increase the number of gray levels in a halftone cell is the susceptibility to printing process variation, which could lead to artifacts such as tone reversal.

In practical applications it has been necessary to impose additional restrictions on multilevel pixels, such that as a dot grows from light to dark, each pixel forming the dot can only grow darker or remain unchanged; the pixel cannot change in the opposite direction. With this limitation, the number of achievable gray levels becomes:

$$\text{Trinary: } N_T = 2p + 1, \qquad (7.27)$$

$$\text{Quarternary : } N_Q = 3p + 1. \qquad (7.28)$$

These numbers, though much smaller than the unrestricted case, are still significantly greater than the binary case.

7.4.2.5 Halftoning, sampling, and aliasing

Because halftoning with ordered dither imposes a periodic pattern on the image, it has many similarities with the sampling process, but also some important differences. For this discussion, we need to carefully examine certain concepts within digital imaging Spatial resolution and aliasing are two key considerations of digital image processes. People often refer to spatial resolution as the highest frequency that may be reproduced. To be more precise, we should separate resolution and sampling rate. Resolution is the highest spatial frequency that can be seen, and half the sampling rate is the highest the system may represent. Aliasing refers to spurious low-frequency components introduced by the digital imaging process. These fundamental sampled system issues are often best understood from a frequency domain perspective. Such is the case in ordered-dither halftoning. The Fourier transform of the halftone image can be expressed in series form, where one term is the Fourier transform of the original image. Resolution, information density and aliasing can be understood by examining the other terms of the series. These terms depend on the halftone dot fill order. Due to nonlinearity of the process, the analysis is quite lengthy and complicated. Here, we present only a qualitative description of the results. The reader is directed to the references for a detailed mathematical analysis.[32–34,38]

Spatial resolution is dependent on original image contrast and filling order of the halftone dot. At a high-contrast limit, the halftone process resolution is equal to the resolution of the original digital image, while at low contrast the resolution is comparable to sampling at the halftone frequency. To understand these limits, consider a light-dark alternating bar pattern. For a high-contrast input, the black and white bars of the original digital image will be thresholded to black and white, respectively, regardless of the spatial frequency. At low contrast, the gray-scale steps of an entire cell are needed to represent the gray difference between a light bar and dark bar. There would be only one pixel changed in the cell representing the light bar when compared to the cell representing the dark bar. The resolution for intermediate-contrast images is dependent upon the filling order of the cell, where dispersed cells exhibit higher spatial resolution and information density than clustered-dot cells for the same cell size.

Aliasing occurs at the (vector) difference frequency between frequencies in the signal and the halftone pattern. Unlike the sampling case, the aliasing for halftones may be absent or at different amplitudes, depending on the dot filling pattern. The maximum contrast of the aliasing is the same as the contrast of the input image for the usual halftone case. However, if dots are used in which bits are moved, rather than simply added, as gray levels change, then the amplitude of aliasing patterns is unlimited.

Adaptive methods have been proposed that are relevant to perceived edge resolution and aliasing. The ARIES (Aliasing Reducing Image Enhancement System) of Roetling[45] forces equality of total gray content of the continuous tone image and halftone image on a halftone cell sized basis to reduce aliasing, then assigns black and white pixels in a prioritized way allowing for reproduction and enhancement of fine detail. A similar approach was described by Pryor et al.[89] Hammerin and Kruse[90] describe a method for aligning elliptical dots in a manner that best reproduces underlying edges. For a given cell, the ellipse orientation is chosen based upon direction-sensitive edge detection filters.[90]

7.4.3 Error diffusion

Thus far, the halftone methods considered in detail were point processes. Here, we describe the neighborhood process of error diffusion, which uses the concept of fixing the total gray content of the image by calculating the brightness error incurred upon binarizing a pixel and incorporating this error in the processing of subsequent pixels. Due to resulting isolated white and isolated black pixels produced by the basic algorithm, the application of error diffusion has been primarily in display technologies. (Some variations of the method do provide some clustering of like pixels, thereby rendering a printable image). The error diffusion method mitigates the trade-off of screen visibility versus gray-level contouring inherent in ordered-dither methods. Smoothly varying gray- scale images as well as sharp discontinuities are well rendered. A pleasant "blue noise" granular structure is observed with the exception of some possible undesirable worm-shaped artifacts. An example of error diffusion was shown in Fig. 7.9, where the pixel resolution is half of that used in the clustered-dot examples so as to render the images printable and show the method at roughly the resolution of typical use. We discuss the basic algorithm introduced by Floyd and Steinberg[61] and some of its modifications, such as the use of other masks, perturbed thresholds, randomly distributed error, edge enhancement, etc. To develop a deeper understanding of the fundamental technique, a one-dimensional frequency modulation model and a spectral analysis of the general technique are presented.

7.4.3.1 Fundamental algorithm

In error diffusion, as a pixel is thresholded at level T to a binary state, the resulting brightness quantization error is distributed in a weighted manner to neighboring pixels that have not yet been processed. Using the original Floyd-Steinberg[61] diffusion mask, Fig. 7.24 shows schematically an image plane, where pixel P is being processed, X denotes a pixel already processed, {A, B, C, D} denote the set of unprocessed pixels that will receive the quantization error and O denotes unprocessed pixels not relevant to the current calculation. The processing of pixels proceeds in a raster fashion, diffusing the error forward and downward. In the original algorithm, error is distributed to {A, B, C, D} weighted by {7/16, 1/16, 5/16, 3/16}, respectively. A schematic of the algorithm was shown as a feedback loop in Fig. 7.8, and is demonstrated graphically for one-dimensional error diffusion in Figs. 7.25(a) and 7.25(b).

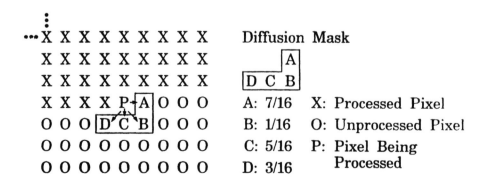

Figure 7.24 Schematic of error-diffusion process operating on an image.

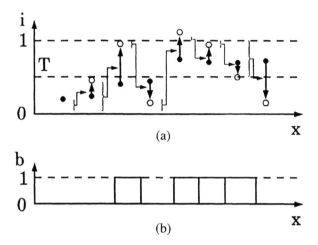

Figure 7.25 Example of error diffusion applied to a one-dimensional signal. (a) Shows the original signal and how the error is propagated forward. (b) Shows the signal after thresholding.

7.4.3.2 Modifications of error diffusion

Many modifications of the original algorithm have been proposed, primarily to eliminate the structured artifacts. First, consider the diffusion mask. The original diffusion weights were derived by trial and error and found to give reasonably good results. Other diffusion masks commonly employed are

$$D = \frac{1}{48} \begin{bmatrix} & & P & 7 & 5 \\ 3 & 5 & 7 & 5 & 3 \\ 1 & 3 & 5 & 3 & 1 \end{bmatrix}, \tag{7.29}$$

$$D = \frac{1}{42} \begin{bmatrix} & & P & 8 & 4 \\ 2 & 4 & 8 & 4 & 2 \\ 1 & 2 & 4 & 2 & 1 \end{bmatrix}, \tag{7.30}$$

which were proposed by Jarvis et al.[91] and Stucki,[92] respectively. It has been shown that the artifacts decrease somewhat with the larger masks, but the directional orientation becomes more pronounced.[71] The mask of Jarvis was also shown to enhance edges strongly, a characteristic of these larger masks.[93]

One key method used to reduce the structured artifacts involves processing the image in other than a common raster fashion. Artifacts are reduced by simply applying error diffusion in a "serpentine raster," where the successive lines are processed in alternating directions. Note that the diffusion mask is flipped as the direction alternates so that the error is always propagated forward onto unprocessed pixels. Other space filling curves can be used to define the order of processing. For example, Witten and Neal[94] processed and diffused error along a Peano curve and Knox[95] provided examples of using a Hilbert curve, both of which greatly reduces structured artifacts but require large memory buffers. Examples of serpentine raster and Hilbert curve are shown in Fig. 7.26.

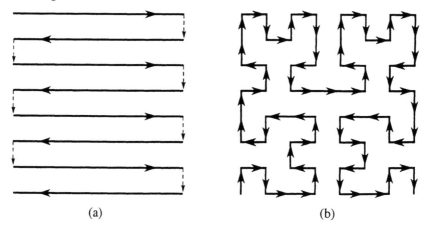

(a)   (b)

Figure 7.26 Examples of (a) serpentine raster, (b) Hilbert curve raster.

Other methods of minimizing artifacts vary the threshold either deterministically or randomly or employ randomization in distributing error and setting thresholds. Billotet-Hoffman and Bryngdahl[47] demonstrated several advantages of utilizing an ordered-dither threshold array in an error diffusion algorithm. The worm artifacts can also be reduced. In addition, more clustered pixel arrangements can be formed, allowing error diffusion to be utilized in the electronic printing process.[96] This scheme provides more gray levels and thus less chance of false gray contours than ordered dither alone. Woo[63] demonstrated that randomly alternating between seven diffusion masks reduced artifacts. At the processing of each pixel, a random number chosen from the set $\{-3, -2, -1, 0, 1, 2, 3\}$, scaled by $1/42$, was added to each coefficient in Stucki's diffusion mask, Eq. 7.30. Deitz[64] reduced artifacts by treating the mask weights as a probability density function. Upon binarization of a given pixel, all the error is distributed to one neighboring pixel. The neighbor is chosen randomly according to the probability density function. Positions of the weights and the threshold could also be randomly varied.

In addition to employing a periodic threshold, several other algorithms are hybrids of error diffusion and ordered dither. Knuth[97] described a dot-diffusion method of halftoning that diffuses quantization error within a cell. Fan[98] described a dot-to-dot diffusion method that diffuses the quantization error from a halftone cell to its neighboring cells in a weighted manner. A schematic of dot-to-dot diffusion for a 45° screen is shown in Fig. 7.27.

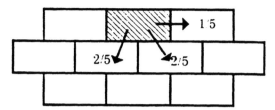

Figure 7.27 Schematic of dot-to-dot diffusion.

Several attempts have been made to improve the symmetry of error diffusion. Peli[99] described a multiresolution approach to halftoning, where accumulated quantization error is uniformly distributed in all directions. Their symmetric form of error diffusion may be implemented using high-speed parallel architectures. Initial results of these new methods show that images can be obtained that are similar to standard clustered-dot or dispersed-dot ordered-dither algorithms. Fan[100] describes a two-pass algorithm that distributes error forward and backward on successive passes.

7.4.3.3 Mathematical description of error diffusion and edge sharpening effects

A one-dimensional model developed by Eschbach and Hauck[65] shows that the error diffusion process can be described as converting the image to a frequency modulated carrier signal with a threshold dependent on the input signal. This would be analogous to an ordered-dither algorithm where the screen frequency (carrier signal) adapts to the signal. A model developed by Eschbach and Knox[72] shows how controlled edge enhancement may be incorporated into the algorithm. Finally, a spectral analysis shows the inherent high-pass filter effect on the texture patterns of error diffusion.[74]

To understand the frequency modulation nature of error diffusion, first consider that error diffusion is a spatially quantized form of pulse density modulation. The binary output of pulse-density modulation can be written

$$b_{PDM}(x) = \text{step}[c(x) - t(x)] \; , \qquad (7.31)$$

where $c(x)$ and $t(x)$ are the carrier and threshold functions, respectively, and $\text{step}(x) = 0$ for $x < 0$, and $\text{step}(x) = 1$ for $x \geq 0$. For pulse density modulation, we may write $c(x) = \cos[\bullet(x)]$, where the derivative of the phase of the cosinusoidal carrier is proportional to the gray-scale image $f(x)$. Solving the differential equation $\partial c / \partial x \propto f(x)$ leads to

$$c(x) = \cos\left[\frac{2\pi}{\Delta f_0} \int_0^x f(x')dx' \right], \qquad (7.32)$$

where f_0 is the maximum intensity of the image and • is the pulse width in the binarized output image. To maintain constant pulse width, the threshold function must be

$$t(x) = \cos\left[\frac{f(x)\pi}{f_0} \right], \qquad (7.33)$$

which upon substitution yields

$$b_{PDM}(x) = \text{step}\left[\cos\left[\left(\frac{2\pi}{\Delta f_0}\right) \int_0^x f(x')dx' \right] - \cos\left[\frac{f(x)\pi}{f_0} \right] \right]. \qquad (7.34)$$

Error diffusion requires spatial quantization within a fixed pixel grid. The effect of the grid can be described by a multiplication by an array of Dirac delta functions and a convolution by a rect function:

$$b_{ED} = b_{PDM}\left(x + \frac{\Delta}{2} \right) \frac{1}{\Delta} \sum rect\left(\frac{x - n\Delta}{\Delta} \right). \qquad (7.35)$$

Figure 7.28 shows the frequency modulation nature of error diffusion, where (a) shows a linear ramp as the input gray signal, (b) and (c) show the carrier frequency and threshold functions, (d) is the thresholded carrier (PDM output) and (e) is the spatially quantized version of (d) (error diffusion).

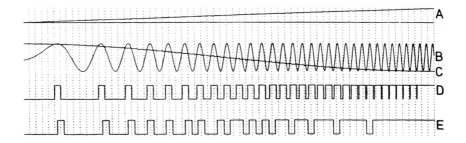

Figure 7.28 Generation of a binary pulse sequence from a one-dimensional gray-wedge input (a) curves, (b) and (c) indicate the carrier and threshold functions, respectively, (d) shows the pulse density modulation output signal, and (e) shows the spatially quantized binary sequence (error diffusion output). Reported with permission from Eschbach and Hauck, Optics Communication (1984).

The error diffusion algorithm may be slightly modified to allow for a controlled degree of edge enhancement. A formalism related to the frequency modulation description given above may be employed to examine the edge enhancement capabilities. The binary output of one-dimensional error diffusion, where all quantization error is passed onto the next pixel only, can be written

$$b_{ED}(n) = \text{step}[\, f(n) - e(n-1) - t_0\,]\ , \qquad (7.36)$$

where $f(n)$ is the gray-scale value of the n'th pixel with the appropriate modified brightness values, $e(n-1)$ is the error incurred upon binarizing the $(n-1)$'th pixel and t_0 is a constant threshold. The cumulative error satisfies the following recursive relationship:

$$e(n) = b(n) - f(n) + e(n-1) \qquad (7.37)$$

and also may be written

$$e(n) = \sum_{l=0}^{n} b(l) - f(l)\,, \qquad (7.38)$$

which upon substitution into Eq. (7.36) yields

$$b_{ED}(n) = \text{step}\left[\, f(n) - \sum_{l=0}^{n-1}[b(l) - f(l)] - t_0\, \right]. \qquad (7.39)$$

In Eq. (7.39) we can see that the sum of the difference between the binary output and the original serves to preserve the overall gray content of the image. We may also compare Eq. (7.39) to Eq. (7.34) to better understand the relationship between error diffusion and frequency modulation. The error sum may be associated with the frequency determining integral, and the slowly varying component $f(n) - t_0$ may be considered an additional phase modulation. Modifying the error diffusion algorithm by generalizing the phase modulation term allows for control over the image microstructure such as enhancement of edges, control of warm artifacts, and control of the fineness of texture.[72,101] In generalized form, the error diffusion process may be written as

$$b_{ED}(n) = \text{step}\left[\, \sum_{l=0}^{n-1}[f(l) - b(l)] - t[f(n)]\, \right], \qquad (7.40)$$

where the threshold is dependent on both $f(n)$ and n. The generalized threshold function allows the process to adapt the local pulse distribution to certain requirements, such as edge enhancement. A simple form for t that allows for controlled edge enhancement capability is

$$t[f(n), n] = t_0 - k \ f(n) , \qquad (7.41)$$

where the amount of edge enhancement increases linearly with the constant k.

We now provide a spatial frequency analysis of the two-dimensional error diffusion process and describe an inherent high-pass filtering effect.[73,93] In two dimensions, the process equation may be written

$$e(m, n) = b \ (m, n) - \left[f(m,n) - \sum_{a_{jk}} e(m-j, n-k) \right], \qquad (7.42)$$

where $a_{jk} \ e(m-j, n-k)$ are the weighted errors from the previous pixels. Fourier transforming yields

$$E(u,v) = B(u,v) - F(u,v) + \left\{ \sum_{a_{jk}} \exp[-i(uj + vk)] \right\} E(u, v), \qquad (7.43)$$

where the capital letters denote Fourier transformed functions of the lower-case functions, u and v are the frequency domain variables and a linear phase term has arisen from shifted errors that were passed onto the neighboring pixels. Equation (7.43) may be re-arranged to yield

$$B(u,v) = F(u,v) + H(u,v) \ E(u, v) , \qquad (7.44)$$

where the filter function $H(u,v)$ is defined as

$$H(u,v) = 1 - \sum_{a_{jk}} \exp[-i(uj + vk)] . \qquad (7.45)$$

We see that the spectrum of the output binary image equals the input image spectrum plus the filtered error term. Figure 7.29 shows a contour plot of $H(u,v)$ for the traditional Floyd-Steinberg four-element mask where the high-amplitude spectral values are denoted by lighter shades. Note that the filter is zero at the origin and one at the corners and therefore has a high-pass filter effect on the error image. This filtering effect may be used to explain inherent high-pass effects of the algorithm and also the blue noise characteristics of the output image. Because quantization artifacts are described by the filtered error

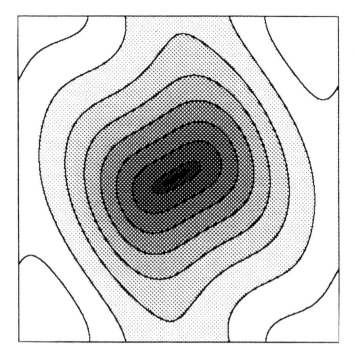

Figure 7.29 Modulus of high-pass filter built into the error diffusion algorithm for the standard four-elements error weights. Zero frequency is located in the center and has zero value. The corners are 1/2 the scan frequency. Where the maximum value of 1.5 occurs. Reprinted with permission from Knox, proceedings of SPIE/IS&T's Symposium on Electronic Imaging Science and Technology (1992).

term $H(u,v)$ $E(u,v)$, the binarized image tends to consist of high-frequency patterns, hence, a blue noise appearance.

One aspect of current research in error diffusion is aimed at understanding and controlling the spatial filtering properties of the process from a linear systems perspective. Other research has been directed toward minimizing worm-like image artifacts. The most interesting development in error diffusion recently has been in the area of color image processing.[75,102,103] For color images, error diffusion can be applied to each of the four separations (C, M, Y, K) independently. Thus for each separation, the pixel value is quantized to 0 or 1 based on the image content in that separation only. This form of error diffusion is sometimes called scalar error diffusion. It has been found that image quality can be improved if the color of each pixel is considered as a vector in a four-dimensional space, with components (c, m, y, k), and vector error diffusion is used. For a binary color printer, the possible output color vectors are (b_1, b_2, b_3, b_4) where the b_i's are binary numbers. In vector error diffusion, the output color vector that is closest to the color vector of a given pixel is chosen as the output for that pixel. A vector color error is calculated and distributed to downstream pixels. Additional criteria can be used to select the output color vector to improve the output uniformity.[103]

In this chapter we have presented encoding methods that are used to reduce the number of quantization levels per pixel in a digital image while maintaining the gray appearance of the image. These techniques are widely employed in the printing and display of digital images. The need for halftoning arises either because the physical processes involved are binary in nature or the processes have been restricted to binary operation for reasons of cost, speed, memory or stability in the presence of process fluctuations. In these applications, the halftoned image is composed ideally of two gray levels, black and white. Spatial integration, plus higher level processing performed by the human visual system, and local area coverage of black and white pixels, provide the appearance of a gray level, or "continuous tone," image.

A brief history of the field was presented, starting with analog methods of halftone image rendering and proceeding to digital techniques. There was a discussion of visual perception, touching only upon the concepts required for the understanding of the halftoning methods. The current methods, ordered dither and error diffusion, were described in detail. For each technique there was a discussion of the methodology as well as issues that include tone reproduction, screen visibility, screen angle, image artifacts and robustness. Current directions of research were discussed.

REFERENCES

1. Eschbach, R., "Recent Progress in Digital Halftoning," *Reprints from IS&T Proceedings 1992-1994*, The Society for Imaging Science and Technology, Springfield, VA, 1994.
2. Jones, P., "Evolution of Halftoning in the United States Patent Literature," *J. Electron. Imaging*, Vol. 3, No. 3, pp. 257-275, July 1994.
3. Roetling, P. G., and Loce, R., "Digital Halftoning," in *Digital Image Processing Methods*, ed. E. Dougherty, Marcel Dekker, NY, 1994.
4. Talbot, W. Fox, "Improvements in the Art of Engraving," British Patent No. 565, 1852.
5. Streifer, W., Goren, R. N., and Marks, L. M., "Analysis and Experimental Study of Ruled Halftone Screens," *Appl. Opt.*, pp. 1299-1317, June 1974.
6. Hepher, M., "A Comparison of Ruled and Vignetted Screens," *Penrose Annual*, Vol. 47, pp. 166-117, 1953.
7. Perry, B. and Medelsohn, M. L., "Picture generation with a standard line printer," *Comm. ACM*, 7(5), pp. 311-313, 1964.
8. Macleod, I. D. G., "Pictorial Output with a Line Printer," *IEEE Transcript on Computers (Short Notes)*, pp. 160-162, Feb. 1970.
9. Schroeder, M. R., "Images from Computers and Microfilm Plotters," *Commun. ACM*, Vol. 12, pp. 95-101, Feb. 1969.

10. Schroeder, M. R., "Images from Computers," *IEEE Spectrum*, Vol. 6, pp. 66-78, Mar. 1969.

11. Knowlton, K., and Harmon, L., "Computer-Produced Greyscales," *Comp. Graph. Image Process.*, pp 1-20, 1972.

12. Arnemann, S. and Tasto, M., "Generating halftone pictures on graphic terminals using run length coding," *Comput. Graphics Image Process.*, 2, pp. 1-11, 1973.

13. Hamill, P., "Line Printer Modification for Better Grey Level Pictures," *Comput. Graph. and Image Proc.*, Vol. 6, pp. 485-491, 1977.

14. Goodall, W. W., "Television by Pulse Code Modulation," *Bell Syst. Technol. J.*, Vol. 30, pp. 33-49, 1951.

15. Roberts, L. G., "Picture Coding Using Pseudo-Random Noise," *IRE Trans. Inf. Theory*, Vol. IT-8, pp. 145-154, Feb. 1962.

16. Schuchman, L., "Dither Signals and Their Effect on Quantization Noise," *IEEE Trans. Commun. Technol.*, Vol. COM-12, pp. 162-164, Dec. 1964.

17. Widrow, B., "Statistical Analysis of Amplitude-Quantized Sampled-Data Systems," *Trans. AIEE (Applications and Industry) pt. II*, Vol. 79, pp. 555-568, Jan. 1961.

18. Thompson, J. E., and Sparkes, J. J., "A Pseduo-Random Quantizer for Television Signals," *Proc. IEEE*, Vol. 55, pp. 353-355, Mar. 1967.

19. Limb, J. O., "Coarse Quantization of Visual Signals," *Australian Telecommun. Res.* Vol. 1, No. 1 and 2, pp. 32-42, Nov. 1967.

20. Limb, J. O., "Design of Dither Waveforms for Quantized Visual Signals," *Bell Sys. Tech. J.*, Vol. 48, pp. 2555-2582, Sept. 1969.

21. Steinberg, E., Easton, R., and Rolleston, R., "Analysis of Random Dither Patterns Using Second-Order Statistics," *IS&T 44th Annual Conference*, St. Paul, May 1991.

22. Mitsa, T., and Parker, K. J., "Digital Halftoning with a Blue Noise Mask," *SPIE/IS&T Symposium on Electronic Imaging Science and Technology*, Feb. 24-Mar. 1, 1991, San Jose.

23. Sullivan, J., Ray, L., and Miller, R., "Design of Minimum Visual Modulation Halftone Patterns," IEEE Trans. Syst. Man Cybernet, Vol. SMC-21, No. 1, pp. 33-38, 1991.

24. Ulichney, R., "The Void-and-Cluster Method for Dither Array Generation," *Human Vision, Visual Processing and Digital Display IV, Proc. SPIE*, Vol. 1913, pp. 332-343, 1993.

25. Delabastita, P. A., "Screening Techniques, Moiré in Four Color Printing," Proceedings of TAGA Conference, pp. 44-65, Apr. 1992.

26. Spaulding, K. E., Miller, R. L., and Schildkraut, J., "Methods for Generating Blue-Noise Dither Matrices for Digital Halftoning," *J. of Electron. Imaging*, Vol. 6, No. 2, pp. 208-230, Apr. 1997.

27. Klensch, R. J., Meyerhofer, D., Walsh, J. J., "Electronically Generated Halftone Pictures," *TAGA Proc.*, Vol. 22, p. 302-320, 1970.

28. Lippel B., Kurland, M., and Marsh, A. H., "Ordered Dither Patterns for Coarse Quantization of Pictures," *Proc. IEEE*, Vol. 59, No 3, pp. 429-431, March 1971.

29. Lippel, B., and Kurland, M., "The Effect of Dither Luminance Quantization of Pictures," *IEEE Trans. Commun.* Vol. COM-19, No. 6, pp. 879-888, 1971.

30. Bayer, B. E., "An Optimum Method for Two-Level Rendition of Continuous-Tone Pictures," *IEEE International Conference on Communications*, Vol. 1, pp. 26-11, 26-15, 1973.

31. Judice, C. N., Jarvis, J. F., and Ninke, W. H., "Using Ordered Dither to Display Continuous Tone Pictures on an AC Plasma Panel," *Proc. of the SID*, Vol. 15 Fourth Quarter, pp. 161-169, 1974.32.

32. Kermisch, D., and Roetling, P. G., "Fourier Spectrum of Halftone Images," *JOSA*, Vol. 65, pp. 716-723, June 1975.

33. Allebach, J. P., and Liu, B., "Analysis of Halftone Dot Profile and Aliasing in the Discrete Binary Representation on Images," *JOSA*, Vol. 67, pp. 1147-1154, 1977.

34. Allebach, J. P., "Aliasing and Quantization in the Efficient Display of Images," *JOSA*, Vol. 69, p. 869-877, 1979.

35. Marquet , M. and Tsujiuchi, J., "Interpretation des aspects particuliers des images obtenues dans une experience de detremage," *Optica Acta*, 8, No. 3, pp. 267-277 1961.

36. Roetling, P. G., "Visual Effects in Binary Display of Continuous-Tone Images," *SID International Symposium Digest of Technical Papers*, Vol. 8, 1977.

37. Roetling, P. G., "Binary Approximation of Continuous-Tone Images," *Photo. Sci. and Eng.*, Vol. 21, No. 2, pp. 60-65, 1977.

38. Roetling, P. G., "Analysis of Detail and Spurious Signals in Halftone Images," *J. Appl. Photo. Eng.*, Vol. 3, No. 12, 1977.

39. Bryngdahl, O., "Halftone Images: Spatial Resolution and Tone Reproduction," *JOSA*, Vol. 68, pp. 416-422, 1978.

40. Ruckdeschel, F. R., Walsh, A. M., Hauser, O. G., and Stephan, C., "Characterizing Halftone Noise: A Technique," *Appl. Opt.*, Vol. 17, Dec. 1978.

41. Matsumoto, S., and Liu, B., "Analytical Fidelity Measures in the Characterization of Halftone Processes," *JOSA*, Vol. 7, No. 10, pp. 1248-1254, 1980.

42. Allebach, J. P., "Binary Display of Images When Spot Size Exceeds Step Size," *Appl. Opt.*, Vol. 119, pp. 2513-2519, 1980.

43. Roetling, P. G., and Holladay, T. M., "Tone Reproduction and Screen Design for Pictorial Electrographic Printing," *J. Appl. Photo. Eng.* Vol. 5, No. 4, pp. 179-182, 1979.

44. Holladay, T. M., "An Optimum Algorithm for Halftone Generation for Displays and Hard Copies," *Proceedings SID*, Vol. 21, No. 2, pp. 185-192, 1980.

45. Roetling, P. G., "Halftone Method for Edge Enhancement and Moire Suppression," *JOSA*, Vol. 66, pp. 985-989, 1976.

46. Allebach, J. P., "Visual Model-Based Algorithms for Halftoning Images," *Proc. SPIE*, Vol. 310, pp. 151-158, 1981.

47. Billotet-Hoffman, C., and Bryngdahl, O., "On the Error Diffusion Technique for Electronic Halftoning," *Proc. SID*, Vol. 24, pp. 253-258, 1983.

48. Riseman, J., Smith, J., d'Entremont, A., and Goldman, C., "An Apparatus for Generating an Image from a Video Signal," U.S. Patent 4,800,442, Jan. 24, 1989.

49. Melnychuck, P., and Shaw, R., "Fourier Spectra of Digital Halftone Images Containing Dot Position Errors," *JOSA*, Vol. 5, pp. 1328-1338, 1988.

50. Haas, D., "Contrast Modulation in Halftone Images Produced by Variation in Scan Line Spacing," *J. Imaging Technol.*, Vol. 5, No. 1, p. 46-55, 1989.

51. Loce, R., and Lama, W., "Halftone Banding in a Xerographic Image Bar Printer," *J. Imaging Technol.*, Vol. 16, No. 1, pp. 6-11, Feb. 1990.

52. Loce, R., Lama, W., and Maltz, M., "Modeling Vibration-induced Halftone Banding in a Xerographic Laser Printer," *J. Electron. Imaging*, Vol. 4, No. 1, pp. 48-61, Jan. 1995.

53. Bloomberg, S. J., and Engledrum, P. G., " Estimation of Color Errors Due to Random Pixel Placement Errors," *J. Imaging Technol.*, Vol. 16, No. 2, pp. 75-79, Apr. 1990.

54. Naing, W., Miyake, Y., Taniguchi, T., and Kubo, S., "An Evaluation of Image Quality of Tri-Level Images Obtained by a New Algorithm," *J. Imaging. Technol.*, Vol. 15, No. 1, Feb. 1989.

55. Lama, W., Feth, S., and Loce, R., "Hybrid (Gray Pixel) Halftone Printing," *J. of Imaging Technol.*, Vol. 15, No. 3, pp. 130-135, June 1989.

56. Auyueng, V., "Raster Output Scanner with Subpixel Addressability," U.S. Patent No. 5,325,216, May 28, 1994.

57. Steinbach, A., and Wong, K. Y., "Moiré Patterns in Scanned Halftone Pictures," *JOSA*, Vol. 72, No. 9, pp. 1190-1198, Sept. 1982.

58. Shu, S. J., Springer, R., and Yeh, C. L., "Moire Factors and Visibility in Scanned and Printed Halftone Images," *Opt. Eng.*, Vol. 28, No. 7, pp. 805-812, July 1989.

59. Inose, H., Yasuda, Y., Murakami, J., "A telemetering system by code modulation - Δ- Σ modulation," *IRE Transcript Space Electron. and Telemetry*, Vol. SET-8, pp. 204-209, 1962.

60. Inose, H., and Yasuda, Y., "A Unity Bit Coding Method by Negative Feedback," *Proc. IEEE*, Nov. 1963, pp. 1524-1535.

61. Floyd, R., and Steinberg, L., "An Adaptive Algorithm for Spatial Grey Scale," *Proc. Soc. Info. Display*, Vol. 17, No. 2, p. 75, 1976, also SID Int. Sym. Digest of Tech. Papers, pp. 36-37.

62. Hale, J. A. G., "Dot Spacing Modulation for the Production of Pseudo Gray Pictures," *Proceedings SID*, Vol. 17, pp. 63-74, 1976.

63. Woo, B., "A survey of halftoning algorithms and investigation of the error diffusion technique," SB thesis, MIT, 1984.

64. Dietz, H., "Randomized Error Distribution," Industrial Associates at Brooklyn Polytechnic Institute, Apr. 1985.

65. Eschbach, R. and Hauck, R., "Analytic Description of the 1-D Error Diffusion Technique for Halftoning," *Opt. Commun.*, Vol. 52, No. 3, pp. 165-168, Dec. 1, 1984.

66. Eschbach, R. and Hauck, R., "A 2-D Pulse Density Modulation by Iteration for Halftoning," *Opt. Commun.*, Vol. 62, No. 5, pp. 300-304, June 1, 1987.

67. Eschbach, R., and Hauck, R., "Binarization Using a Two-Dimensional Pulse Density Modulation," *JOSA – A*, Vol. 4, No. 10, p. 1873, 1987.

68. Eschbach, R., "Pulse Density Modulation on Rastered Media: Combining Pulse Density Modulation and Error Diffusion," *JOSA-A*, Vol. 7, No. 4, pp. 708-716, Apr. 1990.

69. Sullivan, J., and Miller, R., "New Algorithm for Image Halftoning Using a Human Visual Model," *Proc. SPSE's 43rd Annual Conference*, pp. 145-148, Roch. N.Y. May 20-25, 1990.

70. Engledrum, P. G. "Optimum density levels for multilevel halftone printing," *Jrnl. of Imaging Sci.*, Vol. 30, No. 5, pp. 220-222, 1987.

71. Ulichney, R., *Digital Halftoning*, MIT Press, 1987.

72. Eschbach, R., and Knox, K. T., "Error-Diffusion Algorithm with Edge Enhancement," *JOSA*, Vol. 8, No. 12, p. 1844, Dec. 1991.

73. Weisbach, S., Wyrowski, F., and Bryngdahl, O., "Error Diffusion as a Filter in Digital Optics," *Conference of the German Society for Applied Optics*, June 1990.

74. Knox, K. T, "Spectral Analysis of Error Diffusion," *Proc. IS&T Annual Meeting*, St. Paul, May 16, 1991, pp. 448.

75. Kolpatzik, B.W., and Bouman, C.A., "Optimized Error Diffusion for Image Display," *J. Electron. Imaging*, Vol. 1, No. 3, pp. 277-292, 1992.

76. Cornsweet, T. N., *Visual Perception*, New York: Academic Press, 1970.

77. Cohen, R. W., and Gorog, I., "Visual Capacity - an Image Quality Descriptor for Display Evaluation," *RCA Eng.* Vol. 20, No. 3, pp. 72-74, Oct.-Nov. 1974.

78. Roetling, P. G., "Visual Performance and Image Coding," *Proc. SID*, Vol. 17/2, pp. 111-114, 1976.

79. Daly, S., "The Visible Differences Predictor: An Algorithm for the Assessment of Image Fidelity," *Human Vision, Visual Processing, and Digital Display III*, Proc. SPIE, Vol. 1666, pp. 2-15, 1992.

80. Zhang, X., Silverstein, D. A. and Farrell, J. E., "The Color Image Quality Metric S-CIELAB and its Application on Halftone Texture Visibility," IS&T/SPIE Electronic Imaging, 1997.

81. Steinberg, E., Rolleston, R., and Easton, R., "Analysis of Random Dithering Patterns Using Second-Order Statistics," *J. of Electron. Imaging*, pp. 396-404, Oct. 1992.

82. Sullivan, J. R., and Ray, L. A., "Digital Halftoning with Correlated Minimum Visual Modulation Patterns," U.S. Patent No. 5,214,517, assigned to Eastman Kodak Company, 1993.

83. Lin, Q., "Halftone Images Using Special Filters," U.S. Patent No. 5317418, assigned to Hewlett-Packard Company, 1994.

84. Lin, Q., "Halftone Image Formation Using Dither Matrix Generated Based upon Printed Symbol Modules," U.S. Patent No. 5469515, assigned to Hewlett-Packard Company, 1995.

85. Daels, K., and Delabastita, P., "Tone Dependent Phase Modulation in Conventional Halftoning," *Proceedings of IS&Ts 47th Annual Conference*, May 15-20, 1994.

86. Amidror, I., Hersch, R. D., and V. Ostromoukhov, "Spectral Analysis and Minimization of Moiré Patterns in Color Separation," *J. Electron. Imaging*, Vol. 3(2), pp. 295-317, July 1994.87.

87. Fink, P., *Postscript Screening: Adobe Accurate Screens*, Adobe Press, Mountain View, Calif., 1992.

88. Nishikawa, M., "Method of Forming Oblique Dot Pattern," U.S. Patent 4,805,003, Feb 14, 1989.

89. Pryor, R. W., Cinque, G. M., and Rubinstein, A., "Bilevel Displays - a New Approach," *Proceedings SID*, Vol. 19, pp. 127-131, 1978.

90. Hammerin, M., and Krus, B., "Adaptive Screening," *Proceedings of IS&T*.

91. Jarvis, J. F., Judice, C. N., and Ninke, W. H., "A Survey of Techniques for the Display of Continuous-Tone Pictures on Bilevel Displays," *Comput. Graph. Image Process.*, Vol. 5, pp. 13-40, 1976.

92. Stucki, P., MECCA, "A multiple-error correcting computation algorithm for bilevel image hardcopy reproduction," Research Report RZ 1060, IBM Research Laboratory Zurich, Switzerland, 1981.

93. Knox, K. T., "Error Image in Error Diffusion," *Proc SPIE/IS&T's Symposium on Electronic Imaging Science and Technology*, Vol. 1657, San Jose, Feb. 9-14, 1992.

94. Witten, I. H. and Neal, M., "Using Peano curves for bilevel display of continuous-tone images," *IEEE CG&E*, pp. 47-52, 1982.

95. Knox, K. T., "Threshold Modulation in Error Diffusion or Non-Standard Rasters," *Proc. SPIE*, Vol. 2179, pp. 159-169, 1994.

96. Eschbach, R., "Active Binarization Techniques in Printing Applications," presented at *Annual Meeting of DGaO*, June 8-13, 1992, FRG.

97. Knuth, D. E., "Digital Halftones by Dot Diffusion," *ACM Trans. Graph.*, Vol. 6, No. 4, pp. 245-273, Oct. 1987.

98. Fan, Z., "Dot-to-Dot Error Diffusion," Journal of Electronic Imaging, pp. 62-66, 1993.

99. Peli, E., "Multiresolution, error-convergence halftone algorithm," *J. Opt. Soc. Am. A.*, Vol. 8, No. 4, pp. 625-636, 1991.

100. Fan, Z., "Error Diffusion with a more symmetric error distributions," *Proc. SPIE*, Vol. 2179, pp. 150-158, 1994.

101. Levien, R., "Output Dependent Feedback in Error Diffusion Halftoning," *Proceedings of the IS&T*.

102. Kolpatzik, B.W., and Bouman, C.A., "Color Palette Design for Error Diffusion," *Proceedings of IS&Ts 46th Annual Conference*, May 9-14, 1993.

103. Klassen, R.V., and Eschbach, R., "Vector Error Diffusion in a Distorted ColorSpace," *Proceedings of IS&Ts 47th Annual Conference*, May 15-20, 1994.

8

DOCUMENT RECOGNITION

John C. Handley
Xerox Corporation
Webster, New York

The much-heralded age of the paperless society has yet to appear. There is an increasing demand to convert paper documents to electronic versions, not only for archived documents but for recently published ones as well. Document recognition technologies serve a vital role in document processing systems by providing the bridge from printing back to electronic storage and editing. Ironically, the proliferation of word processing software has *increased* the need for easy access to the contents of paper documents. Judging by the explosive growth of the World Wide Web, electronic publishing will be a ubiquitous human activity and document recognition will be an important means to obtain formatted content.

Document recognition is the task of converting the image from a scanned document page into an editable electronic format where characters and other page objects are represented symbolically. Clearly raster images can be edited, but what we are discussing here are documents in the form of word processor files, where at least some of the formatting information is preserved. Documents have two representations, physical and logical. A physical representation is the layout of the rendered document: the characters and graphics on the page. The logical representation of a document is the labeled symbolic representation: character codes and labels for the function of those characters. Most World Wide Web users have some experience with a logical representation. The language used to express the format of a web page is hyper-text markup language (HTML). The intent is to represent the content of the document through image data or text and use tags to label function. The actual physical format seen on a screen or printed out is determined by the web-browser by interpreting the tags. The goal of document recognition is to reverse the process.

There is a wealth of structure in document images. By design, humans easily recognize the form and content of documents, but useful computational

---

The views represented herein are those of the author and are not intended to represent the position of the employer.

methods remain elusive. Ignoring completely the image processing and pattern recognition problems regarding character recognition, adequate document image models do not exist except in specialized domains. A special purpose document recognition system for bank checks or credit card slips has better accuracy than an off-the-shelf software package whose users will expect recognition of magazine pages, financial reports, telephone books and business cards, and so on. A flexible and complete model to fit an arbitrary document image has yet to be formulated. Like natural language, syntactical rules are routinely broken. Even if the existing universe of document styles could be codified, new ones will be produced to stand out from the rest. Without a set of first principles for document creation, document recognition will always be at the mercy of style. Moreover, document creation is no longer limited by technology of metal type and ink printing. Through digital manipulation of documents on desktop personal computers, the range of styles is unbounded.

Document recognition consists of two processes: document image analysis and document image understanding. The former is an image processing function whereby logical objects (e.g., paragraphs, captions, table headings, etc.) on the page image are found and grouped together using image attributes. The latter refers to the analysis of the logical objects and their attributes relative to a model such as a document type expressed in HTML. In real systems it is sometimes difficult to separate the two.

The entire conversion process requires that all symbols on the page be converted from their image representations. For characters, the field of optical character recognition is quite mature. It consists of two major subfields: machine print recognition and handwritten character recognition [1, 2]. The latter is much more difficult than the former because handwritten letters have more inherent variability.

There is recent interest in the field and a number of surveys and special journal editions [3-8]. Rather than duplicate the content of other works, we cover aspects of document recognition not adequately surveyed: tables, forms, and font identification. We still review the major contributions to document analysis and understanding.

8.1 DOCUMENT IMAGE ANALYSIS AND UNDERSTANDING

After a document is scanned into computer memory, the image is processed to remove noise and correct skew [9]. Image-based methods are used to identify objects on the page upon which the document model is built. They can include paragraphs, captions, headings, titles, footers, headers, abstracts, graphics, equations, tables, etc. For example, a digital library system for full-text information retrieval for journal articles may identify titles, authors, and abstracts for indexing [10]. Logical objects are identified from image attributes. Methods for identification include texture analysis for halftone detection and pattern recognition techniques to determine fonts.

For example, from Fig. 8.1, one might want to recover the following tagged format:

```
<chapter>
<chaptitle>Morphological Processing of Binary
Images</chaptitle>
<intro><par>The present...image</par></intro>
<section>
<sectitle>Pixel Regions</sectitle>
<par>Appreciation...</par>
<par>Two pixel...<emphasize>strong neighbors</
emphasize>
... <emphasize>weak neighbors</ emphasize>...
<emphasize>neighbors</ emphasize>...
In this
vein,<reference>neighbors</reference>depict...</par>
<par>When...</par>
<picture name="neighbors" image="neighbors.tif">
<caption>Various...</caption> </picture>
.
.
.
</section>
</chapter>
```

The tagged format represents the logical structure. Different interpreters could place the graphics differently, use different fonts, break lines differently, use different numberings for chapters and pages, and so on.

One must recognize the logical structure from the physical layout. Figure 8.2 represents the geometric layout one could reasonably expect from a document analysis algorithm. If one could identify the string "Morphological Processing of Binary Images" as being in a large font at the top of the page, one could identify it as a chapter title. One could find the four paragraphs using information about the texture of text. The graphic also has a certain texture, and using context, one could infer the text line beneath it to be a caption. The possible arrangements of objects on the page is captured formally in a document structure, perhaps represented as a tree or a grammar.

Document image analysis consists of two steps: document segmentation and block classification (e.g., finding the blocks in Fig. 8.2 and then assigning labels). The first step is to segment the image into functionally uniform regions such as lines of text or halftoned pictures. After segmentation, features are extracted for each block and a classifier is used to estimate its class. The classes used depend on the document processing model and can be as simple as two classes: text/non-text. Segmentation strategies can be bottom-up in which connected components are grouped into larger regions, or top-down, wherein image operations are done at finer and finer detail.

Chapter 3

Morphological Processing of Binary Images

The present chapter covers some morphological algorithms for binary images. These include boundary detection, conditional dilation, curve filling, thinning, segmentation, and restoration. The next chapter explores algorithms employing the hit-or-miss transform. The intent is not to produce a complete compilation of morphological algorithms, but rather to provide a geometrically grounded introduction to some of the most useful techniques. In all cases, it should be recognized that actual use of a particular methodology will often require preprocessing to put the image into a form suitable for application of the algorithm and postprocessing to provide an acceptable output image.

3.1 Pixel Regions

Appreciation of operators concerned with pixel regions requires familiarity with basic concepts having to do with pixel topology. We briefly review the ones necessary for the present text.

Two pixels are said to be **strong neighbors** if they are vertically or horizontally adjacent. They are said to be **weak neighbors** if they are diagonally adjacent. They are simply said to be **neighbors** if they are either strong or weak neighbors. In this vein, Figs. 3.1 (a), (b), and (c) depict the strong-neighbor, weak-neighbor, and neighbor masks, respectively, for the origin.

When concerned with geometric understanding, we often refer to a collection of pixels as a region. A region is said to be **strongly connected** if for any two pixels x and y in the region, there exists a sequence of pixels also in the region such that the first pixel is x, the last is y, and each pixel in the sequence is a strong neighbor of the next. A region is simply said to be **connected** if the same definition applies with

(a)   (b)   (c)

Fig. 3.1. Various neighbor masks for the origin: (a) strong-neighbor mask, (b) weak-neighbor mask, (c) neighbor mask.

Figure 8.1 Binary document image.

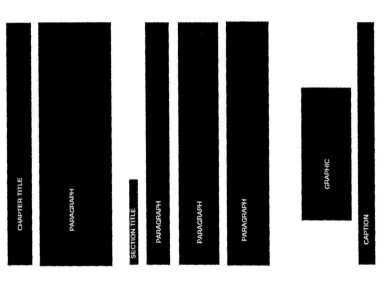

Figure 8.2 Geometric layout.

8.2 RUNLENGTH SMOOTHING ALGORITHM AND VARIANTS

In the oft-cited method of Wong, Casey, and Wahl [11], regions of the page image corresponding to lines of text, line drawings, and graphics are determined. The aspect of the method of interest here is the runlength smoothing algorithm (RLSA) for document segmentation. The first step is to convert the image into horizontal runlength data. White runlengths less than some threshold H are converted to black runlengths. This step is called a horizontal smear. The original image is converted to a vertical runlength representation and white runlengths less than V are converted to black for a vertical smearing. The horizontally and vertically smeared images are "anded" together to produce an image containing large black connected components. As a final step, the "anded" image is smeared horizontally with a smaller threshold H_f. The resulting image has connected components (CC's) corresponding to lines of text, line drawings, and graphics which are further classified into text, line art, and graphics using texture features and pattern recognition algorithms (Fig. 8.3). While the smearing operations may be intuitive, the effect is to classify each pixel into either a member of the foreground or background. White pixels become members of the foreground by being near to black pixels in a way that elicits the structure of the underlying image. (This notion of segmentation by nearness to black pixels is also found in the recursive morphological closing transform of Chen *et al.*[12]. In that transform distance to the nearest black pixel is generalized as the minimum number of consecutive closings by a suitably chosen structuring element needed to reach a white pixel). A black pixel is part of the foreground by default. A white pixel is the intersection of a unique pair of horizontal and vertical white runlengths. If the horizontal runlength is less than H and the vertical runlength is less than V, or if the horizontal runlength is simply less than H_f, classify the white pixel as foreground. (In the original formulation [11], the image sampling resolution is 240 samples per inch, $H = 300$, $V = 500$, and $H_f = 30$). Thus the RLSA is a method to classify background pixels into object or background using the lengths of the containing horizontal and vertical runlengths (h and v, respectively) as features. Indeed, if one does a scatter plot of white pixels by their containing runlengths, it is evident from the clusters that form why these particular H, V, and H_f values were chosen: they partition the h-v plane into rectangular regions corresponding to object and background:

$$\text{Foreground} = \{(h,v): 1 \le h \le H, 1 \le v \le V\} \cup \{(h,v): 1 \le h \le H_f\}.$$

Figures 8.1 and 8.3 show a scanned document binary image and the result of the RLSA. Features are extracted from the regions in the original that correspond to the connected components in the smoothed image. Those

Figure 8.3 Result of RLSA.

features, such as eccentricity and black pixel density, are used to classify the regions into text, horizontal lines, vertical lines, or halftones and graphics. A linear discriminant function was trained for classification.

The RLSA is the basis of other document analysis systems. In the system of Le *et al.* [13], the unmodified RLSA works well on medical journal images. Neural network classifiers are trained on the resultant blocks to distinguish text from nontext, resulting in an accuracy of 99.61%. Seven features are used for each block (the first four are the original ones used in Ref. 11:

1. height
2. eccentricity: the ratio of height to width
3. density: the ratio of black pixels to area
4. mean horizontal black runlength
5. block height times the mean horizontal black runlength
6. eccentricity times the mean horizontal black runlength
7. density times the mean horizontal black runlength.

These seven features capture shape and texture information that allows text and nontext to be differentiated.

The original values of parameters H, V, and H_f are not appropriate for all documents nor for every region within a document. In the method of Fisher *et al.* [14] document images are segmented into text and nontext components using the RLSA algorithm with adaptive H and V thresholds. The parameter V is the interline space as found through a histogram of horizontal projections and H is found as the most populous cell in a histogram of distances between white to black transitions. After smoothing, the boundaries of the resulting CC's are used to analyze the original image. Statistics are gathered for each CC which is then classified as text, nontext, or unknown. If unknown, additional statistics are obtained and more rules are executed. This procedure continues until all CC's are classified.

Chauvet *et al.* [15] use different H and V parameter settings to iteratively segment images of insurance forms. At the first stage, the first two steps of the RLSA are executed (at 200 dpi) with (H=15, V=15) to get major blocks. Within each major block, lines of text are found by running RLSA with (H=30, V=0). Word segmentation is done within each line by RLSA with (H=6, V=0). Finally, characters are isolated within words with (H=0, V=22). It is noted that the first two steps of the RLSA correspond to a morphological closing with an H-length horizontal structuring element "anded" with the result of closing with a vertical structuring element of length V. For classification, texture measurements use the function $CL(H,V)$, the number of 8-connected components remaining after executing RLSA with H and V. There are six classes:
1. halftones
2. words (isolated words, e.g., page numbers)
3. lines (isolated lines of text, e.g., captions and footers)
4. paragraphs (blocks of text)
5. equations
6. graphics (line art, graphs, charts, etc.).

The following features are used to classify blocks:
1. 8-connected component density ($CL(0,0)$/area)
2. symmetry factor: $CL(0,10)-CL(10,0)$
3. horizontal reduction factor: $CL(0,10)/CL(0,0)$
4. vertical reduction factor: $CL(2,0)/CL(0,0)$
5. black pixel density
6. horizontal black runlengths/area
7. vertical black runlengths/area
8. mean horizontal black runlength
9. mean vertical black runlength
10. block area/page area
11. eccentricity (width/height).

Linear discriminant functions obtained from a maximum likelihood classifier yield nearly 100% accuracy on resubstitution data.

8.3 BACKGROUND COVERING

Instead of finding a covering of the black areas of a page to find objects, the dual approach is to find a subcovering of the background and use its complement as a segmentation of the foreground. Baird [16] developed a method by which the white space in a bilevel document image is approximated by a sequence of coverings by white rectangles (see also Baird *et al.* [17, 18]). The largest white rectangle containing each white pixel is computed. This collection of white rectangles produces the inverse image of the document. In segmenting the document image, one would like to merge characters or perhaps lines of text, depending on the granularity desired. This step is a precursor to region classification. Starting with a null subcovering, one adds white rectangles of constrained shapes (elongated vertical and horizontal) until a stopping point is reached. The stopping point depends on the degree of segmentation desired. The result is a subcovering of white space producing black connected components in its complement corresponding to functional objects on the page. The class of documents that can be segmented this way depends on inter-column, inter-line and inter-character statistics, but apparently contains a large number of useful documents.

Within a scanned document image there are a finite number of pixels and hence white rectangles. Let $R = \{r_1,...,r_N\}$ be the unique maximal white rectangles. Note that no rectangle is a subset of another. A subcovering C_n is the union of a subset of R. Let $C = \cup_{i=1}^{N} r_i$ be the entire background. There are 2^N-1 non-null coverings, i.e., increasing sequences $\{C_n\}$ converging to C. The goal of the algorithm is to produce a sequence of subcoverings which terminates in a desired state. Properly chosen sequences elicit greater detail (smaller blocks) in the complement. Sequences are constructed in a greedy manner by selecting elongated rectangles first and avoiding small squares. A stopping rule depends on the percentage of rectangles already in the subcover and the shapes of the ones left. This method relies on the rectilinear arrangement of page objects.

8.4 DOCUMENT DECODING

The document decoding method of Refs.19 through 22 poses character and layout recognition as an inverse problem: What is the most probable original text (or text plus formatting) that gives rise to the observed document image? The problem is modeled as one in communications. The original text, or text with formatting information, is called the *message*. For example, the source may be PostScript or LaTeX documents. An imager renders the message as an image. The image passes through a *channel* that distorts the image and is intended to model printing and scanning degradation. Let M be the original message, Q the rendered image, and Z the observed image. A maximum a posteriori (MAP) estimate of the (unobserved) message is one that maximizes the probability of M given Z:

$$\hat{M} = \max_{M} \arg P\{M \mid Z\}.$$

(All random variables are discrete, so there are no measure-theoretic complications.) The probabilistic structure must be defined and a search method used to find $\hat{M}$. However, the estimated message can contain all of the logical information. The imager model is a Markov source where characters (templates) are placed according to horizontal or vertical displacements and are not allowed to overlap. Transitions are chosen between states randomly, and each transition has attributes $(Q_t, m_t, a_t, \vec{\Delta}_t)$, where Q_t is a template (character), m_t is a message string, a_t is a transition probability, and $\vec{\Delta}_t$ is a vector displacement. A path $\pi = (t_1, ..., t_P)$ defines an imaged document:

$$Q_\pi = \bigcup_{i=1}^{P} Q_{t_i}[\vec{x}_i], \tag{8.1}$$

where $\vec{x}_1 = 0$ and $\vec{x}_{i+1} = \vec{x}_i + \vec{\Delta}_{t_i}$ and a message $M_\pi = m_{t_1}, \cdots, m_{t_P}$. It can be shown that the MAP estimate is found to maximize

$$\Pr\{M, Z\} = \sum_{\pi \mid M_\pi = M} \Pr\{Z \mid Q_\pi\} \Pr\{\pi\}, \tag{8.2}$$

where $\Pr\{Z \mid Q\}$ is the (analytic) channel model. The decoder computes $\hat{M}$ from observed Z by finding the most probable path π using dynamic programming. This method is computationally expensive but improvements are possible [20, 21]. The method requires modeling and training on documents and noise sources. However, this approach captures document recognition as a large numeric optimization problem and is fully general.

8.5 X-Y TREES

A related technique for document analysis and understanding is the method of X-Y trees (Krishnamoorthy *et al.* [23]). This method consists of a hierarchical decomposition of a document image plus an interpretation of each block at each level according to a grammar. (A brief overview of grammars is given in the appendix).

The X-Y tree method recursively partitions a document image using vertical and horizontal projections. One may start with either a vertical or horizontal projection. Assuming a horizontal projection, a histogram of the sums across scanlines is tallied. The histogram is thresholded with those indices below marked by 0's and those above marked with 1's. Contiguous strings of 0's and 1's are called white and black *atoms*, respectively. For example, the first level vertical projection produces the segmentation in Fig. 8.4 where the vertical coordinates of the boxes are determined by vertical projections. Each block in the segmentation corresponds to a black atom. Each block is then projected in the orthogonal direction. Figure 8.5 shows a block of Fig. 8.4 segmented horizontally. This level of segmentation finds words (or characters). As each block represents a node, a tree is constructed alternating between vertical and horizontal segmentations from the document image down to connected components. Each level is intended to represent a different logical level (e.g., pages, columns, lines, words).

After the image has been recursively decomposed, the next step is to interpret the blocks. Each block has a sequence of tokens associated with it. For example, in Fig. 8.4, the atoms have been assigned symbols based upon their height. Large white atoms are assigned 'S', black atoms having height approximately that of text lines are assigned 'p', etc. The sequence of symbols is then processed by a lexical analyzer to group the symbols into tokens (or *molecules*). For example, a sequence of 'R' followed by one or more 'p's, followed by 'R' or 'S' might be assigned a token that could represent a paragraph or caption. A single-token look-ahead parser for context-free grammars parses the molecules and assigns entity labels (nonterminals) to each token. Labels are determined by the allowable patterns admissible by the grammar. If syntax is not legal, an error is reported. Sequences of entities with the same label are merged.

This method is extended to a hierarchy of grammars. A correct parse at one level labels the sub-blocks that compose it. Each of these labels corresponds to a further grammar. The sub-blocks are then parsed according to the grammar determined by its label, and so on, until a complete parse is obtained. A correct parse depends on the correct labeling of its sub-blocks, so the document image must be parsed down to its lowest level. Thus, the document interpretation becomes a search problem. For example, the graphic and its caption at the bottom of the page in Fig. 8.4 may be mislabeled as a section title and a paragraph. Upon parsing the graphic in the horizontal direction, a syntax error occurs, forcing a new parse at the page level and re-parsing at the page level.

This method is similar to document decoding in that it seeks a thorough logical interpretation of a document image, requires document models to be built or estimated, and can be computationally expensive.

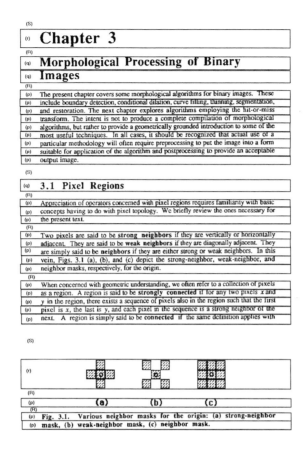

Figure 8.4 First order X-Y segmentation.

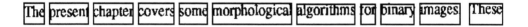

Figure 8.5 Second order X-Y segmentation.

8.6 TABLE RECOGNITION: IDENTIFICATION AND ANALYSIS

The exact notion of a table is difficult to describe. Certainly it should incorporate the idea of symbols arranged into rows and columns. Using horizontal and vertical lines (called ruled lines or rulings) or wide white space, complex two-dimensional arrangements are possible. The notion of a table overlaps that of a form (see the next section).

Tables can represent database or spreadsheets entries where the semantic relations between cells are precise or they can represent a convenient tabulation of data the structure of which is evident only from the text itself. It is difficult to posit a table model for all the structures a reader might what to call a table.

Figures 8.6 through 8.8 show the three varieties of tables: line-less, semi-lined, and fully lined. Cells that are not enclosed by lines are called open. It is difficult to determine the cells of line-less and semi-lined tables because the separators between all the cells are not explicit. This is more evident in large complex tables with multi-line cells.

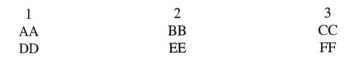

1	2	3
AA	BB	CC
DD	EE	FF

Figure 8.6 Lineless table.

1	2	3
AA	BB	CC
DD	EE	FF

Figure 8.7 Semi-lined table.

1	2	3
AA	BB	CC
DD	EE	FF

Figure 8.8 Fully lined table.

There are two related but distinct problems in table recognition: table identification and table analysis. Table identification is a segmentation problem: how to extract those image components belonging exclusively to a table? One might use text information such as the keyword 'table,' but often the presence of rules and a rectilinear arrangement of textual elements is conclusive. Most table detection methods rely on the presence of a grid of horizontal and vertical rulings but some progress has been made with line-less tables.

Table analysis seeks to uncover the structure of the information implied by its geometric arrangement. The simplest analysis seeks to reconstruct the look of a table using formatting commands without regard to the actual content. For example, it may not matter that a cell is a column header if it appears in the correct place when reconstructed. A more sophisticated approach attempts to determine the semantic relationships among the cells and classify them as either descriptor or data cells (Douglas *et al.* [24]).

The table identification method of Tachikawa [25] extracts runlengths, combining them into connected components and uses large connected components as table candidates. Among these candidate regions, horizontal rulings are extracted by combining runlengths longer than some specified threshold and collecting those with length approximately the width of the connected component. If the number of rulings is greater than some threshold, the region is deemed a table. This procedure finds only fully lined tables. The advantages of this method are speed and the ability to work with runlength compressed data.

After a page image has been segmented, the invention of Ikemure [26] provides a means to determine whether a ruled area is a table or a figure by comparing the number of pixels comprising the horizontal and vertical rulings in the region to the total number of black pixels in a binarized image. If the ratio is sufficiently large, a significant proportion of the pixels belong to rulings and thus the region is a table.

The table analysis of Itonori [27] uses bounding box information of "textual elements" and ruled lines to find cells and assign each a column and row number. It is assumed that the cells are simple and contain characters that can be combined into text blocks by proximity. The table need not be fully lined. First, bounding boxes of characters are merged if they are close together. These merged elements constitute text blocks. The bounding boxes of the text blocks are expanded until they encounter a ruling or the end of a ruling. If a text block is enclosed by lines, that text block is a cell. Otherwise, white "virtual" ruling lines are inferred from vertical and horizontal projection profiles of the text blocks. For example, the vertical projection results in a profile where at each x-position, the sum of the heights of the cells above it is tallied. The profile results in a series of "hills" and "valleys." The beginnings and ends of the hills provide delimiters for the columns. The same procedure is applied to the horizontal projection to obtain rows. The boundaries of the text blocks are expanded to cell boundaries, but are forbidden to overlap. The columns and rows are enumerated and row and column numbers are assigned to each cell. This method requires no image information other than the bounding boxes of connected components.

The sophisticated analysis method of Douglas *et al.* [24] uses natural language processing concepts to represent and analyze tables. They attempt to characterize the information contained within a table, regardless of its form. They posit a canonical representation for tabular information, listing several table transformations with respect to which table information is invariant. Their application is the interpretation of a particular class of tables containing highly structured information in the construction industry. Domain labels are column headings in the canonical representation and a list of n-tuples are values, where n in the number of columns. The data are lines consisting of character bounding boxes and spaces between characters. Alphanumeric characters, recognized by OCR, are tagged. A sequence of characters is content-bearing if it contains at least one alphanumeric character. Column breaks are determined by intersecting vertically overlapping lines. The spaces that survive intersection of all such lines are deemed gaps between columns. Whether the columns are those of a table (rather than columns of text) is determined through a set of rules that use alphanumeric density and the column width relative to the width of the text body being analyzed. Within a column, adjacent lines are merged into cells. Once the cells have been determined and labeled with their unique column/row coordinates, the table is analyzed semantically using recognized characters. Domain knowledge (e.g., construction materials) is used to establish whether a phrase is a domain label or a domain value and whether, based upon the cell's horizontal coordinate, a cell's semantic type is consistent with others in its column.

Laurentini and Viada [28] identify and analyze tables by looking for a network of vertical and horizontal rulings with a consistent arrangement of text blocks (cells). A region is a table if the following analysis runs to completion. First, vertical and horizontal runlengths are used to determine rulings. These rulings are then examined to find a network of crossing and connected rulings. Connected components in the image with the size and aspect ratio of characters are grouped into words and words grouped into phrases to form text blocks. An attempt is made to reconcile the network and arrangement of text blocks so that each text block falls into a lined cell. If more than one text block occupies a cell, horizontal and vertical projections of the blocks are taken. Gaps in the profiles are candidates for virtual rulings (invisible delimiters). If these new virtual rulings along with the old rulings do not separate each text block into a unique cell or the cells are not arranged in a regular rectangular manner, the analysis fails and the region is not a table. Otherwise, the table is reconstructed using an OCR program and the found rulings. This method works with fully lined tables or at least those with a few rulings and well-defined column and row structures.

Hori and Doermann [29] analyze table-form documents which are fully lined. The task is to find all the cells, by which they mean the rectangles that are formed by the rulings and enclose strings of text. Their contribution is the ability to handle degraded documents where characters can overlap rulings. The algorithm operates on two versions of the binary image, one at the original scanned resolution and one at a reduced resolution. In the reduced image, a pixel is black if any pixel in a square region about the corresponding pixel in the original is black. This has the effect of merging broken or dotted rulings. However, this introduces the problem of characters overlapping with lines of the table-form. Inner and outer boxes (bounding white and black areas, respectively) are obtained from the image and classified according to their size and aspect ratio into one of character, cell, table, bar, noise, character hole, and white character island. Inner boxes can be nested in outer boxes. Box coordinates for the original and reduced resolution images are maintained.

Cells are defined to be inner boxes having outer boxes of strings nested inside. Boxes in the original are inspected for characters touching lines, and are separated if found. The boxes in the reduced image are more reliable in the sense that they are formed with broken and dotted lines rendered as solid lines. But they are also more likely to have touching characters. Boxes in the reduced and original images are compared and their differences reconciled. Strings are characters that are nested within the same cell. Character boxes are collected into lines of text. Since the cell coordinates do not match precisely the positions of the rulings, adjustments are made to line up the cells and their neighbors with rulings to avoid gaps and allow spaces for rulings to be drawn between the cells. The result is a collection of bounding boxes corresponding to an ideal version of the scanned table-form.

The method of Hirayama [30] detects and analyses tables that have vertical and horizontal rulings. The first task is to segment a binary document image into regions containing text, tables, and figures. The first step in segmentation is to find the CC's of the runlength-smeared document image. Bounding boxes of the CC's are classified as vertical or horizontal lines, character strings, or other objects according to their heights. Character strings are grouped together to form text regions. The remaining regions are nontext: tables and figures. Tables are required to have horizontal and vertical lines. Lines are grouped together when they intersect, are close and parallel, or their endpoints are close. The regions containing a group of linked lines are called table area candidates. A bounding box of rulings is added to the table region in case some cells are open. Within a table area candidate, all rulings are extended by virtual lines to terminate into the most extreme ruling. The table area is thus segmented into a "lattice" being composed of a grid of rectangles. Next, rectangles that are separated only by virtual lines are joined. The resultant polygons form cells if they are rectangular and enclose only character strings or are empty. Some polygons correspond to cells and others not, but the region as a whole is judged to be a table area if there is at least one nonempty cell and if noncell areas constitute a fraction of the total candidate area. Now in the lattice version of the table, there is a grid of m columns and n rows. The separators between these may be virtual. It is necessary to assign these virtual cells to proper table rows by aligning columns. Alignment is done pairwise from left to right using the well-known string-to-string correction dynamic programming algorithm [31] where the weights for the substitution cost are distances in baselines between two text strings and there is a fixed insertion and deletion cost. For example, in Fig. 8.9 there are three columns and five virtual rows. With the deletion and insertion cost sufficiently high, the alignment algorithm matches string AAAA with DDDD and CCCC with FFFF in the first two columns. The string BBBB is "deleted" and string EEEE is "inserted." The string BBBB is "deleted" and string EEEE is "inserted." That is, BBBB has no match in the second column and EEEE has no match in the first column. For the second and third columns, DDDD matches GGGG, FFFF matches IIII,

EEEE is "deleted" and JJJJ is "inserted." Thus AAAA, DDDD, and GGGG share a row as well as CCCC, FFFF, and IIII. This completes the pair-wise column alignment. Next, the dynamic programming algorithm applied to the previous column, matches HHHH to BBBB. But string JJJJ has no match in any previous column, so a new row is supplied for it. Finally, the remaining unmatched strings are given their own rows. The result is a table with five rows, {AAAA, DDDD, GGGG}, {EEEE}, {BBBB, HHHH}, {CCCC, FFFF, IIII}, and {JJJJ} and three columns {AAAA, BBBB, CCCC}, {DDDD, EEEE, FFFF}, and {GGGG, HHHH, IIII, JJJJ}.

AAAA	DDDD EEEE	GGGG
BBBB		HHHH
CCCC	FFFF	IIII JJJJ

Figure 8.9 Table analysis example.

Graph rewriting techniques can be brought to bear on table identification and analysis [32]. Objects on a page image can be represented as nodes of a graph where links exist between vertically and horizontally adjacent objects. The objects are labeled during a segmentation step as one of {"character," "word," "line," "paragraph," "column structure," "tabular structure," "indexed list," "jagged text," "unformatted text region," "image region,"} where a word is a sequence of characters, a line a sequence of words, a column a sequence of lines, etc. A subgraph of the original graph is rewritten as another (simpler) graph. For example, two adjacent columns are rewritten as a tabular structure. There is a small set of possible rewrite rules. Table identification is the task of starting at a object and choosing rewrite rules until no more rules can be found. If the result is a table, then a table is found, else the procedure starts anew at a different, untouched object. The sequence of productions produces information about the table structure, namely columns and rows. The entire search procedure must be carefully controlled to avoid a combinatorial explosion. Details are found in Ref. 32.

8.7 FORM RECOGNITION

Form recognition methods seek to extract data from fields entered into a form such as tax returns or credit card applications. Forms themselves may be recognized and turned into electronic versions for data entry or printing (Takeda *et al.* [33]). Technologies enabling the conversion of paper forms with filled-in data into electronic databases have enormous economic impact due to the cost savings over keyed data entry. The data may be represented by hand printed or machine printed characters. Due to high accuracy requirements, often a human editor views and corrects the data before storage.

Liebowitz Taylor *et al.* [34] list five steps in the form conversion process: (1) scanning the image, (2) enhancing the image, (3) identifying the form, (4) extracting the regions, and (5) interpreting the data. Step 2 can involve skew correction, thresholding, and noise reduction. If there are multiple form types in the data stream, it is necessary to determine the type of form so that the form image can be registered with a blank form to extract regions (step 4). Finally, the data in each region must be interpreted through OCR. Data fields can be mark-sense (i.e., check-boxes), numeric, alphabetic, or alphanumeric in either hand printed or machine printed characters. Fields might have lexical information associated with them that assist in recognition such as money, addresses, telephone numbers, zip codes, or proper nouns. For example, if a field is numeric, the OCR system would have greater accuracy by restricting the number of classes to ten.

Form identification may be as simple as recognizing a form number or barcode in a fixed position or it may involve matching features against a form database. Before recognition, the form itself is removed from the image. This can be done by printing the form with a special dropout ink and capturing the forms with a scanner equipped with a bulb or filter of the same color. Some scanners come with multiple bulbs or filters to handle a variety of forms. Other methods remove the form from the image algorithmically and reconstruct broken characters.

Liebowitz Taylor *et al.* [34] propose a system to process United States Internal Revenue Service (IRS) forms. There are 160 different forms where the data fields are delimited by horizontal and vertical lines. Filled-out forms are scanned at 75 dpi with 6 bits (64 levels) of gray. Form identification is done by a neural net classifier using geometric features. These features, which are also used to register a blank form image over the original, uses 9 crossings shown in Fig. 8.10. The form image is divided into 9 equal-sized rectangles. Counts for each type of crossing in each region are tallied to form an 81-component feature vector. Registration uses these crossings and compares them to a standard form model database. Registration proceeds by comparing lists of crossing points from the scanned and standard versions and choosing the standard with the highest correlation.

In the method of Wang and Srihari [35] forms are analyzed without an a priori model. Form structure is determined from horizontal and vertical line segments extracted from connected components. Long horizontal and vertical lines are characterized by their horizontal runlength histograms. Each connected component is analyzed to find long lines. Overlapping characters are pruned and short runlengths are merged as necessary. Open or closed boxes

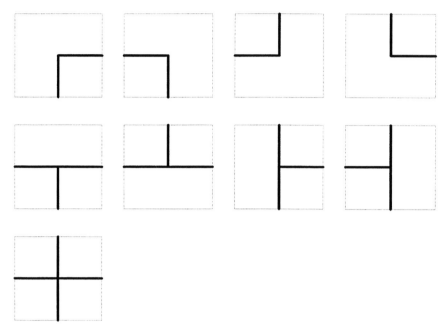

Figure 8.10 Nine crossings used for form identification features.

containing entries are described as a sequence of key-points obtained from the list of lines. Lines have a thickness and thus are black rectangles. The four extreme points of a line are key-points. In addition, there are eight types of concave and convex key-points associated with intersecting lines (Fig. 8.11). Boxes are described by a sequence of these 12 types of key-points. After the horizontal and vertical lines are found, the key-points are determined and used to construct box descriptors. The boxes, represented by key-point sequences, fully describe the form. Horizontal and vertical lines are used to extract the characters (the text map) which are processed by OCR. However, the form removal can result in broken or incomplete characters. These defects are repaired using an arc-filling technique that tries to fill in the missing strokes. When processing is complete, the result is a model of the form represented as a set of boxes and the text that was interpreted within these boxes.

Lin *et al.* [36] supply a method for form identification where a form is represented by graphs which are used to match a form against a model database. After the form is scanned, noise and characters are removed. The resulting horizontal and vertical lines are represented as a graph with lines representing edges and intersection points representing nodes. Smallest cycles in the line graph are found and these represent frames (boxes for entries). Using these labeled frames as nodes, a forest of graphs is formed where nodes are linked if they share a horizontal edge. The forest can be represented as a single string. Similarly, horizontal graphs are formed and represented as a string. Two additional strings are obtained from the sorted list of the vertical and horizontal positions of the lines of the form. These four strings are used to match forms in a model database. Once identified, the original form image can be processed to interpret the fields of interest.

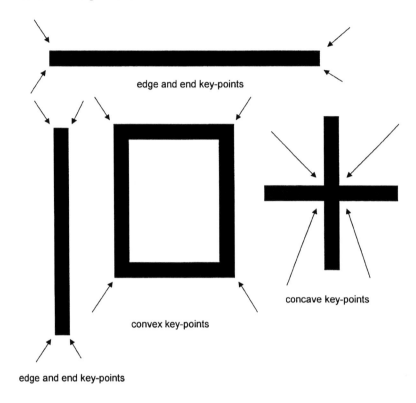

edge and end key-points

concave key-points

convex key-points

edge and end key-points

Figure 8.11 Key-points used to describe form-entry boxes.

Yu and Jain [37] provide methods for learning the structure of a blank form, extracting the data from a filled-in form, and correcting the dropouts caused by line removal. These techniques rely on a generalization of the line adjacency graph used to represent connected components (see Pavlidis [38]). Blocks of adjacent and horizontally overlapping runlengths form nodes in a block adjacency graph. Horizonally overlapping blocks are linked together to express a connected component as a graph. Nodes of graphs are inspected to identify long lines that represent the frames of a form. Upon removal of the form, broken character strokes are interpolated using a block of runlengths.

The method of Garris [39] removes form frames and preserves characters without an explicit form model. This method finds dominant horizontal lines through a Hough transform (see Gonzalez and Woods [40]). A horizontal line is a sequence of short vertical runlengths which are removed from a filled-in form image. Characters are preserved by following a set of carefully crafted rules based on vertical runlength statistics. Damage caused by line removal is repaired by executing another set of carefully crafted rules. The results are shown to be superior to a commerical package and provide a 3% increase in handprinted lower case character recognition accuracy.

8.8 FONT IDENTIFICATION

In document recognition, font identification can be used to help label blocks according to function, to aid in reconstruction of the original look, and to boost the accuracy of OCR by selecting a classifier tuned to a particular font.

To identify the function of a text object (e.g., title) or to perform reconstruction, it is often important to recover information about the fonts used in the original. Such information may be the dominant font used in a page or the font used in a word. A font is represented by a font family (Times Roman, Century Schoolbook, etc.); size (e.g., 10 point); weight (light, normal, or bold); slope (Roman, italic, or slanted); and spacing (proportional or fixed pitch). One might also include underlining or reduce the font family to serif/sans-serif. It is usually assumed that the smallest entity analyzed for font is a word.

8.8.1 Font extraction by morphological methods

Gross shape differences between characters of different fonts can be used to segregate italic, bold, and Roman text. The method of Bloomberg [41] uses morphological operations to remove all but either italic or bold text. For italic text at 300 dpi, a hit-or-miss template is used to detect the left side of italicized characters slanted at about 12 degrees (Fig. 8.12) where '+' is the origin; '0' and '1' indicate miss and hit, respectively; and the remainder are "don't care" pixels. The template is weak enough to mark italic characters in a variety of typefaces. The first stage is to perform the hit-or-miss transform with the above template, followed by a optional dilation with a small vertical structuring element (SE). The italicized text is marked by clusters of pixels, while other text is indicated by isolated points. Next, a closing merges pixels for italicized pixels but merely enlarges isolated marks. After an opening, only the italicized marks remain. This image is called the final seed. The original image is smeared horizontally by a small horizontal SE, causing characters to merge, but not across word boundaries. The smeared original is registered with the final seed. Connected components in the smeared original which cover seeds are italic words.

Bold text is indicated by thick character strokes relative to the Roman text. Iterative thinning operations on the text image remove Roman text before bold text. Each thinning step is an erosion by [1 1] followed by [1 1], where bold indicates the origin. After each thinning step, a test opening is done with a vertical SE meant to preserve down-strokes. The number of pixels in the opened image is stored. If the ratio of the previous number of pixels after opening and the current number exceed a threshold, then the opening has

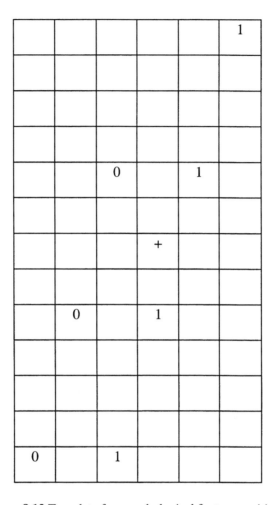

Figure 8.12 Template for morphological font recognition.

removed most of the Roman text and what remains are seeds for the bold text. At this point, the seeds are expanded and used to extract the bold words from the original image as in the italic case.

8.8.2 Projection profiles

Font descriptors can be obtained from projection profiles. In the work of Zramdini and Ingold [42], a font is described by the following properties: font family (Courier, Helvetica, Avant Garde, Times, Palatino, Bookman, or New Century Schoolbook), size in typographic points, weight (light, normal, or bold), slope (roman, slanted, or italic), and spacing mode (proportional or fixed-pitch). A line of text (a word) has three zones, from the lowest point to the baseline (lower zone), from the baseline to the x-height (middle zone) and from the x-height to the topmost point (upper zone) Fig. 8.13. Consider an m by n binary word image (region within the smallest bounding rectangle) with origin at the lower left-hand corner. Let $V[j]$, $0<=j<m$, be the vertical projection. The baseline is estimated as

$$H_{base} = \arg\max_{0<j<m} V[j] - V[j-1] \qquad (8.3)$$

and the x-height by

$$H_{x-height} = \arg\min_{0<j<m} V[j] - V[j-1]. \qquad (8.4)$$

Three features are gathered from the vertical projection:
1. h_1, the height, m
2. h_2, the upper zone height, $m - H_{base}$
3. h_3, the middle zone height, $H_{x-height} - H_{base}$.

The horizontal projection $H[i]$, $i <= 0<n$, is computed only for pixels whose vertical coordinate falls in the estimated middle zone. The fourth feature, dn, is the average of this profile over horizontal coordinates. Finally, the fifth feature, dr, is the variance of the horizontal projection:

$$dr = \frac{1}{n-1} \sum_{i=0}^{n-1} (H[i+1] - H[i])^2. \qquad (8.5)$$

Not all words or lines of text have all three zones. If a text item is short or has only descenders, the feature vector (dr, dn, h_3) is used; if the item has ascenders, four features are used: (dr, dn, h_2, h_3); and if all zones are present, all five features are used for classification. In the experiments presented, a Gaussian maximum-likelihood classifier was trained. Classification accuracy ranged from 84% to 99% for the seven fonts listed above.

Figure 8.13 Zones for font recognition.

Morris [43] used features from the frequency domain for typeface classification. Random text strings from *Wuthering Heights* were rendered in 10-point fonts at 300 dpi on a 512 by 64 pixel grid with an approximately uniform band of white space surrounding the text. Features were obtained from the power spectrum of the binary image samples by averaging power over 38 rectangular regions. Figure 8.14 shows the general pattern for the regions (14 regions shown). Assuming normality, principal component analysis shows that 17 of these features account for 99% of the variance. Using an estimated pooled covariance matrix from a sample of 100 text samples in each of 55 font families, a linear Gaussian maximum-likelihood classifier yields a total error of classification of 6% with individual class accuracies ranging from 88% to 100%. This result is not strictly comparable to that of Zramdini and Ingold because here the characters have a single point size and the text images are rendered but not printed and scanned. Still, together they show that classical statistical pattern recognition techniques can yield good accuracy on a realistic number of font families.

Figure 8.14 Several rectangular masks in the frequency domain; gray shows overlap.

8.9 CONCLUSION

Document recognition draws from a diverse set of technologies including image processing, statistical pattern recognition, signal processing, and formal languages. In spite of the inherent difficulties in document modeling, document recognition is one of the major commercial successes in image analysis and understanding. Expectations are high in this field because we seek to understand highly structured human-produced artifacts, but on the other hand, human production precludes models built from first principles. The result is an appeal to general mathematical techniques such as neural networks and grammars that are not always optimal along with a plethora of ad hoc techniques to fill in the gaps. We have sampled much of the diversity by surveying recent contributions to document image analysis, table recognition, form processing, and font recognition.

8.10 APPENDIX

Central to image understanding is the notion of pattern. In statistical pattern recognition, patterns are interpreted probabilistically. Features, extracted from observed patterns, have a probability distribution conditioned or parameterized by the class of observed patterns. The Bayes decision rule states that the best class yields the feature observation most probable. Alternatively, patterns can be modeled algorithmically by grammars in which a sequence of tokens are combined into a terminal symbol. The act of combination or interpretation is called parsing. A legal parse produces a terminal symbol which acts as a class label. A pattern outside the scope of the model produces a syntax error. Syntactic approaches afford the ability to encode structural knowledge into the model. For example, it is clear that technical journal pages obey a set of rules regarding the relative positions of title, author, abstract, body, and footnotes on a page. These rules can be codified into a page grammar. Upon receiving a string of tokens representing these entities, one can interpret them (perhaps uniquely), and if not, an alternate grammar can be tried. A grammar is basically a one-dimensional notion although one can construct grammars on more general structures such as graphs. In more than one dimension, it is not clear how to obtain the next token. The total ordering of strings makes this concept unambiguous. Most page grammar approaches provide a method for approximating a 2D problem by a set of 1D problems.

A language is a (possibly infinite) set of strings over a finite alphabet. There are at least two ways to compactly represent some languages: a set of rules governing how the strings are produced or an algorithm that determines whether a string belongs to the set. The first case is captured by the notion of grammar. A grammar is defined by a 4-tuple (V_T, V_N, P, S) where V_T is a set of terminals (strings from the alphabet), V_N is a set of nonterminal symbols, P is a set of production or rewrite rules $a \rightarrow b$ where a and b are members of $V = V_N \cup V_T$, and S is a designated nonterminal called the start symbol. The language generated by the grammar is the set of all strings of terminal symbols which can be produced by a finite sequence of rewrites. A special type of grammar is a context-free grammar where productions take the form $A \rightarrow \beta$ where A is a member of V_N and β is a non-null member of V. A string from a context-free grammar can be represented by a tree which has the start symbol as its root and terminal symbols as its leaves. A parser is a mechanism by which it is determined whether a string is a member of a language. Lex and yacc are popular tools for constructing parsers for context-free languages (Levine et al. [44]). Lex is a lexical analyzer that produces terminal symbols and yacc (yet another compiler compiler) is a tool for constructing parsers for context-free languages. In the case of string grammars, a rewrite replaces one string by another.

In extending the notion of grammars to more general structures, the notion of embedding must be clarified. More general definitions are possible, but for the sake of clarity, we deal with the simplest case here. A graph is a nonempty set of vertices V along with a set of edges $E \subset V \times V$. In the case of graph grammars, subgraphs are rewritten according to production rules. However, after a subgraph is rewritten, it must be embedded into the original graph. For string grammars, there is no ambiguity, but for graph grammars, ruled must be specified. This is one reason for the enormous complexity and generality of these structures. One typically applies a set of rewrites in reverse until a start symbol is found. The strategy for doing this is application dependent. The edges and vertices can have labels, called attributes, which can further govern embedding.

REFERENCES

1. S. Mori, C.Y. Suen, and K. Yamamoto, "Historical review of OCR research and development," *Proceedings of the IEEE*, Vol. 80, No. 7, pp. 1029-1058 (1992).
2. G. Nagy, "At the frontiers of OCR," *Proceedings of the IEEE*, Vol. 80, No. 7, pp. 1093-1100 (1992).
3. *International Journal of Imaging Systems and Technology*, Vol. 7 (1996).
4. Special Issue on Optical Character Recognition, *Proceedings of the IEEE*, Vol. 80, No. 7 (1992).
5. *Machine Vision and Applications*, Vol. 5 (1992).
6. Special Issue on Document Image Analysis Systems, *IEEE Computer*, Vol. 25, No. 7 (1992).
7. Special Issue on Document Image Analysis, *International Journal of Pattern Recognition and Artificial Intelligence,* Vol. 8, No. 5 (1994).
8. Y. Y. Tang, S.-W. Lee, and C. Y. Suen, "Automatic document processing: A survey," *Pattern Recognition*, Vol. 29, No. 12, pp. 1931-1952 (1996).
9. A. Amin. S. Fischer, A. F. Parkinson, and R. Shiu, "Comparative study of skew detection algorithms," *Journal of Electronic Imaging*, Vol. 5, No. 4, pp. 443-451 (1996).
10. G. A. Story, L. O'Gorman, D. Fox, L. L. Schaper, and H. V. Jagadish, "The RightPages image-based electronic library for alerting and browsing," *IEEE Computer*, Vol. 25, No. 9, pp. 17-27 (1992).
11. K. Y. Wong, R. G. Casey, and F. M. Wahl, "Document analysis system," *IBM Journal of Research and Development*, Vol. 26, No. 6, pp. 647-656 (1982).
12. S. Chen, R. M. Haralick, and I. T. Phillips, "Extraction of text words in document images based on a statistical characterization," *Journal of Electronic Imaging*, Vol. 5, No. 1, pp. 25-36 (1996).

13. D. X. Le, G. R. Thoma, and H. Wechsler, "Classification of binary document images into textual or nontextual data blocks using neural network models," *Machine Vision and Applications*, Vol. 8, pp. 289-304 (1995).

14. J. L. Fisher, S. C. Hinds, and D. P. D'Amato, "A rule-based system for document image segmentation," *Proc. 10th International Conference on Pattern Recognition*, Atlantic City, New Jersey, IEEE Computer Society Press, pp. 567-572 (1990).

15. P. Chauvet, J. Lopez-Krahe, E. Taflin, and H. Maitre, "System for an intelligent office document analysis, recognition and description," *Signal Processing*, Vol. 32, pp. 161-190 (1993).

16. H. S. Baird, "Background structure in document images," *International Journal of Pattern Recognition and Artificial Intelligence*, Vol. 8, No. 5, pp. 1013-1030 (1994).

17. H. S. Baird, S. E. Jones, and S. J. Fortune, "Image segmentation by shape-directed covers," *Proc. 10th International Conference on Pattern Recognition*, Atlantic City, New Jersey, IEEE Computer Society Press, pp. 820-825 (1990).

18. H. S. Baird, S. J. Fortune, and S. E. Jones, US Patent 5,430,808, "Image segmenting apparatus and methods," AT&T, 4 July 1995.

19. G. E. Kopec and P. A. Chou, "Document image decoding using Markov source models," *IEEE PAMI*, Vol. 16, No. 6, pp. 602-617 (1994).

20. A. C. Kam and G. E. Kopec, "Document image decoding by heuristic search," *IEEE PAMI*, Vol. 18, No. 9, pp. 945-950 (1996).

21. G. E. Kopec, A. C. Kam, and P. A. Chou, "Document decoding using modified branch-and-bound methods," US Patent 5,526,444, Xerox Corporation, 11 June 1996.

22. G. E. Kopec, P. A. Chou, and L. T. Niles, "Automatic training of character templates using a text line image, a text line transcription and a line image source model," US Patent 5,594,809, Xerox Corporation, 14 January 1997.

23. M. Krishnamoorthy, G. Nagy, S. Seth, and M. Viswanathan, "Syntactic segmentation and labeling of digitized pages from technical journals," *IEEE PAMI*, Vol. 15, No. 7, pp. 737-747 (1993).

24. S. Douglas, M. Hurst, and D. Quinn, "Using natural language processing for identifying and interpreting tables in plain text," *Fourth Annual Symposium on Document Analysis and Information Retrieval*, University of Nevata at Las Vegas, pp. 535-546 (1995).

25. M. Tachikawa, "Table region identification method," US Patent 5,048,107, Ricoh Company, 10 Sept. 1991.

26. Y. Ikemure, "Method and apparatus for recognizing table and figures having many lateral and longitudinal lines," US Patent 5,502,777, Matsushita Electric Co, Ltd., 26 March 1996.

27. K. Itonori, "Table structure recognition based on textblock arrangement and ruled line position," *Proc. Second International Conference on Document Analysis and Recognition*, IEEE Computer Society Press, pp. 214-217 (1993).

28. A. Laurentini and P. Viada, "Identifying and understanding tabular material in compound documents," *Proc. International Conference on Pattern Recognition*, IEEE Computer Society Press, pp. 405-409 (1992).

29. O. Hori and D. S. Doermann, "Robust table-form structure analysis based on box-driven reasoning," *Proc. Third International Conference on Document Analysis and Recognition*, IEEE Computer Society Press, Montreal, Canada, pp. 218-221 (1995).

30. Y. Hirayama, "A method for table structure analysis using DP matching," *Proc. Third International Conference on Document Analysis and Recognition*, IEEE Computer Society Press, Montreal, Canada, pp. 583-586 (1995).

31. R. A. Wagner and M. J. Fischer, "The string-to-string correction problem," *Journal of the ACM*, Vol. 21, pp. 168-173 (1974).

32. M. A. Rahgozar and R. Cooperman, "A graph-based table recognition system," *Proc. SPIE* Vol. 2660, pp. 192-203 (1996).

33. H. Takeda, M. Tsuchiya, H. Suzuki, S. Yamada, T. Matsuda, H. Fujise, Y. Kuno, and I. Koai, "Method and system for producing from document image a form display with blank fields and program to input data to the blank fields," US Patent 5,228,100, Hitachi, 13 July 1993.

34. S. Liebowitz Taylor, R. Fritzson, and J. A. Pastor, "Extraction of data from preprinted forms," *Machine Vision and Applications*, Vol. 5, pp. 211-222 (1992).

35. D. Wang and S. N. Srihari, "Analysis of form images," *International Journal of Pattern Recognition*, Vol. 8, No. 5, pp. 1031-1052 (1994).

36. J.-Y. Lin, C.-W. Lee, and Z. Chen, "Identification of business forms using relationships between adjacent frames," *Machine Vision and Applications*, Vol. 9, pp. 56-64 (1996).

37. B. Yu and A. K. Jain, "A generic system for form dropout," *IEEE PAMI*, Vol. 18, No. 11, pp. 1127-1134 (1996).

38. T. Pavlidis, *Algorithms for Graphics and Image Processing*, Computer Science Press, Rockville, MD (1982).

39. M. D. Garris, "Intelligent form removal with character stroke preservation," *Proc. SPIE* Vol. 2660, pp. 321-332 (1996).

40. R. C. Gonzalez and R. E. Woods, *Digital Image Processing*, Addison-Wesley, New York, (1992).

41. D. S. Bloomberg, "Multiresolution morphological analysis of document images," *Proc. SPIE* Vol. 1818, pp. 648-662 (1992).

42. A. Zramdini and R. Ingold, "Optical font recognition from projection profiles," *Electronic Publishing*, Vol. 6, No. 3, pp. 249-260 (1993).

43. R. A. Morris, "Classification of digital typefaces using spectral signatures," *Pattern Recognition*, Vol. 25, No. 8, pp. 869-876 (1992).
44. J. R. Levine, T. Mason, and D. Brown, *Lex & Yacc*, O'Reilly & Associates, Inc., Sabastopol, CA (1992).

9

LEXICON-DRIVEN HANDWRITTEN WORD RECOGNITION

Paul D. Gader
University of Missouri – Columbia
Columbia, Missouri

The focus of this chapter is on a class of methodologies for computer recognition of handwritten words. Handwriting recognition, and more specifically, character recognition, is one of the oldest problems of pattern recognition. In fact, one of the first demonstrations of the Mark I Perceptron neurocomputer of Rosenblatt, performed in the late 1950s, involved character recognition.[1]

There are two classes of handwriting recognition problems: on-line and off-line. In on-line recognition, handwriting is fed directly to the computer through an electronic pad that captures pressure, position and velocity information, such as a personal digital assistant. In off-line recognition, handwriting is produced on paper and then imaged. Off-line applications include automatic sorting of mail with handwritten addresses and automatic check sorting.

There has been significant progress in handwriting recognition since 1990. For example, in that year, the United States Postal Service sponsored a conference on advanced technologies. Handwritten ZIP code recognition was discussed in many papers, but handwritten word recognition was discussed in only one paper.[2] The results were not good, about 63% correct recognition with a dictionary, or lexicon, of only 7 words! The first Frontiers in Handwriting Recognition Workshop was held that year also.[3] One paper was given on handwritten word recognition there as well. The results were 87% with a dictionary of 25 words. However, for the USPS paper, the writers are unconstrained whereas in the second case, there was only one writer. In contrast, current word recognition rates for an unconstrained number of writers are over 90% for dictionaries of size 100 and around 85% for dictionaries of size 1000.[4-9]

One of the most popular methods for handwritten word recognition is the so-called segmentation-based, lexicon-driven method. This is in contrast to speech recognition, in which the hidden Markov model is the most common technique.[10] Hidden Markov models are also used for handwriting recognition and can be combined with the segmentation-based, lexicon-driven method.[11-13] It is not clear which method is preferable for handwriting recognition.

In this chapter, segmentation-based, lexicon-driven handwritten word recognition methods that do not rely on hidden Markov models are discussed. One particular method from this class will be used to discuss the issues involved in deriving such an algorithm. The particular system is under development by the author and his colleagues and has been for several years.[5,6,14-30] Several general issues, not only in handwriting recognition but also in computer vision, are discussed here.

It is assumed that the reader is familiar with the basics of pattern recognition and image processing, including neural networks, although an understanding of these areas is not a requirement for understanding much of the discussion.

9.1 BACKGROUND

9.1.1 What is lexicon-driven handwritten word recognition?

Consider the problem of reading the handwritten word in Fig. 9.1. Although many of the individual characters are illegible, the entire word is easy to read if one is familiar with the English language. This example illustrates the fact that recognition of handwriting requires more than recognition of shapes; it requires knowledge about the language itself. This knowledge is usually referred to as a language model and it comes in many forms. The language places constraints on the sequences of characters that can appear in a word. For example, if we know a word is from the English language, then the word cannot contain the sequence "iiiwwwi."

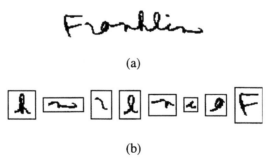

(a)

(b)

Figure 9.1 The handwritten word image of "Franklin" is easy to read in part (a). In part (b), the individual characters of the word are isolated and rearranged. They are difficult to read correctly when viewed separately. This illustrates that handwritten word recognition requires more than recognizing each individual character.

The simplest language model is a lexicon. In this context, the term lexicon refers to a list of all possible words. We shall refer to the words in a lexicon as strings to distinguish them from word images. The lexicon may be tagged or not. If it is tagged, then each word is assigned a number of attributes. For example, a word may be tagged as a noun or a verb or a proper name.

Lexicons are very important in commercial applications such as computer systems that sort checks or mail with handwritten addresses. The number of handwritten words that can appear on a check is quite limited. The words that can appear are likely to be contained in the set

{ zero, one, two, ..., ten, eleven, ..., twenty, thirty, ..., hundred, thousand, million}

that contains 31 words. In addition, other words such as "and" or "no" may appear. Thus, a handwritten word recognition system that is designed to read checks could use a lexicon of approximately 40 words as a language model.

Mail sorting applications are similar. Sorting mail properly requires reading the entire address on a piece of mail. This is a difficult task. The numeric fields, namely, the ZIP code and the street number, are generally easier to read than the rest of the address. If a system can generate several "guesses" concerning the identities of the street number and the ZIP code, then there is a small list of possible street names. This small list constitutes a dynamically generated lexicon that can be used to help recognize the street name and sort the mail properly.

Lexicon-driven handwritten word recognition is a class of methodologies for designing computer programs to read handwritten words that use lexicons as a language model. The inputs to a lexicon-driven word recognition program are a digital image of a handwritten word and a lexicon. The output of such a program is the lexicon itself with one additional feature: each string in the lexicon is assigned a score indicating the degree to which the string represents the word image. The string with the highest score can be taken as the recognition result or the scored lexicon can be used in further processing involving contextual information.

9.1.2 Why is handwritten word recognition difficult?

Handwritten word recognition is difficult because there are ambiguities present in the patterns that can be detected within a handwritten word. The fundamental problems are: (1) patterns that are meant by the writer to represent a single character can appear to represent different or multiple characters, (2) pieces of handwritten characters can look like other characters, and (3) handwritten characters within words are often illegible. These problems are the motivation for many of the techniques presented in this chapter and are therefore now discussed in greater detail.

Character segmentation is the process of isolating the individual characters in a digital image of handwriting. Ideally, a handwritten word recognition system would use a sequential process in which character segmentation was performed prior to character recognition. Unfortunately, as is often true in image pattern recognition, it is difficult to segment without recognizing and it is difficult to recognize without segmenting. Figure 9.2 depicts a classic problem. Both images are equally valid representations of "ll" and "u." The first interpretation requires a segmentation into two segments whereas the second requires only one segment. A system cannot ascertain which hypothetical segmentation is correct without using recognition and context. This leads to the principle of tightly coupled segmentation and recognition.

Figure 9.2 Segmenting these images correctly requires word level context and recognition.

One approach to tightly coupled segmentation and recognition is to generate an "over-segmentation," illustrated in Fig. 9.3. A word image is segmented into primitives. Each primitive is a sub-image of the word image and should represent a piece of a character or a character. If the segmentation of the word into primitives is performed properly, then every true character is either a primitive or a union of primitives. Unions of primitives (including primitives) are called segments. A segmentation is a sequence of segments that uses all primitives of a word image.

A match score can be computed between a segmentation with n segments and a character string with n characters. Each segment is assigned a score indicating how well the segment matches the corresponding character in the string. We call algorithms that assign these scores character class confidence or character class membership algorithms. Typically, character recognition algorithms that return confidence values are used. Confidence values from the individual segments are aggregated to produce a match score. Dynamic programming efficiently computes the best sequence of segments to match lexicon strings.

In Fig. 9.3, the best matches between a word image and the lexicon strings "Richmond" and "Edmund" are shown. The match scores between each segment and the corresponding character are also shown (scaled between 0 and 100). For example, the match between the first segment of the match to "Richmond" and "R" is 0.53. It is interesting to note that the segmentation found by the algorithm to match "Richmond" is not the one most humans would have chosen. This is not unusual and, in this case, is caused by the fact that the character confidence assignment algorithm prefers the "c" without the curl on the top more than it is bothered by the existence of the curl next to the "h." In the latter case, it can be interpreted as a perfectly nice "h" with a speck of noise next to it.

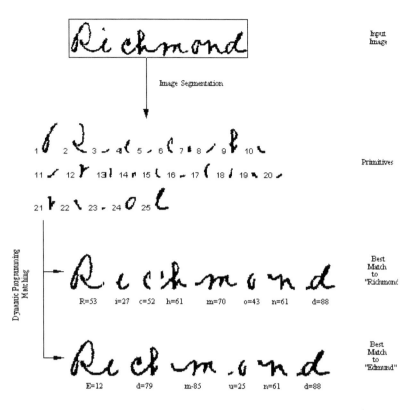

Figure 9.3 Illustration of segmentation-based, lexicon-driven word recognition. The algorithm finds the best assemblies of primitives to match the strings "Richmond" and "Edmund."

This approach provides a framework for solving the problems illustrated in Fig. 9.2 because an over-segmentation algorithm should divide both patterns into two primitives. If a lexicon string contains an "ll," then the primitives can be kept separate to match the pair of characters. If a lexicon string contains a "u," then the union of the primitives can be formed to match the single character.

While over-segmentation provides a basis for a solution to the problem of ambiguity between single and multiple characters, it leaves several to be solved. When matching assemblies of primitives to a lexicon string, the system must process pieces of characters, multiple characters, or single characters with parts of other characters attached, etc. The system must be able to reject such assemblies as non-characters while providing accurate confidence values for assemblies that do represent characters (including characters that are illegible). Another problem not solved by over-segmentation is that a pixel pattern may represent many different characters depending on the context.

An example illustrating these problems is shown in Fig. 9.4. A match made by our system of an image of the word "Cowlesville" to the lexicon string "Avenue" is shown. The second segment, which serves as a piece of a character, could be either "v" or "o." The third segment, which serves as the character "l," could be either "e" or "l." As noted in Fig. 9.2, the fifth segment in Fig. 9.4 could be "u." The first and fourth segments are assemblies of primitives that do not represent characters. However, one version of our system assigns fairly high class membership values for matching to the classes "a" and "n," respectively, which can result in an erroneous overall match. Although one might be quick to point out that the fourth segment should not be assigned a high value, one must remember that a hastily written "n" can easily appear as a sequence of bumps in which the number of bumps may be more than two.

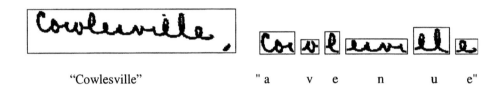

"Cowlesville"   " a   v   e   n   u   e"

Figure 9.4 Character class overlaps and inadequate performance on non-character classes can cause erroneous word matches.

There are multiple considerations to take into account to remedy these problems. One is that a program should not classify characters until at least the word level context is taken into account. Therefore, a character classifier is not really what is needed here. Rather, we need a character class confidence assignment algorithm. Another consideration is that the relationships between adjacent characters are important. For example, the fifth segment should be less likely to be considered as a "u" when viewed next to the sixth segment, which is much smaller. We will discuss methods for incorporating these considerations into word recognizers later in this chapter.

Another difficulty in this type of word recognition approach is estimation of word confidence from the set of character class confidence values for each segment. Typically, the average value is used. However, the average does not perform well in situations when pieces of characters look like characters.

Fig. 9.5 illustrates the point. An image of the word "Juana" is matched against the strings "Juana" and "Lares." When computing the match between the word image and the correct string, low scores are assigned to most of the segments in the dynamic programming derived segmentation because they are not written legibly. When computing the match score between the word image and the incorrect string "Lares," most of the segments are also assigned low scores. In the latter case, however, the segmentation algorithm happens to find a very nice "r" in the segmentation and assigns it a high score. Using the average results in the incorrect string receiving a higher score than the correct string.

Incorrect segmentations may contain segments that have high confidence values. Often, there will only be one or two such segments and so, in the language of robust statistics, the confidence values computed from these segments are outliers.[31, 32] We will show how robust estimators can be used to improve word level confidence estimation.

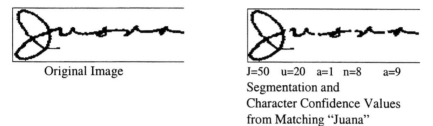

Original Image

J=50   u=20   a=1   n=8   a=9
Segmentation and
Character Confidence Values
from Matching "Juana"

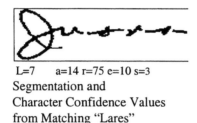

L=7   a=14 r=75 e=10 s=3
Segmentation and
Character Confidence Values
from Matching "Lares"

Figure 9.5 Illustration of how robustness is required in estimation of word level confidence to properly process non-character segments that look like characters.

9.1.3 Overview of typical system

The handwritten word recognition system we are discussing can be represented in a diagram as shown in Fig. 9.6. We have discussed several of the difficulties inherent in building such a system and the motivation for such an approach. In the following sections, we discuss approaches to each module shown in Fig. 9.6. We will provide some guiding principles in each section. For segmentation, we have found that an aggressive splitting algorithm with very little grouping works best. In the case of character class confidence assignment, we have found the accurate representation of all possibilities is more important than character recognition.

These two results are interpreted in the context of the Principle of Least Commitment, stated by David Marr in his book *Vision*.[33] Marr's principle states that irrevocable decisions should be delayed as long as possible. We found that this principle is invaluable, not only in handwriting recognition, but in a variety of computer vision problems.[17, 24, 34, 35]

A discussion of the choices of features for character class confidence assignment, with emphasis on the difficulties involved with interpreting non-characters, is also carried out. The basic dynamic programming algorithm is provided in detail together with a novel generalization that incorporates relationships between adjacent characters. Finally, we discuss using optimal, robust estimation techniques to estimate word level confidence.

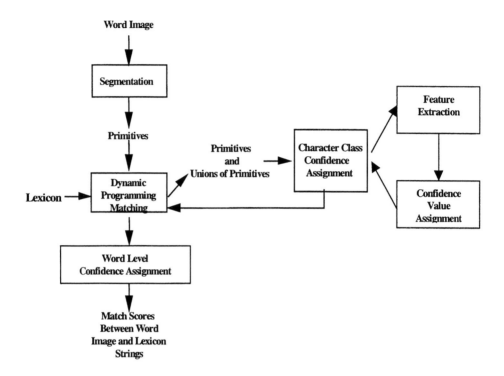

Figure 9.6 Lexicon-driven, segmentation-based handwritten word recognition system diagram.

9.2 SEGMENTATION APPROACHES

The role of the segmentation algorithm is to produce the primitive segments. The lexicon is not generally used at this stage of the algorithm. A thorough review of segmentation algorithms is given in Ref. 36. In our system, the processing is divided into two phases, initial segmentation and character splitting. The initial segmentation phase consists of connected component labeling, connected component grouping, and noise removal.

The purpose of grouping is to create primitives that consist of multiple connected components. The motivation for this process is that many characters, such as the "F" or the "i" in Fig. 9.7, consist of multiple connected components. Grouping is a difficult process. Ideally, an algorithm would detect dots and horizontal line segments (which we refer to as bars) and identify them as connected components that require grouping. The bars from the "F" can be detected using standard image processing techniques. However, in light of the fact that the algorithm is not yet ready to read the word, it is not clear which of the other connected components should be grouped with the bars. There is some possibility that the bottom bar could be part of a "t" and that it should be grouped with the segment that we know is the "r." The point is that we cannot hope even to group properly without reading the word.

This is an example of the Principle of Least Commitment. The bars are grouped later during the dynamic programming matching process. In this way, the optimal grouping can be formed for each string in the lexicon. However, during the grouping, we must ensure that no bar can be considered to be an individual character. We assign each bar a character class confidence value of -∞. This ensures that any segmentation containing an isolated bar will have a very low match score to any string.

Figure 9.7 Grouping simple connected components such as horizontal bars is error-prone if one cannot read the word.

Noise cleaning involves removing stray marks, including dots and punctuation. The dot from the "i" seems informative but it is indistinguishable from a stray dot, such as the one above the "n". Dots are often not near the "i."

After initial segmentation, each initial segment is considered a candidate for splitting into pieces. In our system, this is a heuristic process based on local shape and size and something called the distance transform of the background. Let I denote a binary image and $I(x,y)$ the value of I at location (x,y). A distance transform of the background is an image D with the same domain as I. The value of $D(x,y)$ is the distance from that point to the closest foreground point.

The splitting process is illustrated in Fig. 9.8. The idea is to find points along the top and bottom contour of the segment that are likely to be locations at which characters intersect. These points are determined based on the shape of the contour. Along the top, or north, contour of the segment, valleys are found. Along the bottom, or south, contour of the segment, peaks are found. A starting point is computed near each valley and peak. These points, labeled with the numerals 1 through 8 in Fig. 9.8, serve as starting points for splitting paths. Starting from each point, a path is found from the point through the segment. The paths are constructing iteratively. At each step, the algorithm tries to append another point onto the path. The criterion for selecting the next point is to stay as far away from the segment without going too far "sideways." That is, the algorithm follows the ridges of the distance transform D. Once all the paths have been found, some are removed from consideration as splitting paths. For example, paths that are too close together or that do not actually separate the segment into two pieces are discarded. Full details are provided in Refs. 5 and 21.

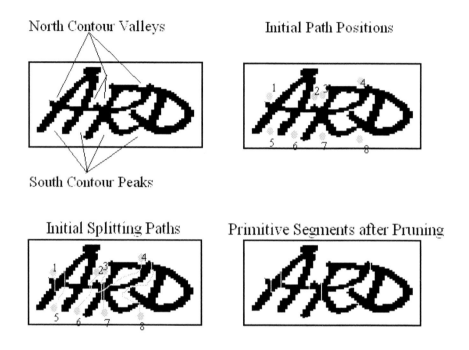

Figure 9.8 Illustration of character splitting process.

Once the primitives are found, unions of primitives are formed. This process is straightforward. Several heuristic rules, such as the existence of large gaps or height-to-width ratio, are used to determine when a union of primitives is either too simple or too complex to serve as a character. An example of legal unions of primitives is shown in Fig. 9.9.

9.3 CHARACTER LEVEL PROCESSING

9.3.1 Features

Features are very important in any pattern recognition application. They define the representation of the patterns for the classifier. In the case of handwritten word recognition, feature representations have several requirements. First, they must characterize character shapes. Second, they must be useful for distinguishing characters from non-characters. Many researchers have focused on the first requirement with not much thought given to the second. Although

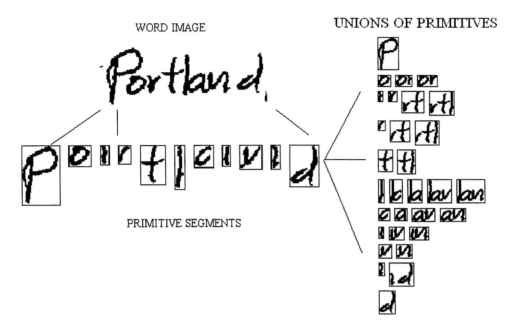

WORD IMAGE

UNIONS OF PRIMITIVES

PRIMITIVE SEGMENTS

Figure 9.9 Example of primitives and all legal unions of primitive segments of a word image.

there are many features proposed for handwritten characters, most features that characterize character shapes will encode the position and direction of edge elements in some form.

In our system, we use two types of features, called the transition features and the bar features. They are fully described in Ref. 5 but we describe them briefly here. The bar features encode directional information from both the foreground and background. First, eight feature images are generated, each corresponding to one of the directions East, Northeast, North, and Northwest, in either the foreground or background. For each of the eight feature images, 15 different sub-image zones are created. Each feature image has an integer value at each location that represents the length of the longest bar which can fit at that point in that direction. The integer values in each zone are summed and the sum is normalized between zero and one. The result is a set of eight feature values for each zone. Since there are 15 zones, the bar feature vector is 120-dimensional.

The transition features encode the location of edge elements (or transitions). A fixed number of edge locations are encoded as fractional distances from boundaries along each row and column. Since the input images have variable dimensions, the locations are down-sampled using local averaging into fixed size arrays. The result is a 100-dimensional feature vector representing positions of edges.

These feature sets perform well for character classification and character confidence assignment, assuming characters are used as inputs to a multi-layer feed-forward neural network (MFNN).[5,23] However, there is an inherent problem with using these types of features in an MFNN in the context of word recognition. The problem is that non-characters can have vastly different representations. Consider the two data clusters shown in Fig. 9.10. A neural network can easily draw a decision boundary between the two clusters. However, samples that are not from the two clusters may come from many different locations in the space. The output of the neural network is difficult to control on such a varied sample.

By contrast, assume that we measure the distance from an unknown input sample to all the training samples. If the input sample is from one of the clusters, the distances to the samples in that cluster should be small and the distances to the other cluster should be relatively large. The key point is that the distance from a sample that is not from either cluster should be relatively large and this information can be used to help distinguish non-characters from characters.

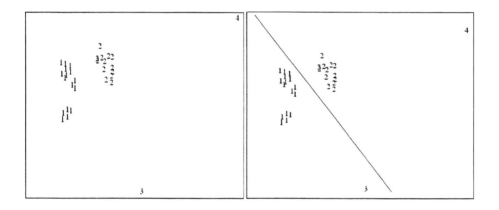

Figure 9.10 A decision boundary can easily be drawn through the classes represented by samples from classes 1 and 2. However, if samples 3 and 4 are presented to the decision boundary based classifier, the outputs may be high or low.

We have implemented this idea, in a more refined form, using Kohonen's Self-Organizing Feature Map (SOFM).[6,30] The nodes in the SOFM serve as prototypical examples of characters. The pattern of distances in the SOFM characterizes characters and provides a uniform representation of non-characters. If an input sample is a non-character, then ideally the distances from all the sample to all the nodes should be large. We use a scaling of the form 1/(1+d) to map these distances to low values. In this way, disparate looking non-characters have a uniformly low pattern of activation across the

SOFM, as illustrated in Fig. 9.11. The scaled values of the SOFM are used as inputs to a character class confidence assignment algorithm, such as an MFNN.

9.3.2 Character confidence assignment

We have shown that word recognition does not require the classical notion of a character classifier. We do not need to assign a class label to an unknown input character. We need a system that can answer the question "To what degree does an unknown input image represent a given character class?"

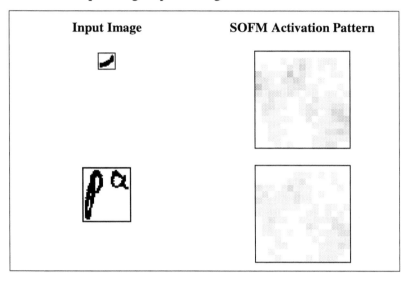

Figure 9.11 The SOFM activation patterns for non-characters provide a more uniform representation of the non-character class than the usual feature sets.

Several methods can be used. In our work, we have used mainly MFNNs. We do not train them in the standard fashion. The standard method for training neural networks for pattern classification is to use class coding. In standard class coded training, the MFNN has one output node for each class. The desired output is 1 for the true class and 0 for all other classes. Rather than use 0 and 1, we use values between 0 and 1, which we refer to as class confidence or class membership values.

Sample outputs of a neural network trained this way are shown in Fig. 9.12. The difficulty lies in determining what the outputs should be for each class. We have used a k-nearest neighbor approach together with prior knowledge of how the character was used. Most of our characters were extracted from handwritten words. Therefore, we know exactly how the character was used in the word. For example, the pattern in Fig. 9.12(c) was actually used as an "R." We always assign a very high desired output for the known class. For the other classes, we look at the k-nearest neighbors of a given sample. We use the percentage of neighbors from each of the classes to define the desired outputs for the other classes. Full details are given in Ref. 23.

This methodology of assigning outputs had a very interesting result. In several independent experiments, character recognition results on isolated characters decreased significantly but word recognition results increased significantly. We again interpret these results in the context of the Principle of Least Commitment. Binary outputs are essentially irrevocable decisions made at the character level. It is clear from the examples that binary class membership decisions should not be made at the character level. We get even better results when we use the Kohonen SOFM as the feature representation, as discussed in Ref. 6.

Similar results should be attainable by accurate estimation of probability density functions or other localization methods. The problem of estimating character class confidence values accurately for handwritten word recognition is an important problem that is not yet fully understood.

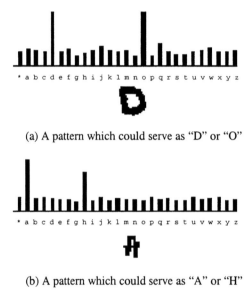

(a) A pattern which could serve as "D" or "O"

(b) A pattern which could serve as "A" or "H"

(c) A pattern which could serve as many characters

Figure 9.12 We have performed experiments with desired outputs such as these. Each pattern is assigned a desired confidence in each class between 0 and 1. The horizontal axis indicates class and the vertical axis indicates degree to which the pattern represents the class. Further research is needed to produce accurate measures of the possibilities that patterns can serve as representatives of each class.

9.4 WORD LEVEL PROCESSING

After a handwritten word is segmented into primitives and unions of primitives and character class confidence values are assigned to the unions of primitives, the lexicon matching process can proceed. We discuss the baseline dynamic programming algorithm. We have extended the algorithm to take into account spatial compatibility between adjacent characters but we refer the reader to Ref. 5 for the extension. We then discuss the problem of estimating word level confidence.

9.4.1 Dynamic programming matching to lexicons

Dynamic programming is a widely used optimization technique. Dynamic programming can be used to find a shortest path through a graph efficiently if the dynamic programming condition holds. The condition can be stated as follows:

Dynamic Programming Condition: Let G be a graph containing nodes A, B, and C. If P is the shortest path from node A to node C and node B is on P, then the portion of the P from A to B is the shortest path from A to B.

The Dynamic Programming Condition may or may not be true for a given problem. If it is true, then it can be used to prune many possibilities. If we start at a node A and find the shortest path from A to some node B, then we no longer need to consider any other paths from A to B when looking for the shortest path to C. Thus, those paths can be pruned, resulting in a vastly more efficient computation.

Dynamic programming is often used in string-matching applications, such as spell checking. String matching can be formulated as a shortest path problem.[37] The lexicon-driven, handwritten word recognition approach can be considered to be a generalized string-matching application. Instead of matching a string to a string, unions of primitives are matched to a string. The dynamic programming algorithm takes the character confidence values for all the unions of primitives from a given word image and a string and returns a value indicating the confidence that the word image represents the string.

The algorithm is implemented using a match matrix approach. Two arrays are formed: the dp array and the path array. The rows of each array correspond to the characters in the string. The columns of each array correspond to primitives. The (i,j) element in the dp array is the value of the best match between the first i characters in the string and the first j primitives. This value may be negative infinity if there is no legal match. If there are w characters and p primitives, then the (w,p) element of the dp array is the value of the best match between all the primitives and all the characters. The (i,j) element in the path array is a pointer back to the previous row indicating which union of primitives produced the best match. A more precise description follows.

Let the primitives of the image be denoted by S_1, S_2,..., S_p. Let the characters in the string be denoted by c_1, c_2, ..., c_w. Let $x(S,c)$ denote the degree to which the segment image S represents character class c. For each pair $m,n \in \{1, 2, ..., p\}$, let

$$S_{mn} = \bigcup_{h=m}^{n} S_h$$

and let Legal (S_{mn}) be a Boolean function that returns TRUE if the union image S_{mn} is "character-like" and FALSE otherwise. The value of Legal (S_{mn}) is computed by evaluating a set of heuristic criteria designed to measure the complexity of S_{mn} as described in Ref. 5.

The values of the dp array and path array are computed as follows:

For i = 1, ..., w

    For j = i, ..., p

1   *IF i = 1, (that is we are matching against the first character) THEN*

2   *IF Legal(S_{1j}) THEN*

3 $dp(1,j) = x(S_{1j}, c_1)$

4   *ELSE*

5 $dp(1,j) = -\infty$

6   *ENDIF*

7   *ELSEIF Legal(S_{ij}) THEN*

8

$$dp\,(i,j) = \max_{k}\,\{dp\,(i-1,k) + x(S_{k+1,j}, C_i) | i \le k \le j\, and\ Legal\,(S_{kj})\, is\ TRUE\,\}$$

9   *ELSE*

10 $dp(i,j) = -\infty$

11   *ENDIF*

12   *ENDIF*

13   *IF dp(i,j) > -∞ THEN*

14

$$j_2 = \arg \max_{k}\,\{dp\,(i-1,k) + x(S_{k+1,j}, C_i) | i \le k \le j\, and\ Legal\,(S_{kj})\, is\ TRUE\,\}$$

15   *path(i,j) = j_2*

16   *ENDIF*

    ENDFOR

ENDFOR

Once the match array has been computed, the best match can be found using the path array.

As an example, consider the image of the word "Richmond" shown in Fig. 9.3. The input image is first segmented into primitives. The best matches to the string "Richmond" is shown. Appendix A contains all the character class confidence on all the unions of segments necessary to perform the matching process using the above algorithm. Table 9.1 is the dp array, call it dp, for matching the image of "Richmond" to the string "Richmond" at the end of the algorithm and Table 9.2 is the path array, call it p.

In Table 9.1, the value $dp(1,1)$ is the confidence that the first primitive is in the character class "R" (i.e., the value of $x(S,c)$, where S is the first primitive and c is the character "R"), $dp(1,2)$ is the confidence that the union of the first two primitives is in the character class "R," and $dp(1,3)$ is the confidence that the union of the first three primitives is in the character class "R," etc. The table indicates that the best way to match unions of primitives to "R" starting with primitive 1 is to take the union of the first three primitives, which is the result most humans would agree on. The membership of the union of the first three primitives in the class "R" is 0.53.

Now consider $dp(2,5)$ and $p(2,5)$. The value $dp(2,5)$ is the maximum value of $\{dp(1,k) + x(S_{k+1,5}, \text{"i"}) \mid k = 1, 2, 3, 4\}$. In this case, it turns out that the maximum value occurs for $k = 3$ so the best way to match between the first 2 characters and the first 5 primitives is to match the first 3 primitives to "R" and the 4th and 5th primitives to "i." The match score is entry (2,5), 0.80, which is the sum of the previous match score 0.53 and the match score of the union of the fourth and fifth primitives (0.27 here). The value $p(2,5)$ is set to 3 to indicate that the second character is matched to primitives 4 and 5. Continuing, the algorithm leads to the best match between all characters and all primitives, entry (8,25). The score 4.58 is normalized to the range [0,1] by dividing by the length of the string, resulting in an overall match score of 0.57 which is the average of the match scores assigned to the segmentation in Fig. 9.3.

To find the segmentation resulting in the best match, the path array is used. Starting at the (8,25) entry, we follow the path array backwards. The fact that $p(8,25) = 22$ indicates that the 8th character, "d," is matched to the union of the 23rd, 24th, and 25th primitive segments. The fact that $p(7,p(8,25)) = p(7,22) = 18$ indicates that the 7th character, "n," is matched to the union of the 19th, 20th, 21st, and 22nd primitive segments, and so on.

As mentioned previously, we have extended the algorithm to include relationships between adjacent segments. In a graph theoretic sense, it is interesting to note that since the extension incorporates information about the compatibility of the nodes, the dynamic programming condition no longer holds. The shortest path between A and B depends on which node comes after B. Therefore, dynamic programming no longer provides an optimal solution when the adjacency information is included. However, when the adjacency information is included, the recognition results are in fact significantly higher. This interesting result points out that optimal solutions are not always the best solutions. Every optimal solution is relative to some criterion function. A sub-optimal solution to one criterion function may be better than an optimal solution to a different criterion function.

	1	2	3	4	5	6	7	8	9	10	11	12	13	14	15	16	17	18	19	20	21	22	23	24	25
"R"	11	38	**53**	45	44	0	0	0	0	0	0	0	0	0	0	0	0	0	0	0	0	0	0	0	0
"i"	0	23	27	71	**80**	52	53	54	52	51	55	0	0	0	0	0	0	0	0	0	0	0	0	0	0
"c"	0	0	0	38	90	**132**	108	108	86	86	87	63	63	63	63	62	65	65	0	0	0	0	0	0	0
"h"	0	0	0	0	42	95	144	147	157	**193**	189	166	162	141	145	121	100	95	85	77	76	0	0	0	0
"m"	0	0	0	0	0	50	104	154	156	167	202	208	219	227	261	**263**	272	258	178	177	216	231	0	0	0
"o"	0	0	0	0	0	0	74	114	169	163	179	206	241	261	236	236	279	**309**	312	290	281	280	281	0	0
"n"	0	0	0	0	0	0	0	83	122	178	181	190	223	261	269	269	275	285	322	337	352	**370**	356	321	0
"d"	0	0	0	0	0	0	0	0	0	0	0	0	0	0	0	0	0	0	0	0	0	0	0	0	**458**

Table 9.1 Dynamic programming array for matching the image of Richmond to the string "Richmond."

	1	2	3	4	5	6	7	8	9	10	11	12	13	14	15	16	17	18	19	20	21	22	23	24	25
"R"	0	0	**0**	0	0	0	0	1	8	8	6	9	11	11	12	12	13	16	15	15	19	20	20	19	20
"i"	0	1	1	3	**3**	5	5	5	5	5	5	5	7	7	14	10	16	16	16	15	15	19	16	21	19
"c"	0	0	2	3	3	**5**	5	5	5	5	5	7	7	7	8	11	11	11	12	14	14	16	21	19	21
"h"	0	0	0	3	4	5	6	6	6	**6**	6	6	6	8	8	10	11	11	15	15	15	16	16	17	21
"m"	0	0	0	0	4	5	6	7	8	9	10	10	10	11	11	**10**	16	16	12	13	15	15	16	21	19
"o"	0	0	0	0	0	5	6	7	8	9	10	11	12	13	14	14	**16**	17	16	15	17	17	17	19	21
"n"	0	0	0	0	0	0	6	7	8	9	9	9	11	13	14	14	13	17	18	18	17	**18**	18	19	23
"d"	0	0	0	0	0	0	0	7	8	9	10	10	11	12	13	13	16	17	17	17	20	20	20	22	**22**

Table 9.2 Dynamic programming path array for match of Richmond to "Richmond."

9.4.2 Estimation of word-level confidence from character level confidence

The dynamic programming algorithm uses the average character confidence value of a segmentation as the value of a match between a segmentation and a character string. As shown earlier, the average is not always desirable because it is not robust. In this section, we show how a method based on linear combination of order statistics (LOS) operators can be used.

The word robust is often used. In statistics, it is precisely defined. We do not go into precise definitions here, but we try to give an intuitive idea. Suppose that one is trying to estimate a parameter of population from sample data. Ideally, the sample data are all obtained from the underlying population. However, sometimes our data are corrupted by samples from a different population. For example, suppose we are collecting data on the weights of all the people in a given geographical area. Assume that the data are collected by asking people to provide their weight. Assume further that some people lie about their weight. Then, their samples are not actually from the true distribution. A robust estimator will not be biased by a percentage of measurements that are not truly from the distribution. If one person out of every 10 says that he/she weighs 1000 pounds, the average value will probably be affected. In contrast, the median value will not be affected. Thus, the median is robust to conditions under which 10% of the samples are from another distribution. In fact, the median is robust if up to (but not as many as) 50% of the values are from another distribution.

There are many types of robust operators.[38] The LOS operators include robust and non-robust operators. The max, min, median, and mean are all examples of LOS operators. Of these, only the median is robust. We now define LOS operators more precisely.

Let $x = (x_1, x_2, ..., x_n)$ be a vector of real numbers. The ith order statistic of x is the ith smallest element of x. We denote the ith order statistic as $x_{(i)}$ where $x_{(1)} \leq x_{(2)} \leq ... \leq x_{(n)}$. If $n = 2k-1$ then the kth order statistic is the median.

Let $w = (w_1, w_2, ..., w_n)$ be a vector of real numbers constrained so that:

$$\sum_{i=1}^{n} w_i = 1 \qquad \text{and} \qquad 0 \leq w_i \leq 1 \qquad \text{for } i = 1, 2, ..., n.$$

The LOS operator on x with weight vector w is defined as

$$\text{LOS}(x, w) = \sum_{i=1}^{n} w_i x_{(i)}.$$

In our application, the vector x represents the character class confidence values for the segments in a segmentation and LOS(x, w) is the word confidence. If all the weights are $1/n$, then the LOS operator is the sample mean. If $w = (1,0,0,...,0)$, then the LOS operator represents the min. The α-trimmed mean is another example of a robust LOS operator. In this case, the first and last α% of the weights are 0, and the rest of the weights are

$$w_i = \cfrac{1}{n - 2\left(\cfrac{\alpha}{100}\right)n} \quad .$$

For example, if $n = 8$ and $\alpha = 25$, then the non-zero weights are 0.25 and the α-trimmed mean is

$$\frac{1}{4}\sum_{i=3}^{6} x_{(i)} .$$

This operator is robust since two of the samples can be from another distribution and not affect the result. (More specifically, if we let two of the samples go to $+\infty$ or $-\infty$, there is no effect on the output).

We now discuss how LOS operators can be used to improve handwritten word recognition. In our discussion, we once again ignore the character adjacency considerations since they make the discussion more complex and do not contribute to the conceptual understanding of the use of the LOS operators.

9.4.2.1 Baseline dynamic programming algorithm revisited

Let I be a word image and let $L = \{L_1, L_2, ..., L_T\}$ be a set of words or strings. In the baseline dynamic programming algorithm, a match between a string $L_m \in L$ and I is computed by maximizing the match score between segmentations of I and the individual characters of the word W, where the match score is computed using an average. We use the LOS to recompute the match score between L_m and I.

Let $L_m = c_1, c_2, ..., c_n$ where c_i is the ith character in L_m. Assume the word image I is segmented into n segments. Our character recognition algorithms provide confidence values that the ith segment represents the ith character in the string. Denote these values by $x = (x_{c1}, x_{c2}, ..., x_{cn})$. The baseline system computes the match score between the segmentation and string by averaging these confidence values. More precisely, let the primitives of I be $S_1, S_2, ..., S_p$. Denote the set of all legal unions of primitives by

$$U_I = \{ S_{jk} : \quad S_{jk} = \bigcup_{h=j}^{k} S_h , \quad j \leq k, \quad \text{and} \quad \bigcup_{h=j}^{k} S_h \text{ is legal} \}.$$

A segmentation of length n of I is a sequence of elements of U_I of the form

$$S = S_{k_0+1,k_1}, \; S_{k_1+1,k_2}, \; ..., \; S_{k_{n-1}+1, \, k_n} \qquad \text{where } \; k_0 = 0 \text{ and } \; k_n = p.$$

A segmentation of length n must use all primitive segments. Let S_n denote the set of all segmentations of I of length n. A match score between a string L_m and a segmentation $S \in S_n$ can be computed as

$$\text{segmatch} \, (\, L_m , S) = \frac{1}{n} \sum_{i=1}^{n} x_{c_i} (S_{k_{i-1}+1, \, ki}) \, .$$

The dynamic programming algorithm efficiently computes the best value of segmatch, that is:

$$\text{best\_match} \, (L_m, I) = \max\{ \, \text{segmatch}(L_m , S) : S \in S_n \, \}.$$

9.4.2.2 Robust matching algorithm

LOS operators can be used in place of the average to compute the match between a given segmentation and a string. We would like to find the best match over all possible segmentations, that is, we would like to find

$$\text{LOS\_match}(L_m, I) = \max_{S \in S_n} \{ \sum_{i=1}^{n} w_i x_{(c_i)}(S_{ki - 1 + 1, \, ki}) \} \, .$$

However, this optimization is difficult to implement directly using dynamic programming because of the need to sort the confidence values. A sub-optimal approach did not perform better than the present method.[39] The present method computes seg_match using the baseline method and then recomputes the match scores for the highest ranked lexicon strings using LOS operators. Since the correct string is almost always among the highest ranked strings, this method is successful. A different LOS operator is required for each different length string. The algorithm is:

Algorithm: Estimating Word Level Confidence Using LOS Operators

Inputs: Word Image, I

        Lexicon of Candidate Strings, $L = \{L_1, L_2, ..., L_T\}$

        Set of weight vectors $\mathbf{w}_2, \mathbf{w}_3, ..., \mathbf{w}_{mlen}$ of lengths 2, 3, ..., mlen satisfying the LOS constraints

Outputs: Lexicon of Candidate Strings Tagged with Word Confidence Scores,

        $L^* = \{ \, (L_1, wconf_1), (L_2, wconf_2), ..., (L_T, wconf_T) \, \}$

Step1:   For each string $L_m \in L$

           Let $conf_m$ = best_match (L_m, I)

                    n_m = number of characters in L_m

$$S_{o,m} =$$
$$\mathrm{argmax}_{s}(\mathrm{best\_match}\ (L_m, I)) = \mathrm{argmax}_{s}(\max\{\ \mathrm{segmatch}(L_m, S): S \in S_n\})$$

Step2:   Create list of pairs sorted by conf_m:

$$\{\ (L_{(1)}, \mathrm{conf}_{(1)}), (L_{(2)}, \mathrm{conf}_{(2)}), ..., (\ L_{(T)}, \mathrm{conf}_{(T)})\ \}\ \text{where}\ \mathrm{conf}_{(1)} \geq \mathrm{conf}_{(2)} \geq ... \geq \mathrm{conf}_{(T)}.$$

Step3:   Choose the top N ranked strings, $\{\ (L_{(1)}, \mathrm{conf}_{(1)}), (L_{(2)}, \mathrm{conf}_{(2)}), ..., (\ L_{(N)}, \mathrm{conf}_{(N)})\ \}$

Step4:   For $m = 1, 2, ..., N$

$$\mathbf{x}(S_{o,(m)}) = \quad \text{vector of character scores of the segments of } S_{o,(m)} \text{ corresponding}$$
to $L_{(m)}$

$$\mathrm{wconf}_m = \quad \mathrm{LOS}(w_{nm}, \mathbf{x}(S_{o,(m)}))$$

Step5:   For $m = N+1, ..., T$

$$\mathrm{wconf}_m = \quad \mathrm{conf}_m$$

Step6:   Return $\{\ (L_{(1)}, \mathrm{wconf}_{(1)}), (L_{(2)}, \mathrm{wconf}_{(2)}), ..., (\ L_{(T)}, \mathrm{wconf}_{(T)})\ \}$

The string with the highest ranking $L_{(1)}$ can be taken as the recognition result. The success of the method depends upon the weight vectors used in the LOS operators. For example, in Ref. 18 we found that weights of the form

$$w_i = \frac{w_1}{(1+p)^i} \qquad \text{for } i = 1, 2, ..., n \qquad \text{and} \qquad p > 0$$

(the exact form is a little more complicated and can be found in the referenced paper) provided improved results over the average. As can be seen, the weights decrease exponentially as a function of the subscript. Thus, small values are assigned larger weights than low values. These weights do not precisely define robust operators but they are "closer" to robust operators than the average.

For example, consider the problem illustrated in Fig. 9.5. The outlier, the segment that is not a character but looks like an "r," has the maximum value. However, it has the smallest weight and therefore contributes less to the word confidence. Thus, these kinds of operators contribute to a solution of the problem caused by non-characters that look like characters.

The above weights were found through a suboptimal investigation. However, optimal LOS weights can be found through the use of quadratic programming. The problem of finding the LOS operator that minimizes the mean square error between the LOS of the input data and the desired outputs can be formulated as:

$$\text{Minimize} \quad \sum_{i=1}^{T}\left[LOS(\mathbf{x}_i, \mathbf{w}) - y_i\right]^2 \quad = \quad \sum_{i=1}^{T}\left[\sum_{j=1}^{n} w_j x_{i(j)} - y_i\right]^2,$$

subject to the constraints

$$\sum_{i=1}^{n} w_i = 1 \qquad \text{and} \qquad 0 \le w_i \le 1 \text{ for } i = 1, 2, \ldots, n.$$

This is a constrained optimization problem with a quadratic objective function and linear constraints. It can be solved using standard optimization methods.

Although optimality is guaranteed, we observe that there are many possible optimal solutions. Some are much better than others. There are several methods for computing optimal LOS operators for the handwritten word recognition application because a variety of ways exist to define the desired outputs and to choose training lexicons. Our experimental results show that a wide range of performance results (ranging anywhere from 46% correct to approximately 80% correct on the same image data and testing lexicons[4] can be achieved with different, but reasonable, choices of these optimization parameters and training set construction methodologies. Thus, even though optimal solutions are available, heuristics are used to define the training methodology.

9.5 EXPERIMENTAL RESULTS

The system we have described has evolved over the course of several years and many sets of experiments have been performed at different stages of the development. Generally, results are presented in pairs; we present the results of the penultimate system and the newest version of the system in order to show the effect of an algorithmic modification on the system. We present the most recent pair of results.

For our penultimate system, we use the dynamic programming algorithm that includes spatial compatibility between adjacent characters. The feature sets used are the Kohonen Self-Organizing Feature Maps constructed from bar and transition features. The character confidence assignment algorithms are multi-layer feed-forward neural networks with desired outputs established by the k-nearest neighbor-based algorithm. For the newest version of the system, we use the system with all the above-mentioned modules and the additional module which is the estimation of the word-level confidence values using optimal weights found using quadratic programming.

Experimental results are very dependent upon the data sets. In Table 9.3, we show results for a test set of 1317 handwritten word images. Three different sets of lexicons are used, but the images are the same. In the first row, results are shown for a set of lexicons called lex100. There is a different lexicon in lex100 for each word image. Each lexicon has length 100. In the second and third rows, results are shown for sets of lexicons called lex2 and lex3. There are different lexicons in each of these sets for each word image also. However, for these lexicons, the lengths vary. For lex2, the average length of a lexicon is 100 whereas in lex3 the average length is 1000. The lexicons were generated by other groups algorithmically, as described in the references.

Table 9.3 Summary of Testing Results.

	Recognition Rate of Baseline System	Recognition Rate of Optimal LOS
Lex100-Lexicons of Size 100	84%	86%
Lex2-Lexicons of Average Size 100	91%	91%
Lex3-Lexicons of Average Size 1000	81%	84%

9.6 CONCLUSION

We presented an in-depth view of lexicon-driven, segmentation-based handwritten word recognition by focusing on a representative system from among several developed. The difficulties in handwritten word recognition were illustrated with real examples. Several approaches to solving the problems associated with the difficulties were described.

We provided detailed numerical examples to allow the reader to work through the dynamic programming algorithm on real data. We showed how the Principle of Least Commitment plays a major role in the development of algorithm components for handwritten word recognition. We pointed out how the notion of an optimal algorithm can be misleading; heuristics still play a role in the choice of sample data, optimization parameters, and criterion functions. We provided experimental results and pointed out that the recognition rates quoted are actually highly dependent on data sets.

We hope the reader has attained an understanding of how lexicon-driven, segmentation-based handwritten word recognition systems work. There is still need for significant research before generic handwriting recognition programs will become available.

9.7 ACKNOWLEDGMENTS

Many people have contributed in many ways to the ideas presented in this chapter. They have become too numerous to list in an acknowledgment section; many of them are included in the references as co-authors on papers. Those who have contributed include students and faculty members at the University of Missouri, research staff of the Environmental Research Institute of Michigan (ERIM), the staff of Arthur D. Little Inc, and of the U.S. Postal Service.

REFERENCES

1. R. Hecht-Nielson, *Neurocomputing*, Addison-Wesley (1990).

2. J. Favata and S. Srihari, "Recognition of Handwritten Words for Address Reading," *Proceedings, U.S. Postal Service Advanced Technology Conference*, Washington, D.C., pp. 192-205, (1990).

3.   J. C. Simon and O. Baret, "Handwriting Recognition as an Application of Regularities and Singularities in Line Pictures," *Proceedings, First Int'l Workshop Frontiers in Handwriting Recognition*, Montreal, pp. 23-39 (1990).

4.   W. Chen, P. D. Gader, and H. Shi, "Improved Dynamic Programming-Based Handwritten Word Recognition Using Optimal Order Statistics," *Proceedings, SPIE Conf. Statistical and Stochastic Methods in Image Processing II*, San Diego (1997).

5.   P. D. Gader, M. Mohamed, and J. H. Chiang, "Handwritten Word Recognition with Character and Inter-Character Neural Networks," *IEEE Trans. Sys. Man Cybernetics*, vol. 27, no. 1, pp. 158-165 (1996).

6.   J.-H. Chiang and P. D. Gader, "Hybrid Fuzzy-Neural Systems in Handwritten Word Recognition," *IEEE Trans. Fuzzy Sys.*, vol. 5, no. 4, pp. 497-508 (1997).

7.   M. Shridhar, G. Houles, and F. Kimura, "Handwritten Word Recognition Using Lexicon Free and Lexicon Directed Word Recognition Algorithms," *Proceedings, Int'l. Conf. Document Analysis Recognition (ICDAR '97)*, Ulm, Germany (1997).

8.   G. Kim and V. Govindaraju, "Handwritten Word Recognition for Real-Time Applications," *Proceedings, Third International Conf. Document Anal. Recognition (ICDAR '95)*, Montreal, pp. 24-28 (1995).

9.   G. Dzuba, A. Filatov, and A. Volgunin, "Handwritten ZIP Code Recognition," *Proceedings, Int'l. Conf. Document Analysis Recognition (ICDAR '97)*, Ulm, Germany (1997).

10.   L. Rabiner, "A Tutorial on Hidden Markov Models and Selected Applications in Speech Recognition," *Proc. IEEE*, vol. 77, no. 2, pp. 257-286, (1989).

11.   M. Chen and A. Kundu, "An Alternative to Variable Duration HMM in Handwritten Word Recognition," *Proceedings, Third International Workshop on Frontiers in Handwriting Recognition,* Buffalo, NY, pp. 82-92 (1993).

12.   M. Y. Chen, A. Kundu, and J. Zhou, "Off-line Handwritten Word Recognition Using Hidden Markov Model Type Stochastic Network," *IEEE Trans Patt. Anal. Mach. Intell.*, vol. 16, no. 5, pp. 481-496 (1994).

13.   Y. Bengio, Y. LeCun, and D. Henderson, "Globally Trained Handwritten Word Recognizer Using Spatial Representation, Space Displacement Neural Networks and Hidden Markov Models," in *Advances in Neural Information Processing Systems*, vol. 6, J. Cowan and G. Tesauro, Eds. (1993).

14. F. Stentiford, "Automatic Feature Design for Optical Character Recognition Using an Evolutionary Search Procedure," *IEEE Trans. Patt. Anal. Mach. Intell.*, vol. 7, pp. 349-355 (1985).

15. M. Mohamed and P. D. Gader, "Handwritten Word Recognition Using Segmentation-Free Hidden Markov Modeling and Segmentation-Based Dynamic Programming Techniques," *IEEE Trans. Patt. Anal. Mach. Intel.*, vol. 18, no. 5, pp. 548-554 (1995).

16. M. A. Mohamed, *Handwritten Word Recognition using Generalized Hidden Markov Models*, University of Missouri-Columbia, Ph.D. Thesis (1995).

17. P. D. Gader, J. M. Keller, R. Krishnapuram, J. H. Chiang, and M. A. Mohamed, "Neural and Fuzzy Methods in Handwriting Recognition," *Computer*, vol. 30, no. 2, pp. 79-86 (1996).

18. P. D. Gader, M. A. Mohamed, and J. M. Keller, "Dynamic Programming Based Handwritten Word Recognition Using the Choquet Fuzzy Integral as the Match Function," *J. of Electron. Imaging*, vol. 5, no. 1, pp. 15-24 (1996).

19. P. D. Gader, M. Mohamed, and J. Keller, "Fusion of Handwritten Word Classifiers," *Patt. Recog. Lett.*, vol. 17, no. 6, pp. 577-584 (1996).

20. P. D. Gader, M. A. Mohamed, and J. M. Keller, "Applications of Fuzzy Integrals to Handwriting Recognition," *Proceedings, SPIE Conf. Appl. of Fuzzy Logic Technology II*, Orlando FL (1995).

21. P. D. Gader, M. P. Whalen, M. J. Ganzberger, and D. Hepp, "Handprinted Word Recognition on a NIST Data Set," *Mach.Vis. App.*, vol. 8, pp. 31-40 (1995).

22. P. D. Gader and M. A. Mohamed, "Multiple Classifier Fusion for Handwritten Word Recognition," *Proceedings, IEEE Conference Systems, Man, Cybernetics*, Vancouver (1995).

23. P. D. Gader, M. A. Mohamed, and J.-H. Chiang, "Comparison of Crisp and Fuzzy Character Neural Networks in Handwritten Word Recognition," *IEEE Trans. Fuzzy Syst.*, vol. 3, no. 3, pp. 357-364 (1995).

24. P. D. Gader and J. M. Keller, "Applications of Fuzzy Set Theory to Handwriting Recognition," *Proceedings, Third IEEE Int'l. Conf. Fuzzy Systems*, Orlando, FL, pp. 910-917 (1994).

25. P. D. Gader, A. M. Gillies, and D. Hepp, "Handwritten Character Recognition," in *Digital Image Processing Methods*, E. Dougherty, Ed. New York, NY, Marcel Dekker, pp. 223 - 261 (1994).

26. P. D. Gader, M. Mohamed, and J. Chiang, "Segmentation-Based Handprinted Word Recognition," *Proceedings, U.S. Postal Service Advanced Technology Conference*, Washington, DC, pp. 215 - 225, Nov. (1992).

27. P. D. Gader, M. Mohamed, and J. Chiang, "Comparison of Crisp and Fuzzy Character Networks in Handwritten Word Recognition," *Proceedings, North American Fuzzy Information Processing Society Conference*, Puerto Vallarta, Mexico, pp. 257-266, (1992).

28. P. D. Gader, M. Mohamed, and J. Chiang, "Fuzzy and Crisp Handwritten Alphabetic Character Recognition Using Neural Networks," *Proceedings, Artificial Neural Networks in Engineering*, St. Louis, MO (1992).

29. P. Gader and M. Whalen, "Advanced Research in Handwritten ZIP Code Recognition," U.S. Postal Service Office of Advanced Technology, Washington D.C. (1990).

30. J. Chiang, *Hybrid Fuzzy Neural Systems for Robust Handwritten Word Recognition*, University of Missouri-Columbia, Ph.D. Thesis (1995).

31. P. J. Huber, *Robust Statistics*, New York, John Wiley, (1981).

32. P. J. Rousseeuw and A. M. Leroy, *Robust Regression and Outlier Detection*, New York, John Wiley (1987).

33. D. Marr, *Vision*, San Francisco, CA, W. H. Freeman and Company (1982).

34. J. M. Keller and P. D. Gader, "Fuzzy Logic and the Principle of Least Committment in Computer Vision," *Proceedings, IEEE Conf. Systems, Man, Cybernetics*, Vancouver, in press.

35. J. M. Keller, P. D. Gader, and C. W. Caldwell, "The Principle of Least Commitment in the Analysis of Chromosome Images," *Proceedings, SPIE Conf. Appl. of Fuzzy Logic Technology II*, Orlando FL (1995).

36. R. G. Casey and E. Lecolinet, "A Survey of Methods and Strategies in Character Segmentation," *IEEE Trans. Patt. Anal. Mach. Intell.*, vol. 18, no. 7, pp. 690-707 (1996).

37. D. Sankoff and J. Kruskal, *Time Warps, String Edits, and Macromolecules: The Theory and Practice of Sequence Comparison*, Reading, MA, Addison-Wesley (1983).

38. S. Kassam and H. Poor, "Robust Techniques for Signal Processing: A Survey," *Proc. IEEE*, vol. 73, no. 3, pp. 433-481 (1985).

39. H. Shi and P. D. Gader, "Lexicon-Driven Handwritten Word Recognition Using Choquet Fuzzy Integral," *Proceedings, IEEE Conf. Systems, Man, and Cybernetics*, Beijing, China, pp. 412-417 (1996).

Appendix

This appendix provides upper and lower case character class confidence scores that can be used to work the dynamic programming example for the string "Richmond" shown in Fig. 3.   The first table shows lower case character confidence values for selected character classes for unions of primitives from the Richmond word image starting from primitive 2 (the second character).

Start and End Indices of Primitive Unions		Lower Case Character Classes Appearing in "Richmond"						
Start	**End**	**c**	**d**	**h**	**i**	**m**	**n**	**o**
2	2	10	44	9	12	10	7	8
2	3	7	73	14	16	9	10	8
2	4	8	45	24	12	11	10	8
2	5	5	44	41	12	10	8	6
2	6	5	5	59	17	12	5	5
2	7	5	3	59	19	15	6	6
2	8	5	3	55	17	20	8	4
3	4	13	27	9	21	9	7	13
3	5	23	32	16	28	12	8	13
4	4	11	8	8	18	7	8	16
4	5	63	10	16	27	10	9	8
5	5	2	20	4	32	9	15	7
6	6	52	12	5	8	8	8	8
6	7	28	11	6	9	8	12	10
6	8	28	7	7	10	8	13	62
6	9	6	57	8	8	7	22	5
6	10	6	66	27	7	9	13	5
6	11	7	13	42	11	19	8	6
6	13	6	6	52	15	18	8	6
6	14	7	3	53	16	20	8	7
6	16	8	3	55	15	19	8	5
7	7	4	5	12	82	9	12	24
7	8	7	4	15	16	7	8	9
7	9	9	16	25	9	5	15	8
7	10	8	26	61	7	10	9	11
7	11	6	23	57	11	12	13	10
7	12	8	8	34	9	11	8	11
7	13	8	5	30	9	12	10	9
7	14	8	5	26	9	14	11	9
7	16	8	6	40	10	15	10	7
8	8	5	13	4	21	10	9	10
8	9	14	19	22	12	11	10	10
8	10	15	21	49	9	11	12	12
8	11	8	17	58	12	12	8	12
8	12	10	8	43	11	10	6	9
8	13	10	5	34	12	12	7	8
8	14	10	5	31	12	14	8	8
8	16	10	6	46	13	14	7	6
8	19	5	7	48	15	17	7	5
8	20	4	7	44	16	18	9	6
9	9	8	8	14	10	9	8	15
9	10	7	10	75	6	10	10	7
9	11	11	10	76	13	8	11	9
9	12	9	4	45	9	10	11	9
9	13	8	3	37	10	12	11	8
9	14	9	3	33	10	15	11	8
9	15	9	3	37	11	19	10	8
9	16	9	3	46	12	19	10	7
9	19	5	6	47	17	17	11	5
9	20	5	5	41	18	19	13	6
10	10	13	8	5	74	10	9	7
10	11	8	9	11	14	7	12	8
10	12	7	9	16	15	9	21	11
10	13	7	8	15	11	26	20	11
10	14	8	8	15	10	37	22	9
10	15	6	8	17	10	56	19	5
10	16	6	7	33	9	64	12	4
10	17	9	13	7	9	80	17	10
10	19	10	13	7	9	79	14	10
10	20	10	12	6	9	77	12	9
11	11	6	16	8	27	9	8	12
11	12	10	9	12	13	15	21	23
11	13	9	8	16	11	26	35	12
11	14	9	7	15	8	33	43	8
11	15	9	7	22	7	67	29	5
11	16	8	8	35	8	70	21	4
11	17	10	12	8	9	79	21	9
11	19	11	13	8	8	75	16	7
11	20	12	11	7	9	71	16	8
12	12	7	8	4	13	10	9	4

		c	d	h	i	m	n	o
12	13	4	6	19	7	22	44	25
12	14	6	7	15	7	38	38	18
12	15	7	8	17	8	72	33	6
12	16	7	6	30	9	60	31	5
12	17	10	10	13	6	54	37	6
12	18	10	11	8	7	69	22	5
12	19	11	11	6	8	58	15	7
12	20	10	8	5	10	49	11	11
13	13	3	10	11	40	8	10	33
13	14	6	15	8	19	11	12	25
13	15	11	10	12	11	18	58	8
13	16	20	11	18	31	13	29	9
13	17	8	9	18	9	14	28	7
13	18	8	9	11	10	24	14	6
13	19	7	8	7	10	12	10	11
13	20	6	7	6	10	15	9	15
13	21	8	8	7	9	41	8	16
14	14	4	9	9	42	38	20	42
14	15	13	9	6	25	13	18	7
14	16	24	9	15	54	11	11	10
14	17	9	9	25	9	7	34	7
14	18	8	9	12	9	14	17	7
14	19	7	8	7	9	11	11	11
14	20	7	6	7	10	15	10	15
14	21	9	7	8	9	51	8	17
14	22	8	10	8	10	77	8	12
15	15	5	5	18	77	9	8	9
15	16	16	4	39	38	8	8	9
15	17	8	8	33	12	7	14	5
15	18	9	8	10	10	12	11	6
15	19	8	9	8	10	10	13	11
15	20	9	5	9	11	16	12	18
15	21	10	8	11	10	61	8	15
15	22	8	7	11	9	80	9	12
16	17	14	11	12	22	7	16	12
16	18	17	9	24	18	10	9	21
16	19	14	8	22	10	11	8	22
16	20	20	12	14	8	11	11	29
16	21	17	7	13	8	71	25	15
16	22	10	10	14	10	86	18	12
16	23	11	11	14	10	88	19	11
17	17	9	6	12	47	7	6	16
17	18	16	8	10	13	9	7	46
17	19	11	9	12	9	6	10	49
17	20	10	12	6	11	19	8	16
17	21	10	7	10	7	58	56	13
17	22	10	9	10	8	88	25	9
17	23	10	9	11	9	89	24	9
17	25	4	28	13	9	36	41	5
18	18	3	7	5	37	7	6	25
18	19	12	9	12	9	8	10	11
18	20	12	12	6	9	11	9	7
18	21	9	9	7	8	20	73	9
18	22	9	8	9	9	72	41	8
18	23	9	8	9	9	75	47	9
18	24	7	10	5	10	73	9	15
19	19	6	8	7	68	11	13	14
19	20	6	21	9	12	6	28	8
19	21	8	10	7	12	13	12	6
19	22	7	4	18	12	16	61	5
19	23	5	3	54	11	19	47	5
19	24	4	9	15	10	57	6	6
19	25	8	9	7	9	33	16	6
20	20	1	2	1	14	10	4	4
20	21	7	9	8	15	13	18	5
20	22	5	3	42	11	17	52	5
20	23	4	2	58	12	23	43	4
20	24	7	7	24	6	39	9	5
20	25	7	13	7	7	24	14	6
21	21	3	8	24	28	9	15	9
21	22	3	4	68	8	18	34	4
21	23	4	4	67	11	23	30	6
21	24	7	8	11	7	16	13	7
21	25	8	18	7	8	17	14	6
22	22	12	15	8	59	11	13	12
22	23	15	5	15	46	10	11	15
22	24	9	10	9	9	11	8	8
22	25	14	63	11	9	10	10	10
23	23	1	2	3	20	9	5	10
23	24	13	15	14	8	9	7	43
23	25	16	88	12	7	9	7	9
24	24	10	9	12	9	7	8	62
24	25	19	83	11	8	10	6	9
25	25	48	8	9	8	9	9	7

The second table shows   upper case character confidence values for "R" for those unions of primitives from the Richmond word image that begin with primitive.   (thereby matching against the first character only)

i	j	R
1	1	11
1	2	38
1	3	53
1	4	45
1	5	44

10

SCANNING

Ying-wei Lin
Joseph P. Taillie
Leon C. Williams
Xerox Corporation
Webster, New York

10.1. INTRODUCTION

This chapter discusses the process of converting a hardcopy image into a digital image using a scanner. The components of a scanner are described in some detail: illumination source, optics, scanning mechanical components, image sensor, and electronics. Special attention is paid to the factors that influence the quality of the scanner output. A number of image processing functions often integrated with scanners are also described.

Although many digital images are generated directly with application software on computers, it is often necessary to convert existing hardcopy images into digital form. For example, a photo must be converted to a digital form in order to include it in a publication, send it to a distant location through a network, or archive it on a digital storage medium. Even to make a copy of a grayscale or color image, modern color photocopiers first convert it to a digital image, process it appropriately, and then print it. Typically, the original is a standard sheet of paper, but it could be an overhead transparency, a 35 mm slide, or an engineering drawing. It can also be any specialized kind of document; for example, a rare manuscript or an insurance claim form.

Various types of *scanners* are employed in the digitization of images [1, 2]. Among the simplest of scanners are those found in facsimile (FAX) machines. The images transmitted via FAX machines are typically black and white text and line drawings and the transmission is through low speed telephone lines. Thus, the scanners employ a low sampling rate (~ 200 *spots per inch* or 200 spi), generate digital values of only two levels (1 or 0), and are slow (~ 1 page/min). Scanners for digitizing pictures and photos for inclusion in publications, graphic arts applications, etc., need to preserve the tone scale of the original images, hence the output of the scanner usually has 256 or more

---

The views represented herein are those of the authors and are not intended to represent the position of their employer.

gray levels per color, and the sampling rate is moderate, at 300–600 spi. The speed can be relatively slow because scanning is usually an off-line activity. For Optical Character Recognition (OCR), the sampling rate and number of gray levels need to be sufficient to enable the OCR software to resolve small differences among characters: 300 spi and 256 gray levels per pixel are common. In this setting, productivity is important so the scanning process must be relatively fast. Finally, for office document applications such as copying, either local or remote, and for scanning books and pictures for archiving, the quality and speed demands are both high, so 400 spi and up and 256 gray levels per pixel are commonly used. Speed must be high (~ 10 pages/min) for productivity reasons.

Scanners used in an office environment are the primary focus of this chapter. Office scanners are of three basic types: platen scanners (Fig. 10.1), where the document remains stationary while it is scanned; Constant Velocity Transport (CVT) scanners where the document is fed one sheet at a time through a stationary scanning station; and hand held scanners that can assume a wide variety of configurations. There are also film scanners where the original image is on a photographic film and drum scanners that offer high uniformity, low noise, variable sampling rate, etc., and are used in the graphic arts industry. In order to discuss the basic principles of digitization in a document image processing setting, we focus on the first two types of scanners.

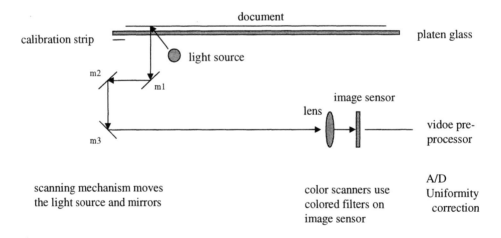

Figure 10.1 Platen scanner schematic.

Platen scanners and CVT scanners consist of several basic components:
1. *optical and mechanical systems*, including a mechanism for supporting the paper original while it is being scanned, a light source for illuminating the paper original, lenses to focus the document onto the image sensor, and a mechanism for moving the

paper document relative to the image sensor. These topics are discussed in Section 10.3.

2. *an image sensor*, for converting light into analog electrical signals. The image sensor is discussed in Section 10.4.

3. *electronics that include analog-to-digital conversion and non-uniformity correction.* These two topics are discussed in Section 10.5.

Section 10.2 provides an overview of the scanning and digitization process. The basic components of scanners are discussed in Sections 10.3 – 10.5. In Section 10.6 we discuss the factors that affect the scanned image quality. Finally, in Section 10.7, we describe a number of image processing functions, mainly to enhance the scanned image, that are often integrated with office scanners.

10.2 FUNDAMENTALS OF THE SCANNING AND DIGITIZATION PROCESS

In this section we present an overview of the scanning process, followed by a discussion of sampling and quantization.

10.2.1 Overview

A schematic illustration of the scanning process is shown in Fig. 10.2. Starting from the beginning of the sequence of events, there is a light source (not shown in the figure) that illuminates a narrow strip of the paper original document. An optical system, represented by a lens in the figure, focuses the light reflected from the original onto the image sensor. Each small area, or picture element (*pixel*), on the original along the narrow strip is focused on a corresponding light sensitive element on the image sensor. The image sensor is a linear array of light sensitive elements; each element generates an electric signal that is proportional to the amount of light collected over its spatial *aperture* and over a short period of time referred to as the *integration time*. The output from each sensor element is therefore proportional to the light reflected from its corresponding pixel on the original; it is a representation of the reflectance of that small area. In this way, one line on the original image is converted to electrical signals contained in the image sensor. The electrical signal from each sensor element is then converted to a digital value via an *Analog to Digital Converter* (ADC or A/D converter) and sent downstream for further processing. In the meantime, the electrical signal from each sensor element is cleared, the paper is advanced one line (relative to the image sensor), and the process described above is repeated for the next line on the original. Thus, line by line the reflectance information on the original is converted to digital values. The digital image is the resulting two-dimensional array of digital values, each value representing the reflectance of a pixel on the original image. The direction that is parallel to the image sensor is called the *fastscan* direction, while the direction of the paper movement is called the *slowscan* direction.

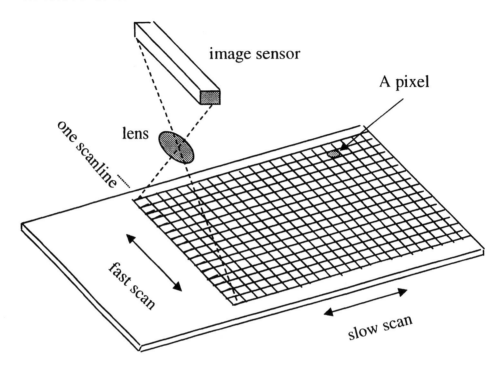

Figure 10.2 Scanning process.

The sampling rate in the fastscan direction is related to the number of elements on the image sensor and the width of the scan. Thus for a 400 spi scanner with a scan width of 12 inches, the number of active sensor elements on the image sensor is $400 \times 12 = 4800$. The optics are usually set at a fixed reduction ratio, hence the basic sampling rate of the scanner in the fastscan direction is fixed. Magnification and reduction in the fastscan direction is commonly achieved through electronic scaling, an image processing operation, rather than through optical means. The sampling rate in the slowscan direction, in contrast, is mainly controlled by the speed of scanning carriage movement for platen scanners and the speed of paper movement for CVT scanners. This speed, and consequently the sampling rate in the slowscan direction, is often variable.

10.2.2 Scanning and sampling

As a result of the scanning process, a two-dimensional image is represented by a two-dimensional array of numbers, each number being a *sample* of the image value at a discrete location. Let $C(x, y)$ be the reflectance value of the original image at position (x, y), where x and y are continuous spatial variables and C possesses continuous values. The value of the image sample at a discrete location (i, j) is denoted by $S(i, j)$. Because we represent the original continuous image with a finite number of samples, in general, it is inevitable that image detail can be lost. For the scanned image to be a faithful representation of the original image, the scanning rate must be sufficiently high, such that important details of the image are not lost in the scanning process. A key theorem in digital signal and image processing is the *Whittaker-Shannon sampling theorem*, which gives the sampling rate requirement for preventing information loss via the sampling process. For simplicity, let us consider the one-dimensional case. For a practical desktop scanner, the sampling theorem may be stated as follows. For a sampling rate of f_s and signal frequency of f, all f that satisfies

$$f < f_s/2 \qquad\qquad (10.1)$$

can be completely recovered from the samples with suitable interpolation. The variable $f_s/2$ is called the *Nyquist frequency* of the sampling system. Therefore, the "sufficiently high" sampling rate depends on the frequency of the signal that we wish to sample. Extension of this theorem to two-dimensional images is straightforward and can be found in standard textbooks on digital image processing [3]. For a scanner with sampling rate of 400 spi, its Nyquist frequency is 200 cycles/inch, or roughly 8 cycles/mm. Signals with frequency content above the Nyquist frequency will appear as, and are indistinguishable from, low frequency signals. This sampling defect is called *aliasing* and should be avoided wherever possible.

Aliasing can be avoided by limiting the high frequency content in an image or by increasing the sampling rate. Limiting the frequency content can be accomplished by applying a low pass filter to the image to blur out high frequency signals prior to sampling. It is clear that filtering should not be excessive, because in the extreme case one could get an image that appears too blurry. The image would be free of any aliasing defect but would be unacceptable. In any real scanner there are a number of places where blurring of the image naturally occurs, therefore aliasing of image information is typically reduced to some extent by the scanning process itself.

10.2.3 Quantization

To complete the image digitization process, the sensor output, which is the sampled image $S(i, j)$, is converted to digital values, with *n bits per pixel* (*bpp*), by the ADC. The process of converting a continuous, analog value to a discrete, digital value is called *quantization*. We represent the resulting quantized digital image by $D(i, j)$. The ADCs employed in scanners are typically linear, resulting in uniform quantization of the signal range. For example, with an 8 bit linear ADC, the input signal range (from the lowest value to the highest value) is divided into 256 equally spaced, discrete reflectance levels. From the standpoint of image representation or image coding, equal spacing in reflectance is not the most efficient way of using the 8 bits per pixel or 256 levels, since, from human visual system studies, we find that the levels need to be more closely spaced in the low signal (dark) region and can be spaced farther apart in high signal (light) regions. Achieving optimal non-uniform quantization is not trivial, and the usual practice is to use a linear ADC to quantize the image to more bits per pixel, say 10 bpp, then use a look up table to condense the number of levels in the light region in a manner such that there remain 256 levels, appropriately spaced. Hence, given that the human visual system can distinguish roughly 200 levels, with the levels optimally spaced, a digital image can be coded with 8 bits per pixel, which is a convenient size from the perspective of available hardware.

The required number of gray levels is somewhat image dependent. For high contrast images such as text and line art and for images with a great deal of detail, the number of gray levels can be quite small, and often two levels (black and white) are sufficient. On the other hand, for images with large areas of smooth gradations, more gray levels are required. If quantization is too coarse, say 6 bits per pixel or less, then the digital image, when it is displayed or printed, may have *false contours*. For high quality color document scanners, it is found that more than 8 bits per pixel per color is needed. A number of high quality color document scanners employ 10 bits per pixel per color.

10.3 OPTICAL AND MECHANICAL SYSTEMS

Below we describe common methods of supporting the paper original, the illumination system, optics, and the mechanical system used for scanning [4].

10.3.1 Paper original support

The original paper document must be held in place while it is scanned. For the platen scanner, the original is held between a flat piece of glass, called a platen, and a backing sheet, which is often composed of white plastic or elastomer. Sometimes low-polish aluminum is used to enhance the contrast at the document edges. The platen is normally stationary, but it may also move to provide the scanning motion. While this simplifies the imaging and illumination optics, it results in a scanner footprint that is quite a bit larger than the scanner box itself. Most systems have some means of controlling the position of the original on the platen. This often takes the form of registration guides around the imaging area.

In a CVT scanner, where a single sheet at a time is fed to the scanner, the paper support function is often performed by a backing-roller that is part of the feeding mechanism. A platen glass, if it exists, is mainly used as a paper baffle or simply as a seal to prevent dirt from entering the optics. In some cases, plastic may be used instead of glass if there is little abrasive contact with the moving sheets. Registration guides are normally part of the feeder mechanism.

10.3.2 Illumination system

A lamp is used to illuminate the original. There are two main choices for office scanners: tungsten halogen and mercury fluorescent. Light Emitting Diode (LED) light sources are sometimes used for low speed scanners.

10.3.2.1 Halogen lamp

The halogen lamp consists of a series of wound tungsten filament segments arranged in a line inside a quartz tube. It is incandescent and operates at modest color temperature (3000 K) and maintains its output fairly well until filament failure. Its main drawback is that it is inefficient; it takes a lot of power (wattage) to provide the needed exposure energy. At this color temperature, most of the radiant output is in the infrared. A corollary to the efficiency problem is that the lamp produces a large amount of heat, which is often difficult to dissipate.

Since most scanners use silicon detectors, it is usually necessary to filter out the infrared radiation; thereby, preventing its entry further into the imaging system. This longer wavelength can result in significant MTF degradation if it is allowed to fall onto the sensor elements. In color scanners, filtration is even more important because colorants on the original tend to be transparent in the infrared. A brightly colored original looks uniformly "white" at 850 mm. A further complication can result with some scanning systems if the lamp assembly is allowed to remain stationary for long periods of time. The heat from the lamp may be sufficient to heat the platen to a hazardous level or damage the original, so cooling and filtration are sometimes used.

The tungsten filaments themselves are quite small in cross-section. To increase the amount of light directed toward the original, an elliptical cylinder reflector is employed. The filament is at a focus of the ellipse and an open end of the ellipse allows the light to pass onto the original, focusing the filament onto the line that will be imaged. A further increase in efficiency can be gained by adding an opposed reflector. The irradiance at the platen surface is governed by the total angle subtended by the "image" of the filament in the reflectors(s).

As stated above, the filament is segmented into discrete sections. This provides the flexibility to make the segments different lengths. By choosing the lengths appropriately, the illumination profile on the original can be tailored to specific needs. This allows correction of end falloff due to the limited length of the lamp, and thus keeps its length as short as possible. Segmenting can also be used to correct for edge falloff in the imaging system. Segmented filaments do produce a small amount of illumination non-uniformity themselves; a method of dealing with that potential problem is described below.

10.3.2.2 Fluorescent lamp

A common alternative to a halogen lamp is the aperture fluorescent lamp, a low pressure mercury discharge lamp whose output is produced by UV excitation of visible-light-emitting phosphors. Fluorescent lamps are much more efficient than halogen lamps, particularly the aperture lamps that confine output to a narrow window, requiring far less power and cooling. System costs are usually slightly lower as well. But fluorescent lamps suffer from a serious problem – the output is strongly temperature sensitive because it depends on the vapor pressure of the mercury in the discharge. Starting the discharge in cold environments can normally be handled by power supply design, however, once started, the output rises gradually, passes through a peak at about 100°F, and begins falling until equilibrium is reached. What's worse, because the output depends on local mercury pressure, the transient is not the same along the length of the lamp, causing non-uniformity.

Methods have been developed for dealing with the transient characteristics of fluorescent lamps, ranging from closed loop thermal control, to light-feedback, to patience. In the first case, temperature is controlled on the lamp wall to force the mercury to condense in a single spot. The mercury temperature is maintained at or near optimum either using passive or active heatsinks. With light feedback, mercury is again collected in one cold spot, and the output is regulated by varying the current. Patience, of course, means waiting until the lamp reaches equilibrium, sometimes requiring that the lamp remain on all of the time. Each method has some disadvantages in terms of cost or customer satisfaction but fluorescent lamps have enjoyed reasonable success, due in part to the ability of scanners to correct for variations through image processing. Multiple-lamp fluorescent systems have been used, combined with color filters, to build a color scanner using a monochrome sensing device.

Because the fluorescent lamp is continuous along its length, its output cannot be profiled. To minimize the end falloff, the lamp is allowed to extend well beyond the edges of the platen. While this is a system design tradeoff, it usually amounts to many centimeters, forcing the frame and covers of the scanner to extend well beyond the maximum document size. In the desktop market this is a major issue. Some designers accept the end falloff in return for a small footprint, reasoning that they can correct electronically. As we see below, there is a penalty for this correction.

Within the last several years, there has been an increase in the use of fluorescent lamps based not on mercury but on xenon. These lamps have the virtue that they are nearly independent of temperature, meaning that they turn on instantly and remain stable for long periods. However, because the phosphor is excited by shorter wavelengths, the efficiency is much lower than for a mercury lamp, so the application range is limited to fairly low scan rates.

10.3.2.3 LED array

Another linear illumination technology that is commonly used for low end applications is the LED array. A series of emitting diodes are arranged in a line and lenslets are used to aim the light at a line on the original. Not only are these arrays stable but they can also be profiled by modulating current and/or duty cycle. They are limited in output energy and in the color selection that is available, though both of these are active areas of research [2].

10.3.2.4 Power supply

A non-optical part of the illumination system that has optical consequences is the power supply. Nearly all scanners use very short integration, or exposure, times (<1 msec). The AC supply used in room lighting would produce obvious banding corresponding to the 60 cycles/sec modulation of the power line. Tungsten halogen lamps are normally operated with DC supplies, though the electronic filtering of the power signal need not be carefully controlled because the thermal time constant of the filament is usually long enough to suppress some modulation. Fluorescent lamps, on the other hand, can resolve power supply variations fairly well so their power supplies are most often high frequency electronic ballasts. The frequency of operation is dependent on the actual integration time, though most are well above the 20 kHz audio threshold. The condition that must be met is that there must be many cycles of the lamp modulation within an integration time so that the ripple is averaged out.

10.3.3 Optical system

There are really two main optical architectures for document scanning in common use: the single reduction lens and the page-width lens array. The optical system for wide body scanners is an unusual case that needs special consideration. Reduction systems are used in applications having CCD array sensors while lens arrays are used with contact sensors. We discuss these two sensor groups in detail below, but, for the present, note that contact sensors are roughly the size of a page while CCDs are much smaller than a page.

10.3.3.1 Reduction system

In reduction systems, a line on the original is de-magnified (less than unit magnification) so that its image fits onto the sensor. The size of the original and size of the sensor determine the required magnification. The magnification, together with the pixel pitch of the sensor, determine the sampling rate in the fastscan direction. The choice of focal length is a compromise among the following:

        object-to-image distance – package size

        field angle – optical aberrations

        effective F/# – radiometric efficiency, optical aberrations

        lens diameter – package size.

An important characteristic of a reduction optical system is the *Depth of Focus* (DOF), that is, the sensitivity to small changes in the object distance. DOF is primarily dependent upon the angle subtended by the lens aperture from the object. Because of the long object conjugate, these systems are fairly tolerant, allowing the scanning of books, wrinkled documents, etc. The image conjugate is quite short, however, so it is necessary to control the distance to the image sensor very precisely. This requires very precise optical alignment and a robust assembly.

Optical system designs usually fall into two main categories: *single carriage* and *full-rate/half-rate* (FR/HR). As the name suggests, the single-carriage mounts all components together in an assembly. This includes the illumination system, imaging optics, and sensor, along with some electronics. The main advantage of the single-carriage design is that the assembly can be built and aligned as a unit that can be installed into the machine. This saves considerable time and alignment tooling during assembly of the machine and it allows for simple replacement in the field.

In order to package the imaging optics into a single assembly, it is necessary to fold the optical path multiple times. Because each fold reduces radiometric throughput and MTF by a constant factor, there is normally a tradeoff between the volume of the assembly and the number of folds. Most common are fairly large packages employing three folds, for example, the geometry (Fig. 10.3). Some very clever schemes have been devised for reducing volume. Figure 10.4 shows one scheme using only three reflecting components to get six folds. The first reflector is a complicated, trapezoidal mirror, while the second and third are simple plane mirrors. While this gives a very compact system, the trapezoidal mirror is difficult and expensive to manufacture.

The full-rate/half-rate system (Fig. 10.1 and 10.5) is a classic scanner from the days of analog copiers. It uses three moving mirrors, in two groups, and a stationary lens. The first mirror, mounted on the illumination assembly, folds

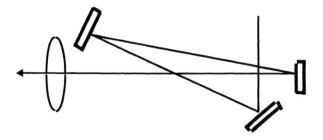

Figure 10.3 Optical path with three folds.

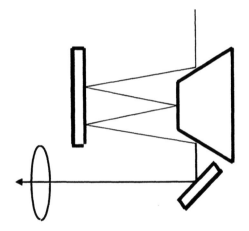

Figure 10.4 An optical assembly using only three reflecting components to get six folds.

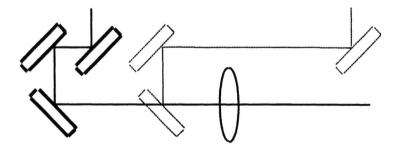

Figure 10.5 Full-rate/half-rate optical system.

the beam through $90°$. The second and third mirrors are mounted together, each folding $90°$ so as to reverse the direction of the beam. As the first mirror scans at the "full rate," the mirror cluster scans at half that rate, since the change in path is doubled during its motion. Care must be taken to allow the first mirror to pass over the lens for the most compact geometries.

The full-rate/half-rate system is a very efficient use of the space inside the box however it introduces relative motion between the optical components, increasing the likelihood of focus and alignment errors, especially as the scan proceeds. Of particular concern is the included angle in the mirror cluster. If it is not exactly $90°$, the object point moves along the surface of the original at a rate slightly different from the carriage motion, resulting in an apparent change in magnification.

10.3.3.2 Full-page width systems

Full-page width systems almost universally employ GRadient INdex (GRIN) lens arrays to form the image. These arrays are composed of a series of gradient index rod lenses (lenslets), normally about 1 mm diameter, arranged in a line. Each lenslet forms an erect, unit magnification image of a patch of the original and the images from adjacent lenslets overlap to form the composite image of a narrow line on the sensor. Because several lenslets contribute to each image point, the radiometric efficiency can be quite high, corresponding to a large effective aperture. At one-to-one conjugates, this equates to a small DOF, which is the principal drawback to lens arrays.

Another problem with arrays is the non-uniform illumination signature introduced by the overlapping of the images. Each lenslet has a distribution of image brightness. Depending on the exact position in the image, the sum of all contributions is slightly different, causing a fast-scan modulation in illumination intensity. This modulation can be controlled to some level by design but it cannot be eliminated.

In practice, there are a small number of GRIN lens types available, however, each can be fabricated to different lengths to accommodate a range of total conjugate lengths, efficiencies, etc. The different types suffer from more or less chromatic aberration so care must be taken in choosing the lens for a color system.

As an alternative to the GRIN lens array, there has been some work performed on arrays of conventional lenses (Fig. 10.6). Again, the overlapping of images requires unit, erect imaging in each sub-lens. The common arrangement for this sort of system uses a singlet objective to form an inverted image, a singlet relay lens to erect that image, and a singlet field lens between these two to bend the imaging bundles around toward the erector. The lenses can be formed in sheets by injection mold and are mounted in precise register. There has been some work on folding the lens assembly using roof mirrors to

shorten the overall length (Fig. 10.7). The mirrors in this case are also formed
by injection mold.

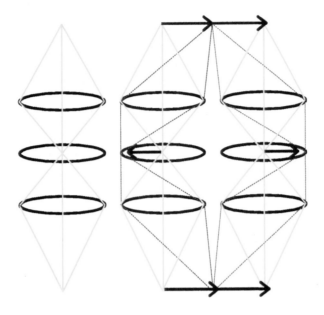

Figure 10.6 Arrays of conventional lenses.

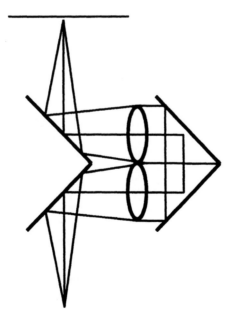

Figure 10.7 Lens assembly with folding by using roof mirrors.

10.3.3.3 Wide body systems

Some of the more interesting imaging geometries are found in designing wide body scanners for the engineering market. At even medium resolution (400 spi), a line can be composed of up to 15,000 pixels, well beyond the range of common image sensors. Often, parallel imaging systems are used to divide the line up into multiple zones. In the naïve approach (Fig. 10.8), the original is treated as a series of sheets butted together, each imaged by its own lens. While this works well in principle, the stitching together of the sub-images is dependent on the proper alignment of the individual lenses. Of course, the magnification must be precisely set for each and they must also be aligned to each other. In addition, the original must be located accurately in the plane of focus since the extreme rays at the stitch point must overlap. While all this can be achieved, it is difficult and expensive to control in a production machine.

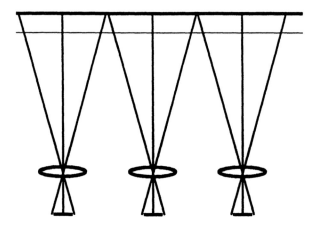

Figure 10.8 Imaging for wide-body scanners, the naïve approach.

Beam splitters have been used to divide the image formed by a single lens (Fig. 10.9), in a spatial sense or in an amplitude sense, into two image lines. Then, multiple sensors can be used to capture portions of the image. In the case of amplitude division, there are two complete images at one-half intensity so the sensors can effectively overlap each other, allowing for some electronic correction at the stitch points. With spatial division, each image is segmented by the beam splitter and can also take advantage of some image overlap. Again, both require careful alignment at the precision of the image plane pixel size, which is usually tens of micrometers. Amplitude division is radiometrically inefficient, while spatial division requires careful positioning of the beam splitter.

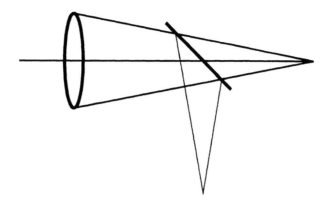

Figure 10.9 Imaging for wide-body scanners, using a beam splitter.

10.3.4 Mechanical system

The scanning operation requires relative motion between the original image and the sensor. It is usually the case that only one motion needs to be provided since linear arrays perform the "scanning" in the fastscan direction. The motion scanning mechanism is a major contributor to the overall design because of the influence on image quality, audible noise, and power consumption. Common types include lead screw and cable/pulley. Dampers at various places in the scanning mechanism are sometimes provided to reduce mechanical vibration. To increase throughput rate, scanning in some scanners is performed in both directions, so that after scanning the first page, the scanning mechanism does not need to fly back to home position to start scanning the next page. Clearly, this makes the downstream image processing a little bit more complicated.

We have already described several of the scanning options, either as part of the imaging system or the original support. Generally, though, we can achieve the required motion in one of several ways. For the *moving-original* method the obvious example is the sheet fed CVT scanner. Moving platen desktop scanners also fall into this category. The single-carriage architecture for the reduction system is an example of *moving the optics and sensor*. The full-page width system is another example. Here, all of the optics move as a group along with the sensor. It could also be thought of as a "not-moving" document, the complement to a moving document scanner.

As we have seen with the FR/HR system, the movement can be virtual. The document, lens, and sensor all remain stationary but the original is made to appear to move with respect to the others by the mirror motions. A final option is to have *no relative motion*. Both the main scan and the sub-scan are supplied by a 2D image sensor. This is not used in document scanners at present because of cost, resolution, and throughput restrictions.

10.4 IMAGE SENSOR

The image sensor is at the heart of the scanning device. By far, the most common sensor used in scanned imaging is the linear *Charge Coupled Device* (CCD) array [5]. This is a slight misrepresentation since many linear arrays are not really CCD sensors. They are photodiode array (PDA) sensors with CCD readout registers. (CCDs can be made photosensitive themselves but normally they are not photosensitive in these types of devices.) These sensors are used exclusively for single-lens systems and in the parallel architectures discussed above. For scanners using 1:1 optics, the image sensor is often a *contact sensor* or *full-width array sensor*. For color scanners, image sensors with color filters are often used. The different types of image sensors are discussed below, followed by a discussion of different methods of acquiring and representing the color information of an image.

10.4.1 CCD sensors

CCDs are integrated circuit devices fabricated in silicon using standard processing methods. Typical CCDs for document scanning are lines of photosites in the range from 2500 to 7500 elements long. Each photosite is roughly square, between 7 μm and 50 μm on a side, depending on the sensor. When exposed to light, the photosite converts radiant energy into electric charge that is stored in the photosite. At intervals of about 1 ms, the charge is transferred into an adjacent shift register and then to a serial output line (Fig. 10.10).

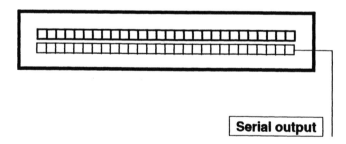

Figure 10.10 CCD image sensor, monochrome.

In use, the sensors are normally integrated into an assembly that can be used to control the optical alignment. There are a variety of schemes for doing this, including permanently bonding the sensor and lens to a common housing or mounting the sensor to an assembly that is pin registered to the lens housing. Careful thought is required in order to simultaneously control and maintain the focus, magnification, skew, and registration of the image while allowing for field service without adjustment. The lens is moved along its axis to control the magnification while the sensor is adjusted to maintain focus. The sensor vertical and horizontal positions control slow and fast scan registrations. Skew and tip (linear change in focus across the image) are removed by rotating the CCD in its plane and about a vertical line, respectively. Once each of these adjustments are made to the module, a single setting may be performed in the scanner to set the object conjugate by moving the entire assembly.

10.4.2 Contact sensors

The other main class of image sensor in common use is the *contact sensor*, sometimes called *full width array* image sensor, of which there are two principal types: composite and monolithic. In the former, CCDs are arranged on a common substrate to build up long sensors out of a series of short ones (Fig. 10.11). In some cases, the individual chips are arranged in a line, leaving a small gap between neighbors. While this is the preferred arrangement, it often produces stitching errors at the gap, though interpolation can be used to minimize the effect. This technique has been used to make 400 spi devices covering page sizes up to A3 (300 mm).

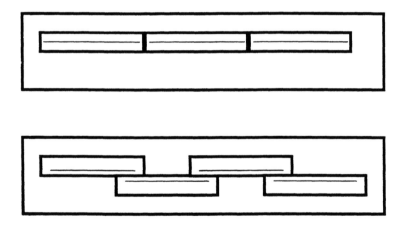

Figure 10.11 Methods of building long image sensors out of shorter pieces. Top, inline; bottom, staggered.

Another arrangement of CCDs is possible using two mirror-image chips. The sensing line is formed close to one edge so that a nearly continuous line can be formed by placing opposite-sense chips facing one another. Of course, the line is not really continuous, so buffering must be used to delay the signal from one line until the optical image has advanced to the other line. In one case, the buffers are analog shift registers built into each chip. The amount of delay can be adjusted by controlling the clocks that operate the shift registers.

Long, monolithic image sensors are fabricated in amorphous silicon, especially for low end applications such as facsimile. These devices are vacuum-deposited on a large substrate, giving a continuous array of photosites. Readout is accomplished by serially addressing each photosite, detecting the accumulated charge, and resetting. Resolution capability and sensitivity are in line with that of multi-CCD devices though there has been little movement toward the higher end.

10.4.3 Color CCD sensors

The picture gets a little more complicated when we consider color sensing. We can simply replicate the linear sensor three (or more) times on the same substrate and apply the appropriate color filters to each row (Fig. 10.12). In fact, this is what is commonly done with CCDs [6]. But there is a problem; we are now reading the image at three different places at any one time. As with the mirror-image sensor, we need to buffer the signals from the first two channels until the third color is read and then recombine the three in register. While easy to do in principle, it requires a lot of memory with a cost to achieve.

The situation becomes even more complex when one considers magnification. In many scanners, magnification in the slowscan direction is handled by varying the scan speed while keeping the line time constant. With a multi-line sensor, this means that the number of lines to be buffered changes depending on the magnification. And if the magnification is such that the required delay time is not an integral multiple of the line time, a color registration error will result. While slight, the error can lead to color fringing on the borders of image structures, particularly text and halftones.

Another method to extend image sensor usage to color is to interlace the color photosites in R-G-B order while maintaining a single line (Fig. 10.13). Since the three-color samples share the space of one pixel, the colors are slightly out of register (color fringing) and the images are each undersampled, which, we saw, could result in aliasing. These problems can be reduced by tipping the photosites so that, when scanned, they sweep out an area that is approximately rectangular and of width equal to the sample pitch. This type of color-interlaced sensor requires more care in the filter deposition since each photosite gets a unique filter.

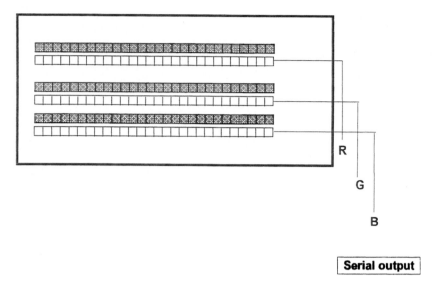

Figure 10.12 Color CCD image sensor, with R, G, B sensors on different lines.

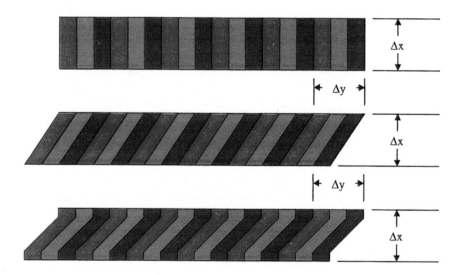

Figure 10.13 Color CCD image sensor, with R, G, B photosites in the same line, but interlaced. Top, straight photosites; middle, tilted photosites; bottom, another example of tilted photosites.

10.4.4   Color separation

Color information can be acquired in four basic temporal modes: page sequential, line sequential, pixel sequential, and simultaneous. We have already discussed the pixel sequential case with the color-interlaced device. The page sequential system is straightforward. The imaging system is used to scan the original in three successive passes, a color change being made between each pass. Color changes can be made, for example, by inserting filters into the optical path or by switching lamps. The virtue of page sequential scanning is that only one imaging path is needed. In most applications, however, two pages worth of image buffer are required to allow three-color image processing but this is only a problem when the buffer is not needed for other reasons. Calibration and correction must also be performed three times and, depending on the scanning mechanics, there can be difficulties with color-to-color registration.

Line sequential color scanning requires an altogether different schema. Here, a line of each color is captured and stored and then the scan mechanics is advanced by one line, where the capture is repeated. This requires that the carriages can stop and start at scan-line resolution so the scan speed is limited. Of course, it is possible to use a line-interlace scheme similar to that used for pixel-interlaced color sensors. Color change can be performed by changing filters but timing with the scan mechanism must be precisely controlled. Three flashing lamps can also be used for the color change, but here again timing and the lamp response time can be issues.

The ideal situation is one where we can read all three colors at each pixel at the same time. This is almost possible using the three-line color sensor discussed above but the three lines do not coincide on the original so we have to use line buffers. What is needed is a technique to bring the lines into coincidence. Color beam splitters can provide this capability with some increase in complexity. Using brute force, a three-way beam splitter can be used to break the image into three separate images that are sampled by discrete CCDs, a technique used in studio video equipment. But even for modest sensor size, prisms get unwieldy and costs increase. We must also consider the cost of three CCDs, three sets of video electronics, and a difficult optical alignment.

A much more elegant method for performing color separation uses the three-line color sensor along with a multi-layer beam splitter [7]. Here, a matched set of interference filter stacks are used to simultaneously break the images into three colored images and displace them spatially relative to each other by the amount needed.

10.5 ELECTRONICS

In this section we describe two key functions performed by the scanner electronics: A/D conversion and non-uniformity correction.

10.5.1 A/D conversion

An analog-to-digital converter is required to transform the analog voltage of each pixel into a digital representation. The basic method for performing this conversion is to compare each voltage against a number of fractional weights of a fixed reference voltage (V_{ref}) to determine the digital value. The two most common types of ADCs are successive approximation converters and flash converters. A selection of which type of converter to be used may be made on the basis of speed of conversion and cost. Typically a successive approximation converter requires only one analog comparator and takes n clock cycles to perform each conversion where n is the number of output bits of the conversion. In contrast to this, the flash converter requires 2^n-1 comparators but requires only 1 clock cycle per conversion.

The major components of a successive approximation converter are shown in Fig. 10.14. A digital-to-analog converter is used to generate the fractional values of the reference voltage. For each conversion, the successive approximation register generates a sequence of trial digital values. Based on the result of each comparison, the successive approximation register value is refined. A binary search technique is used to minimize the number of comparisons. At the start of the conversion, the value of one-half of V_{ref} is used. This corresponds to the most significant bit of the n-bit output being set to 1 with the remaining bits set to 0. If the input to the successive approximation register from the comparator is 1, this indicates that the pixel voltage V_{pixel} is greater than ½ V_{ref}. If the input is 0, it is less than ½ V_{ref}. Based on the result of this comparison, the most significant bit of the n-bit output is set to its final value and the range of possible V_{pixel} voltages is divided in half. Likewise, the cycle is repeated for the next most significant bit, n-1, of the successive approximation register. The second cycle reduces the range of possible V_{pixel} voltages to one-fourth of the total V_{ref} range. Each cycle further subdivides the range by two until all bits have been set and the conversion is complete. The advantage of the successive approximation method is cost. It requires only one analog voltage comparator to determine the digital value. This cost advantage is at the expense of speed. It takes n clock cycles to generate each digital value. It should also be noted that the V_{pixel} voltage should remain constant during the n clock cycles to ensure an accurate conversion. This places additional requirements on the analog sample-and-hold prior to the ADC device.

The major components of a flash converter are shown in Fig. 10.15. In contrast to the successive approximation converter, the flash converter utilizes 2^n-1 comparators that operate simultaneously to determine range of the input voltage V_{pixel}. The resistor divider network values are chosen to provide each comparator with a fixed fraction of the V_{ref} space, one least significant bit apart.

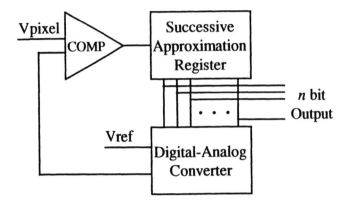

Figure 10.14 Successive approximation of ADC.

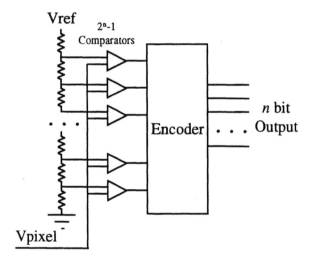

Figure 10.15 Flash ADC.

When the pixel voltage is applied, all comparator outputs whose input reference is less than V_{pixel} will be set to 1. Likewise, all comparator outputs whose input reference is greater than V_{pixel} will be set to 0. Thus, the location where the comparator outputs transition from 1 to 0 identifies the digital value corresponding to the analog voltage V_{pixel}. The function of the encoder is to find this transition and convert the 2^n-1 input bits into the final n output bits. The advantage of the flash converter is speed. It takes only one clock cycle to convert each pixel into a digital value. This speed is at the expense of additional circuitry required to operate 2^n-1 analog voltage comparators in parallel.

The process of analog-to-digital conversion is a *quantization* of the continuous range of analog voltages into a discrete set of numbers representing those voltages. Since this is a lossy conversion, the error associated is sometimes called *quantization error*. The number of bits in a linear ADC process determines the maximum *signal-to-noise ratio (SNR)*, which can be thought of as the maximum signal amplitude divided by the quantization step.

Bits	Maximum Signal-to-Noise Ratio	Dynamic Range, dB
8	256	48.2
10	1024	60.2
12	4096	72.2

10.5.2 Non-uniformity correction

In Section 10.3.2 we described the problem of illumination non-uniformity. We discussed lamp non-uniformity caused by end effects and lamp filament signature. There are also other sources of variation that we consider, including the sensitivity of the sensor, aging of the lamp, local variations in lens transmittance ($\cos^4\theta$ falloff or lenslet signature), etc. Normally, a calibration and correction are needed to remove the effects of these variations. This post-processing capability is likely the most important distinction between a design for electronic capture and one for optical copying. Once the optical image is recorded in a typical copier, for example, there is little one can do to "fix" it.

In the calibration step, we record the response of the scanner to a known input. We then calculate the *electronic gain* required at each pixel which, when applied, makes the corrected response of all pixels the same. Note that these gain values are different for each pixel, and they compensate for the non-uniformity in the fastscan direction from all sources: exposure, optics and sensor elements. These individual pixel gains, for all the pixels in a scan-line, are stored in memory. During scanning, these stored corrections are retrieved and applied to the respective pixel's live video. One example of the electronics used for pixel-to-pixel non-uniformity correction is shown in Fig. 10.16.

Since the same correction is used for all scan-lines, this correction technique does not compensate for variations of light intensity in the slowscan direction. For this purpose, additional sensor elements are provided at the end of the image sensor to detect the light intensity and to provide input to an *Automatic Gain Control* (AGC) circuit.

The quality of the calibration is largely determined by the quality of the known input. Ordinarily, this is a highly uniform target of fixed reflectance, often white. The target is permanently mounted to the platen so as to protect it

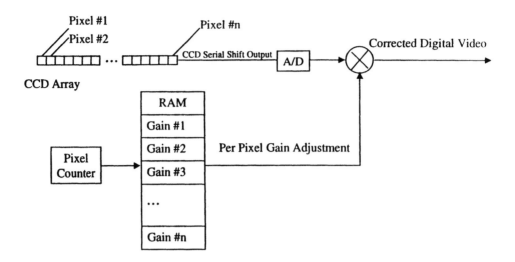

Figure 10.16 Pixel gain correction.

from dirt, scratches, exposure, etc. Should it be damaged or discolored, every scan will show an artifact that is complementary to the damage. In line scanning, even a very small defect can result in a streak across the entire image.

This calibration capability, though powerful, has two limitations. First, the applied gain, even if it were perfect, would amplify the noise along with the signal. Normally, in the shot-noise limit, the signal-to-noise ratio decreases as the signal decreases. A gain of more than about four significantly increases the image noise and related artifacts. Second, the applied gain is not perfect. It is not possible to be too specific here because of implementation differences but, since the stored values are digital, there are quantization errors and noise to take into account. Even if the gain itself is analog, analog-to-digital conversion (and vice versa) is used to measure the response in determining the gain correction, hence some quantization noise is inevitable.

In addition to gain correction, some scanners also provide "offset" correction that compensates for the non-uniformity in the dark current from different sensor elements, that is, the sensor output when there is no light. This non-uniformity is small in well-designed sensors and is sometimes overlooked.

10.6 FACTORS AFFECTING SCANNED IMAGE QUALITY

Besides sampling rate and number of bits per pixel, the imaging quality of a scanner is determined mainly by the following factors: spatial frequency response, noise level, scanning artifacts, color accuracy, and uniformity.

The spatial frequency response of a scanner determines the sharpness of the scanned image. The frequency response is often analytically described by a

Modulation Transfer Function, or MTF, and it is generally different for the fastscan and slowscan directions. Digital filters in the image processing step can be used to increase or reduce the sharpness of the image if desired. Scanner MTF is discussed more fully in Section 10.6.1.

The scanner can introduce noise into the digitized image. A major source of noise is the image sensor. In fact, an image sensor can yield a varying response to a constant level of light input. The electronic components from the image sensor to the ADC also can add noise to the scanner output signal. After the scanned image is converted to digital form, a slight amount of noise can also be introduced in image processing due to the finite precision of the various operations. Scanned image noise is discussed in Section 10.6.2.

A major source of scanning artifacts is the motion quality of the mechanical system. Motion quality issues are discussed further in Section 10.6.3.

For color scanners, the color filters employed play a key role in determining the accuracy of the output color. Color filters are considered in Section 10.6.4.

When a uniform image is scanned the output signal should be constant. To achieve this goal, a uniform light source should be used, but this alone does not guarantee uniform output because the sensors and optical system also have inherent non-uniformities. To obtain uniform output, an electronic non-uniformity correction is provided in high quality scanners. Non-uniformity correction was discussed in Section 10.5.2.

10.6.1 Scanner MTF

It is inevitable that the scanning process will blur an image to some extent. The first source of image blurring is the lens used in the optical path. Here the quality of the optics used as well as the depth of focus are important factors. For optical imaging systems there is a fundamental limit to the MTF that can be achieved, the diffraction limit. High quality optics can approach this limit quite closely.

The image sensor also blurs the image. The main source here is the finite aperture of the sensor elements. If the aperture length in the x direction is l_x, then periodic signals with frequency equal to or greater than $1/l_x$ will be severely attenuated. For a 400 spi scanner, the sensor aperture size (on the image plane) is approximately 1/400 inch in dimension, the aperture MTF has a value of about 60% at 200 cycles/in., zero at 400 cycles/in., and small values above 400 cycles/in.

The second source of blurring in the sensor is lateral diffusion of the charge generated by the input light. For light of long wavelength, the absorption in the sensor is inefficient, hence the light needs to penetrate deeper into the sensor before it is absorbed and charge generated. Hence the charge has a long diffusion length and has more chance to migrate laterally. The result is that the sensor MTF depends strongly on the light wavelength; for light in the infrared region, typical CCD sensors have poor MTFs, hence it is sometimes important to provide image scanners with IR filters to reduce this effect.

A third source of blurring in the sensor is in the CCD, which serves as an analog shift register. The *charge transport efficiency* (*CTE*) is close to, but not exactly, 100%; the inefficiency may be small, but since the number of transport stages is large in typical sensors, the residual charge left behind in each stage has a significant effect on the final result. For this reason, image sensors often have multiple CCDs for charge transport to reduce the number of stages. In the bilinear configuration, for example, there are two CCDs, one to transport the charge of even numbered pixels and one for odd numbered pixels, and the number of stages in each CCD is reduced by a factor of 2. Like the lateral diffusion effect, the CTE effect reduces MTF only in the fastscan direction.

Taking the factors discussed above, the MTF of the scanner, in the fastscan direction, is given by

$$MTF_f = MTF_o \; MTF_a \; MTF_d \; MTF_t , \tag{10.2}$$

where MTF_o is the MTF of the optics, MTF_a is the MTF due to the finite aperture of the sensor, MTF_d is the MTF due to charge diffusion in the sensor, and MTF_t is the MTF due to the charge transfer inefficiency in the CCD.

The mechanical motion of the scan carriage also contributes to image blurring. This is due to the fact that the scan carriage is in constant motion while light is being collected in the image sensors during the integration time. Such motion blur only affects the MTF in the slowscan direction. Therefore for the slowscan direction, the scanner MTF is given by

$$MTF_s = MTF_o \, MTF_a MTF_m , \tag{10.3}$$

where MTF_m is the MTF due to the motion of the scanning carriage. A typical MTF curve in the slowscan direction for a 400 spi scanner is shown in Fig. 10.17.

Because of the blurring from various sources, the aliasing problem mentioned above is reduced by the scanner itself without extra effort. To make scanned images as sharp as possible, it is preferable not to deliberately blur the image any further, and accept any residual aliasing problems.

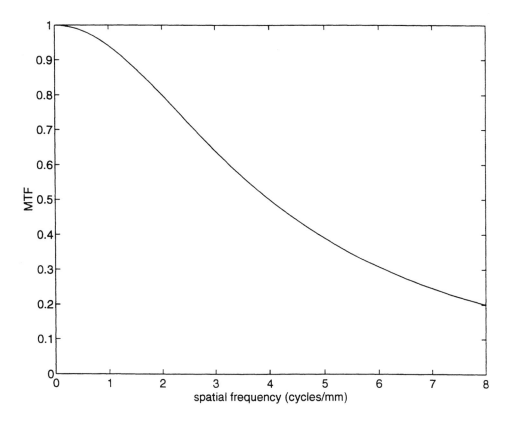

Figure 10.17 Typical MTF of a 400 spi scanner.

10.6.2   Scanned image noise

In the ideal situation, when a uniform input document is scanned, all pixels in the digital image should have the same value. In real scanners the pixel values will vary to some degree, both from place to place (spatial variation) and from time to time (temporal variation). Some of the variations are of a deterministic nature, hence can be detected and corrected. Non-uniformity correction is one such example (Section 10.5.2). Other variations are of a random nature and are referred to as *scanned image noise*.

Scanned image noise is generated from many sources, including the image sensor and other electronics components such as the ADC. From the image sensor, there are two sources of noise:

1. Photon shot noise. This is proportional to the square root of the number of electrons generated, hence it is proportional to the square root of the sensor output signal.

2. CCD and amplifier noise. This is independent of the signal level, but depends on the design and manufacturing process of the image sensor.

The sensor output signal level, as well as the level of the two types of noise, are plotted against light intensity in Fig. 10.18. Notice that the signal-to-noise ratio (S/N) increases with light intensity. Also, at low light level, the signal independent noise (from the CCD and amplifier) dominates, whereas at high light level the signal dependent noise (photon shot noise) dominates. The curve that shows the sensor output signal level as a function of light intensity is called the sensor *Photo Response Curve*. It has a linear region and a saturation region. To achieve suitable quantization steps, the sensor must be operated in the linear region.

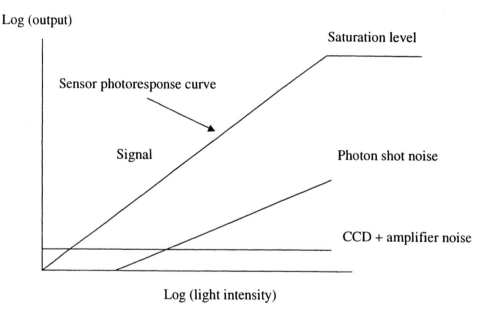

Figure 10.18 Sensor signal and noise output as a function of input light intensity.

In addition to the noise generated in the sensor, other analog electronic components can also generate noise of a magnitude comparable to the amplifier noise mentioned above. Low noise design must be employed to keep such noise to a minimum.

When the image sensor output is quantized, we can regard the resulting digital image as the original image plus a noise component:

$$D(i, j) = S(i, j) + e(i, j), \tag{10.4}$$

where $e(i, j)$ is the difference between the sampled image before quantization, $S(i, j)$, and after quantization, $D(i, j)$. For images with pixel values that are largely random, $e(i, j)$ can be treated as a random variable, and its root mean square value is given by

$$\mathrm{rms}(e) = q / \sqrt{12} \; , \tag{10.5}$$

where q is the quantization step size. As a random noise, the magnitude of this component is much smaller than a quantization step and plays an insignificant role. However, for uniform images that change slowly with position, the image pixel values are highly correlated and the effect of quantization cannot be treated as random noise. As discussed above in connection with quantization, the undesirable effect is false gray level contours and the quantization step must be small enough such that no objectionable contours occur.

After the image has been quantized and converted to digital form, additional noise can be introduced into the image if image processing is performed. For typical image processing operations, finite precision calculations are used and rounding or truncation of the intermediate or final result is inevitable. Errors or noise introduced in such calculations can in general be simulated and sufficient precision can usually be achieved with careful algorithm design. Further discussions of image noise and other aspects of image quality can be found in Ref. [8].

10.6.3 Motion quality

The scan carriage motion quality has a direct influence on the scanned image quality, and for this reason, it has received a great deal of attention [8]. Slight mechanical vibrations in the scanning process can generate quite distracting artifacts in images with periodic components, such as halftones [9]. A goal of the mechanical system design is to keep the vibration in the mechanical system to a level low enough to force all image defects to be below the visual threshold. The scan carriage speed, as a function of time, can be analyzed in terms of its frequency components. For an ideal system, there are no frequency components besides the constant speed; in reality some other frequency components will be present. The tolerable level of these components depends on their frequency as well as their amplitude. Higher amplitude can be tolerated at high frequencies and low frequencies where our visual system is not as sensitive, but in mid- frequencies (corresponding to spatial image disturbances around 1 cycle/mm) the mechanical system must meet tight tolerances in order to generate artifact-free scanned images.

10.6.4 Color accuracy

For a color scanner, the color output must be an accurate representation of the input color. In this section we describe the two most important factors that influence the accuracy of the output color: the color filter and the color calibration procedure.

10.6.4.1 Color filters

Color filters are used for separating the image into its three-color components and it is important to carefully select and optimize the spectral characteristics for each of the R, G, and B filters [10,11]. For an arbitrary color on a document, let r denote its value recorded through the R filter, g through the G filter, and b through the B filter. Ideally, a color scanner should record colors as the human eye sees them. That is, when the eye sees two colors as the same, the two recorded (r, g, b) values should be the same; and when the eye sees them as different, then the two recorded (r, g, b) values should be different. The spectral responses of such ideal color filters are well established. Unfortunately, real filters cannot generate such spectral responses; they can only approximate these ideal responses. As a result, errors are introduced in the scanned colors as represented by real device (r, g, b) values. A particular consequence of the scanner color error is that some colors that look quite different to the human eye can have very close or identical (r, g, b) values. This phenomenon is called *metamerism*. It must be pointed out that even though this problem exists in all real color scanners, current high quality color scanners can produce quite acceptable results in the vast majority of cases.

A way to improve the color accuracy of a color scanner, should it be desired, is to use more than three color filters, for example, using four-color filters. Such color scanners obviously are more costly to manufacture, and the optimal selection of the four-color filters is an interesting research topic.

10.6.4.2 Scanner color calibration

When using the same input document, different scanners will have different (r, g, b) outputs due to differences in light source and color filters, among other factors. To make the scanner output unambiguous, color scanners must be calibrated. The calibration procedure usually consists of scanning some industry standard test targets that have numerous uniform patches, each having a color value specified in a standard color space, such as the CIE Lab (L, a, b) color space. The scanner output (r, g, b) value of each patch can then be compared with its known (L, a, b) value and a transformation function from (r, g, b) to (L, a, b) constructed. Due to cost constraints, such transformation functions are often simple approximations. One such approximation uses a single 3×3 matrix for the linear transformation from *(r, g, b)* to a standard linear color space, coupled with one-dimensional look up tables to perform the non-linear transformations to *(L, a, b)*. The 3×3 matrix coefficients are determined from experimental data, by minimizing the color error over the whole color space. The color accuracy that can be achieved can be quite respectable for the simple methods, and ways of improving the color accuracy are continually being proposed. [12]

10.6.5 Other factors

The list of factors that affect scanned image quality presented above is long, but not exhaustive. An interesting addition is the so-called *integrating cavity effect*. The incident light intensity on the document at a point depends to some extent on the average reflectance of the narrow strip of the document that is being scanned. Compare the case where the narrow strip of the document is mostly white, with the case where the narrow strip is mostly dark. In the latter case, less light is scattered from areas surrounding the point in question and the illumination level there is lower than that in the former case. Consequently, a small white area in a dark surround can appear darker. This effect varies with the illumination system design; it is image dependent and is difficult to correct.

10.7 IMAGE PROCESSING

The scanned image is eventually used in some way; often it is sent to a printer to generate some form of hardcopy output. Image processing is required to make a print of a scanned image, which can be performed by software on a computer or with dedicated hardware. In this section we will describe a few common image processing functions aimed at preparing the image for printing: scaling, background compensation, edge enhancement, de-screening, and image segmentation. We do not discuss here the important topic of halftoning or other binarization techniques, since it is covered in other chapters.

10.7.1 Scaling

We often need to scale an image, that is, to change its size, in order to fit it on a printed page. Since the resolution of printers is usually fixed, say 300 spi, to change the size of an image on the print we must change the *sampling rate* of the image before it is sent to the printer. For example, if we change the sampling rate of an image from 300 spi to 600 spi, when printed on a 300 spi printer, the size of the 600 spi image will be twice that of the original 300 spi image.

The sampling rate of an image in the slowscan direction can sometimes be altered directly in the scanner. The fastscan sampling rate of a scanner is fixed, therefore the sampling rate needs to be changed after the image has been sampled. This operation is called *resampling*. Conceptually, resampling is a two step process. First reconstruct a continuous image from its samples using some form of interpolation method, then resample this reconstructed, continuous image at the desired rate.

The quality of resampling depends on the interpolation method (Fig. 10.19). The simplest kind of interpolation is the nearest-neighbor method, which approximates the continuous image with a series of steps. This method does not yield acceptable quality for typical office applications. An improved method is linear interpolation, which approximates the continuous image with straight lines connecting adjacent sample points. This is the most popular method; it delivers acceptable image quality and is still simple enough to implement in a straightforward, cost-effective manner.

For reduction of an image, the new sampling rate is lower than the original sampling rate. Here one must filter out high frequency components in the sampled image with a digital low pass filter before resampling at a lower rate; otherwise, serious aliasing defects could appear in the resampled image.

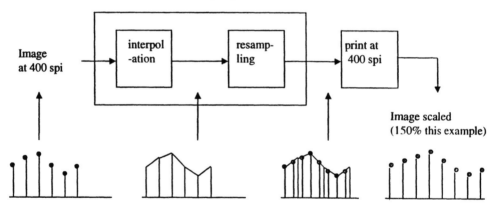

Figure 10.19 Interpolation and resampling.

10.7.2 Background compensation

If the original document has a background that is not white, for example a faded newspaper, the scanned output, when printed, will have a dirty background, and it may not be what we want. Many scanners can detect the background level of the document and make adjustments to the output such that the background on the print looks clean and the contrast is increased for the text etc. The adjustment is usually performed with a look up table that implements shift and linear stretching of the pixel values based on the detected background level (Fig. 10.20).

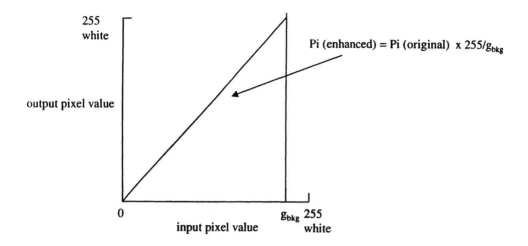

Figure 10.20 Background compensation.

To detect the background level, there are two general approaches. One approach examines a narrow strip at the top of the document and assumes that it is representative of the whole document. Safeguards are provided to handle documents that have a dark border at the top. Using this approach, the background detection and compensation can be performed in one pass. Another approach is to scan the whole page first, construct a histogram of the gray levels of the page, and extract the background level from the histogram. This approach is more robust, but requires the construction of a histogram and it is necessary to either rescan the image or store the scanned image in a buffer in order to perform background compensation.

10.7.3 Edge enhancement

We mentioned above in connection with the scanner MTF that all scanners blur images to some extent. High frequency details of images, such as fine strokes in a text document or fine lines in a drawing, are particularly susceptible to this form of degradation. If uncorrected, image details can easily degrade to unacceptable limits. This is especially important for multiple generation copies, where the "original" is itself a copy. To improve the sharpness of an image, digital filters are used and such filters are usually called *edge enhancement filters*. Note that the frequency components exceeding the Nyquist frequency ($f_s/2$) cannot be recovered, as discussed above, so edge enhancement filters attempt to increase the amplitude of frequency components below the Nyquist frequency. Since our visual system is more sensitive to frequencies in the range of 1 to 2 cycles/mm or about 25 to 50 cycles/in., which is well below the Nyquist frequency of office scanners, it is therefore possible to design enhancement filters that make an image appear sharper.

Digital filters can be applied to an image either in the frequency domain or in the spatial domain. In the first case, the image must be transformed to the frequency domain, its frequency spectrum multiplied by the frequency response of the digital filter, and the resulting spectrum transformed back to the spatial domain. Due to the large size of typical desktop scanner images, this approach is not efficient and is not used in scanners. A much simpler approach is to filter an image in the spatial domain. A discussion of this approach will take us too far afield, so we present here a heuristic discussion of this subject.

Let $D(i, j)$ be the original image, and let $D_{ave}(i, j)$ be a smoothed image where each pixel is an average of the pixels of $D(i, j)$ in a small neighborhood surrounding the position (i, j). Since in $D_{ave}(i, j)$ the image *detail* is smoothed out, the difference between $D(i, j)$ and $D_{ave}(i, j)$ represents the image detail, which is what we want to enhance. Therefore, we multiply the image detail, $D(i, j) - D_{ave}(i, j)$, by a factor k and add it back to the original image, resulting in an image, $D_{enh}(i, j)$, with enhanced details:

$$D_{enh}(i, j) = D(i, j) + k(D(i, j) - D_{ave}(i, j)) . \qquad (10.6)$$

If in the above equation we let $D(i, j)$ denote the value of a pixel at location (i, j) instead of the whole image, then this equation gives us a procedure to process the original image, pixel by pixel, to obtain the enhanced image. This operation is an example of *filtering in the spatial domain.* For this filter, we can adjust the degree of enhancement by changing the parameter k; to obtain more enhancements we use a higher value of k. The value of k needs to be optimized; enhancement that is too aggressive may result in an image that appears unnatural. Another limiting factor is image noise, which is enhanced along with image details. We can also use a different method of averaging to get enhancement in different frequency bands. Frequency domain techniques are often employed to achieve more complex filtering.

10.7.4 Descreening

Original images that are halftones, such as pictures in a magazine or newspaper, require special image processing if they are to be printed on office printers or by a digital copier. A commonly employed image processing method uses a digital filter to remove the periodic halftone pattern in the scanned image before printing; this method is called *descreening*. In order to understand the necessity of this step, we need to review briefly the electronic printing process.

Most printers, including lithography and laser printers, are binary in nature; a place either has ink or no ink in the case of lithography, and a spot is either on or off in the case of a laser printer. To represent intermediate gray levels in an image, regularly spaced dots of varying sizes are generated. If the dots are fine enough, they are not objectionable. Pictures in magazines are usually printed at around 133 dots/in., while newspapers use around 85 dots/in. Most office laser printers also generate halftone dots with a frequency in this range.

If the original image is a halftone (which contains periodic patterns) and the scanned image is sent to the printer directly, the regular dots generated by the printer often form very objectionable interference patterns, or moiré, with the periodic patterns in the scanned image. To prevent this defect from occurring, a digital filter can be used to blur the scanned image first, so as to eliminate its periodic halftone patterns, thereby generating a smooth image that can then be printed free of artifacts.

The design of the descreening filter requires special consideration. It must blur the image just enough to avoid moiré, but not too much to lose important image details. Proper design of the descreening filter is not a simple task, and frequency domain analysis is required.

There are other alternatives to handle halftone originals, each having its strong and weak points. One method is *error diffusion*, which preserves image details but requires more computation. Prints made with it tend to be noisy and show more variations from print to print and from day to day.

10.7.5 Image segmentation

From the discussion presented above, we see that text and line art images require edge enhancement, while halftone images need blurring. These two image processing treatments are incompatible. It so happens that a great deal of documents, such as pages from magazines, have both types of image on the same page. When such pages are scanned, no single treatment can yield optimal results. Of course, the scanned image can be displayed on a computer screen, and a user can manually separate the page into different pieces and process each piece differently, but this can be very cumbersome. This problem led to the development of automatic image segmentation methods [13, 14], where an image is analyzed and each pixel or group of pixels is classified into one of several image types. Each type is treated with a proper image processing method optimized for that image type, all without user intervention.

Image segmentation methods are usually rather complex, because they need to handle the multitude of cases that can arise in the real world. We illustrate the concept by considering the simple case of a black and white document with only a high frequency halftone and text [14]. In this case we only need to detect the presence of a halftone, and an area that is not halftone is text by default. A block diagram showing this method is given in Fig. 10.21.

For halftone detection, the image is divided into small blocks, and the pixels in each block are analyzed to detect the presence of periodic patterns in that block. Results from this detection step are then combined and analyzed, to include information from neighboring blocks, before a final classification is made for a given block. The image-type classification result is then used to select one of several treatments: edge enhancement and thresholding if it is text, and descreening and re-screening if the image is halftone. Since this selection is made block-by-block, it can handle complicated pages with many halftone and text areas. An improved method that can handle color halftone originals is described in Ref. [15].

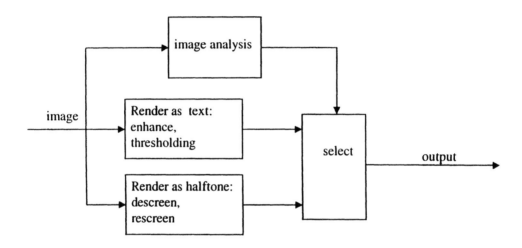

Figure 10.21 Automatic image segmentation method as discussed in Ref. [14].

10.8. CONCLUSION

In this chapter we described the main components of document scanners: illumination, optics, mechanical system, image sensor, and electronics. Each component plays a role in determining, among other things, the quality of the scanned image. We have discussed the key factors of the scanner that have a direct influence on the scanned image quality: resolution, quantization level, MTF, noise level, machine vibration, and uniformity. We can summarize the requirements on a "good" scanner as follows. Spatial resolution should be high enough for the intended application. It must preserve image detail. Quantization resolution should be fine enough to avoid false contours. The MTF of the optics should be close to the diffraction limit, and the sensor MTF should be high by possessing low levels of lateral diffusion and high transport efficiency. Adequate signal (light) level is required to achieve high S/N, and a good sensor and analog electronics design are required to reduce sign-independent noise. Vibration amplitude must result in image defects that are below visual threshold. For uniformity, a stable illumination system, high quality calibration strip, and accurate calibration procedure are required.

We described several image processing functions that are often used with scanners: scaling, background compensation, edge enhancement, descreening of halftoned originals, and image segmentation of composite images to achieve optimal printed image quality.

REFERENCES

1. H. S. Hou, *Digital Document Processing*, Wiley, New York, 1983. Chapter 2. (This comprehensive book, though somewhat dated, is still a good reference.)

2. H. C. Marz and R. L. Nielson, ed., *Cameras, Scanners, and Image Acquisition Systems, Proceedings of SPIE*, Vol. 1901, 1993.

3. W. K. Pratt, *Digital Image Processing*, Second Ed., Wiley, New York, 1991. Chapter 4.

4. J. P. Taillie, "Scanners for Document Capture," OSA Annual Meeting, Rochester, NY, October 21, 1996.

5. A. J. P. Theuwissen, *Solid State Imaging with Charge Coupled Devices*, Kluwer, Dordrecht (Netherland), 1995.

6. *CCD Linear Image Sensor Data Book*, Toshiba Corporation, 1993, pp. 170-183.

7. K. D. Vincent, "Color Imager Utilizing Novel Trichromatic Beamsplitter and Photosensor," U.S. Patent 4,709,144, 1987.

8. D. R. Lehmbeck and J. C. Urbach, "Scanned Image Quality," in *Optical Scanning*, G. F. Marshall, ed., Marcel Dekker, New York, 1991.

9. G. Wolberg and R. P. Loce, "Restoration of Images Scanned in the Presence of Vibrations," *Journal of Electronic Imaging*, 5(1), 1996, pp. 50-65.

10. J. A. C. Yule, *Principles of Color Reproduction*, Wiley, New York, 1967. Chapter 6.

11. G. Sharma and H. J. Trussell, "Figures of Merit for Color Scanners," *IEEE Trans. Image Proc.*, 6(7), 1997, pp. 990-1001.

12. G. Sharma and H. J. Trussell, "Digital Color Imaging," *IEEE Transactions on Image Processing*, 6(7), 1997, pp. 901-932.

13. J. C. Stoffel and J. F. Moreland, "A Survey of Electronic Techniques for Pictorial Reproduction," *IEEE Transactions on Communications*, 29(12), 1981, pp. 1898-1925.

14. J. C. Stoffel, "Automatic Multimode (Continuous Tone, Halftone, Linecopy) Reproduction," U.S. Patent 4,194,221, 1980.

15. J. N. Shiau and B. Farrell, "Automatic Image Segmentation Using Local Area Maximum and Minimum Image Signals," U.S. Patent 5,293,430, 1994.

11

HARDWARE ARCHITECTURE FOR IMAGE PROCESSING

Divyendu Sinha
CUNY/City College of Staten Island
New York, NY

Edward R. Dougherty
Texas A&M University
College Station, TX

In this chapter we consider computer hardware architectures that have been utilized in commercially available CPUs to speed execution. Practical image processing is computer-system dependent and achieving efficient implementation depends on an understanding of the computational environment, whether to take advantage of features within an existing system, to augment that system with additional hardware or peripherals, to network the system, or to purchase a new system more suitable to the task at hand. In real-time environments, appropriate computational hardware is essential. In addition to general issues concerning parallel processing, the chapter treats in detail pipelining for instructions and convolution, multiple-processor systems and scheduling, and two commercial image processing systems. The commercial systems have been chosen to illustrate the capabilities available in today's market while at the same time presenting two different architectures. It is assumed that the reader is familiar with the basics of sequential computing and image processing (for instance, see Ref. 1).

11.1 PARALLELISM

Sequential computing in the context of the ordinary fetch-execute, CPU-memory von Neuman model leads to a waste of time in situations where computation can be executed in parallel. Many imaging tasks cannot be accomplished for real-world applications unless potential parallelism is exploited. There are three basic ways to reduce execution time via concurrent processing: exploitation of control parallelism, program-segment parallelism, and data parallelism.

Control parallelism refers to the situation where two or more instructions can be executed concurrently on a single processor. An example is pipeline implementation of the fetch-execute cycle in an overlapped manner. To effectively achieve control parallelism, the execution order of instructions should be predictable at run-time. Owing to the presence of conditional branch statements (needed to implement loops and IF–THEN–ELSE statements), prediction is difficult. Pipelining and multiple functional units exploit control parallelism.

Program-segment parallelism refers to the situation where different segments of a program are simultaneously executed on different processors. It offers a high potential for concurrency. It is common for two or more segments to work on the same data and this leads to frequent transfer of data from one processor to another. Communication overhead is a major issue in segment selection.

Data parallelism refers to the situation where virtually the same operation is performed on multiple data by many processors. It offers the maximum potential for concurrency. The major difficulty is the synchronization overhead required to coordinate the various processors' activities. SIMD and MIMD machines exploit data parallelism.

It is important to recognize that exploitation of parallelism is not a hardware-only issue. The hardware in conjunction with the compiler can provide parallelism. Without appropriate compiler support, hardware capability will not be fully utilized. On Intel's Pentium-based personal computers, many of the Pentium's innovations for exploiting parallelism are not being utilized by current software. This is due to a single factor: lack of a compiler that can generate code to effectively utilize the Pentium's hardware.

11.1.1 Classification based on instruction and data streams

For the purpose of parallel processing, computer architectures are classified according to whether there are single or multiple instruction streams and whether there are single or multiple data streams. The standard sequential von Neumann architecture is a *single-instruction, single-data (SISD)* system. An SISD system has a single CPU processing a single instruction and a single datum at a time. Communication is synchronized between the CPU and memory. Instructions and data are stored together in memory and there is a nonconcurrent fetch-execute cycle. The system will most likely include an I/O processor; however, if the I/O processor does not possess independent computational facilities, then it does not enhance the speed of actual data processing, so the system is functionally an SISD system. The basic von Neumann architecture is the paradigm of an SISD system. Various modifications exist, for instance, use of a cache (see Section 11.1.3).

Pipelining achieves parallelism via concurrent execution of two or more instructions on a single datum. Such concurrent execution represents a *multiple-instruction, single data* (*MISD*) machine.

In a *single-instruction, multiple-data* (*SIMD*) machine, a single instruction applies to a number of different data. Typically, the architecture is based on a multitude of simple processing elements (PEs) that execute the same (single) instruction on different data. A *systolic array* is archetypical for an SIMD system. It consists of a set of interconnected cells, each capable of performing a simple operation. Data are used at the various cells and are pumped from cell to cell. Information flows in a pipelined fashion and outside communication occurs only at boundary cells. Systolic architectures are designed for real-time implementation of *compute-bound* operations. For these, the number of computations is large in comparison to the number of I/O instructions. Convolution, morphological operations, and matrix multiplication are examples of compute-bound operations.

A machine that achieves concurrency for both instructions and data is called a *multiple-instruction, multiple-data* (*MIMD*) machine. MIMD computation requires a large number of processing elements capable of executing more than a single instruction. Examples are dataflow processors and transputers. In *dataflow* machines, flow is controlled by the availability of operands: an instruction executes when all of its operands are available, either as stored data or as outputs from previously executed instructions. Flow in a sequential machine is maintained by a program counter that results in processing following a predesigned path; dataflow processing is nonsequential and the computation stream proceeds along a program graph absent of predetermined control. *Transputers* are self-sufficient, multiple-instruction-set von Neumann processors. Each instruction set includes directives to send data or receive data via ports connected to other transputers.

11.1.2 Classification based on instruction set

In a *complex instruction set computer* (*CISC*), instructions are designed to facilitate direct mapping of high-level language constructs into processor-level instructions. Designers hoped that this would lead to more efficient code generation by compilers. While this is true, experience in the last two decades has shown that only a few instructions are executed most of the time. Reference 2 suggests that about 25% of the instructions of a large instruction set are used 95% of the time. If this is true, then 75% of the hardware used in implementing instructions is severely underutilized. The frequently employed instructions are simple ones and require less processing time; the infrequently used instructions require extensive microcode support and take considerable processing time.

Based on these observations, a movement began in the 1980s to design *reduced instruction set computers* (*RISC*). RISC architectures are characterized by relatively small numbers of macro-instructions and addressing modes. Execution time is speeded by the resulting simplification of decoding and macro-instruction execution. The space freed by eliminating instructions is used to provide more general-purpose registers (GPRs) in the CPU and to enhance performance of instruction execution by pipelining the fetch-execute cycles of various instructions. To speed performance even further, most RISC machines have split cache; CISC machines usually have a single cache for data and instructions. Most RISC architectures include pipelining (and therefore exist in conjunction with parallel design) and split cache. But pipelining and split cache are not necessary characteristics of RISC architectures. For example, Intel's P6 processor has pipelining and split cache even though it is a CISC processor.

The main distinguishing characteristics of RISC architectures are:

- Fixed-length instructions, leading to simplified instruction decoding and operand fetching.
- Small instruction set, which only implements frequently used operations.
- Extremely limited addressing modes. All arithmetic and logic operations require the destination and one of the sources to be registers. The second source can be a register or a memory operand.
- Large number of GPRs (in excess of 32) to facilitate limited addressing modes.
- Less clock cycles needed per instruction (CPI). CISC architectures have CPI from 2 to 15 (depending upon instructions); RISC architectures have an average CPI less than 1.5.
- High degree of superpipelining (see Section 11.2.5).

11.1.3 Cache

Although not a parallel construct, cache is important to enhanced performance and is often used in conjunction with parallelism to achieve reduced execution times.

Since data and programs are stored in RAM, execution efficiency can be increased substantially by employing RAM with small access time. The major drawback of such an approach is the prohibitive cost of small-access-time RAM. Cache presents a compromise between these two extreme ends. *Cache* is a midrange storage facility between the CPU and main memory that has very small access time (Fig. 11.1). Register transfers between cache and CPU are not as fast as between CPU registers but are much faster than between CPU registers and memory.

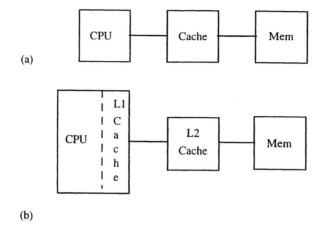

Figure 11.1 Cache: (a) single cache; (b) split-level cache.

Frequently accessed portions of program and data are copied into cache, the aim being to speed up access time for frequently accessed data and instructions. When the CPU needs to access memory, the cache is examined. If the desired location is stored in the cache (called *cache hit*), then the contents of the location can be transferred to or from the CPU extremely fast. If the location is not found in the cache (called *cache miss*), then main memory is accessed. When cache misses occur frequently, performance degrades as time is wasted searching the cache. Performance degradation can be minimized by reducing the occurrences of cache misses. Some scheme must be used to keep frequently accessed words in the cache. Two heuristics are employed: *temporal locality* (an instruction or datum referenced in the recent past will likely be referenced in the near future) and *spatial locality* (data and instructions adjacent to recently referenced data and instructions will likely be referenced in the near future). Should there be too much branching, these principles may not hold; in practice, they work quite well.

When using cache, *consistency* between main memory and cache must be maintained. When data are written to cache, the corresponding main memory location must be updated. But when? Two schemes are widely used: write-through and write-back. In *write-through cache*, if a cache location is modified, contents of the corresponding main memory location are updated immediately. In *write-back cache*, the contents of the corresponding main memory location are updated only when the modified cache location is to be removed from cache (to make room for new data or an instruction). Since a location is modified numerous times during execution, write-back cache provides better performance. On the downside, write-back cache is more expensive because of the complex logic needed to implement the delayed updating of main memory.

On most recent machines, cache is built into the CPU itself. Such a cache is often called *L1 cache, level 1 cache*, or *primary cache*. When additional cache is included [Fig. 11.1(b)], then this cache is known as *L2 cache, level 2 cache*, or *secondary cache*. Access time for primary cache is extremely small (comparable to that of registers). L1 cache is usually 16K bytes and L2 cache size ranges from 128K to 512K bytes. The size of L1 cache affects the performance much more than the size of L2 cache.

Most CPUs have a single primary cache shared between data and instructions. This arrangement has the drawback that the cache may be filled with data (in the case of a short loop that operates on many data) or instructions (in the case of a long loop). To avoid this situation, some processors have separate primary caches for data (*DCache*) and instructions (*ICache*). This arrangement has the added advantage that ICache contents need not be saved back into memory.

11.2 PIPELINED PROCESSORS

A common technique to increase processor performance is to employ pipelines. Most current CPUs have multiple pipelines to speed up different tasks. After discussing general design philosophy, we study the use of pipelining in the instruction fetch and execute cycle and in the ALU. We also discuss design of pipeline processors to perform convolution and morphological gradient. Finally, we mention what is meant by superpipelining.

11.2.1 General principles

In a pipeline, several processors are arranged in stages (*segments*) like an assembly line (Fig. 11.2). To accomplish an operation, micro-operations (μOPs) are concurrently performed in the various segments. The principle is straightforward: rather than use a given number of pulses to perform an operation that must be performed repeatedly, arrange the μOPs so that (after the pipe is full) one full operation is completed at each pulse. At each stage a distinctive μOP is performed and the final result is obtained after all stages have been visited. Concurrent computation is facilitated by associating a register (called a *latch*) with each segment in the pipeline, so that, relative to data, segments are isolated and can concurrently process different data. Pipelining necessitates decomposition of macro-instructions into sequences of more primitive μOPs. These primitive μOPs are employed in a concurrent manner by overlapping steps in the instruction cycle.

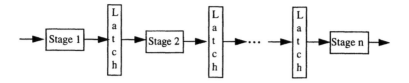

Figure 11.2 An n-stage pipeline.

The stages are designed so that data spend approximately an equal amount of time at each stage. The maximum time spent at any stage is known as the (*pipeline*) *clock cycle*. A pipeline with n stages can process its task on m data in $n + m - 1$ clock cycles. The number of stages and the purpose of each stage varies with processors. A few general design principles are available for pipelines. To process m data points, the *speedup factor* of an n-stage pipeline with respect to a nonpipelined processor is given by

$$S_n(m) = \frac{n}{1 + \dfrac{n-1}{m}} \qquad (11.1)$$

The speedup factor approaches the maximum value n only as $m \to \infty$. For maximum performance, a pipeline must be continuously supplied with data. This is a difficult task and considerable amount of chip area is usually devoted to the logic concerning this aspect of pipelines.

For a given data stream, the higher the number of stages, the greater the speedup. For example, operating on 32 data, a twelve-stage pipeline has speedup factor 8.93, whereas a seven-stage pipeline has speedup factor 5.89. High-performance CPUs have large numbers of stages. For example, to improve upon the performance of the Pentium (P5) chip, one of the "enhancements" Intel introduced in its Pentium PRO (P6) chip is a twelve-stage instruction pipeline (P5 has a five-stage pipeline). On the downside, the cost of implementation and complexity of controlling different stages increases with the increase in number of stages. The graph of performance/cost ratio versus number of stages is almost bell-shaped. Depending on the manufacturing technology used, one can pick an optimal value for the number of stages from this graph.

11.2.2 Instruction pipelining

Recent processors speed up the fetch-execute cycle by overlaying the execution of one instruction with the fetching of the next instruction. Consider the five-stage fetch-execute-store cycle shown in Table 11.1. Each of the five stages is a μOP. Five pipeline segments can perform these five μOPs for different operations in parallel. Control circuitry for each stage works in parallel and the outcome of stage i is passed as input to stage $i + 1$. Table 11.2 provides a pulse-by-pulse timing diagram in which sequence overlap is explicit. From the fifth pulse on, a computed result is stored in memory during each pulse. The sequence can be made more complicated for move involved architectures, such as those having a cache or CPU registers for temporary storage of repeatedly used operands.

Stage	Description
F	Fetch instruction from memory.
D	Decode instruction.
L	Access memory and load datum.
E	Execute instruction.
S	Store result in memory.

Table 11. 1 Five-stage fetch-execute cycle.

Pulse →	1	2	3	4	5	6	7
Stage 1	F(1)	F(2)	F(3)	F(4)	F(5)	F(6)	F(7)
Stage 2		D(1)	D(2)	D(3)	D(4)	D(5)	D(6)
Stage 3			L(1)	L(2)	L(3)	L(4)	L(5)
Stage 4				E(1)	E(2)	E(3)	E(4)
Stage 5					S(1)	S(2)	S(3)

Table 11.2 Timing diagram for a five-stage fetch-execute-store cycle.

Table 11.2 depicts the ideal situation; in a less-than-ideal situation, data and resource dependencies and conditional branch instructions may cause the pipeline to *stall*. A compiler written for a pipelined processor can avoid stalling by scheduling instructions in an appropriate order while still maintaining the integrity of computations. The aim is to keep the data flowing through the pipeline. This technique, though very powerful, is extremely difficult to implement for large programs. Most compilers implement this technique on pieces of code.

Pipelined machines fetch the next instruction before complete execution of the previous instruction. *Instruction prefetching* is based on the spatial locality principle: since instructions are executed sequentially, it is highly likely that the instruction to be executed next is stored immediately following the current instruction in memory. While the current instruction is being executed, cache is searched to check if the next instruction is there. If not, the instruction is fetched from memory and stored in cache.

Should the previous instruction have been a conditional or unconditional branch, then the next instruction fetch could come from an unexpected location. The moment an unconditional branch is decoded, the pipeline must be flushed of prefetched instructions and the new instructions should be fetched. In case of conditional branches, the task is a little harder. Only after the statement executes can we be sure whether the pipeline needs to be flushed. *Branch prediction* is a technique that boosts performance by guessing whether a branch will be taken or not. Branch prediction attempts to infer the proper next instruction address, knowing only the current one based on branch history during program execution. Most computers employ an associative memory, called *branch target buffer* (*BTB*), to keep track of statistics on different branches. After the branch instruction has been executed, we can find out if the prediction was correct. If the prediction was correct, then the flow through the pipeline continues, otherwise instructions must be flushed from the pipeline and fetching restarted from the correct target address.

11.2.3 Case study: convolution

Perhaps the most used digital imaging operation is *windowed convolution*. It is used for many tasks, including noise suppression and edge detection, and is defined via a mask g of numerical weights over some finite window. The window is translated across a digital image pixel by pixel and at each pixel the arithmetic sum of products between the mask weights and the corresponding image pixels in the translated window is taken. For image $f = f(i, j)$ and $(2m + 1) \times (2m + 1)$ mask $g = g(i, j)$, $i, j = -m, -m + 1, ..., m$, where we use an odd-sized square window, the convolution image $h = g * f$ is defined at pixel (i, j) by

$$h(i, j) = \sum_{r=-m}^{m} \sum_{s=-m}^{m} g(r, s) f(i + r, j + s).$$ (11.2)

Convolution is a linear operator and is spatially invariant because it operates in the same manner at each pixel. In principle, an image is defined over the entire Cartesian grid and $h(i, j)$ is well-defined since the double sum can be computed for all window positions over the image; in practice, the image exists within a finite frame and the definition of $h(i, j)$ needs to be adjusted when the translated window extends outside the frame. We take the straightforward approach of defining the image to be zero outside its frame.

When implemented on a SISD machine, convolution is computationally intensive because of the many memory transfers of data and mask weights. Owing to its central role in image processing, extensive effort has gone into developing efficient implementations on various computer architectures. Computationally, it is often used as a benchmark for measuring the efficiency of a particular computer design relative to image processing. We illustrate the manner in which pipelining can speed it up.

Consider 3-point convolution of a digital signal x_1, x_2, x_3,... by the mask with weights a_1, a_2, and a_3. A suitable pipeline architecture is shown in Fig. 11.3, where MULT(a_i) means multiply the input from the register by a_i. At each clock pulse, three successive signal values are read into registers R0, R1, and R2. From the fourth clock pulse onwards, the desired convolution value,

$$c_k = a_1 x_k + a_2 x_{k+1} + a_3 x_{k+2} ,$$
(11.3)

is in register R8 and is sent onto the bus at the next pulse. Table 11.3 shows the contents of each register after each clock pulse. If produced in hardware on a chip, the pipeline of Fig. 11.3 becomes a special-purpose processor designed to accomplish a given task.

Pulse →	1	2	3	4	5	6
R0	x_1	x_2	x_3	x_4	x_5	x_6
R1	x_2	x_3	x_4	x_5	x_6	x_7
R2	x_3	x_4	x_5	x_6	x_7	x_8
R3		x_3	x_4	x_5	x_6	x_7
R4		$a_1 x_1$	$a_1 x_2$	$a_1 x_3$	$a_1 x_4$	$a_1 x_5$
R5		$a_2 x_2$	$a_2 x_3$	$a_2 x_4$	$a_2 x_5$	$a_2 x_6$
R6			$a_1 x_1 + a_2 x_2$	$a_1 x_2 + a_2 x_3$	$a_1 x_3 + a_2 x_4$	$a_1 x_4 + a_2 x_5$
R7			$a_3 x_3$	$a_3 x_4$	$a_3 x_5$	$a_3 x_6$
R8				c_1	c_2	c_3

Table 11.3 Pulse-wise register contents for convolution pipeline.

For real-time imaging many dedicated processors are hardwired. For instance, if a general-purpose computer requires a convolution computation, the operation can be performed on dedicated hardware and not by the CPUs of the main computer. Indeed, entire algorithms, including the discrete cosine transform, convolution, and morphological operations, are performed on special-purpose parallel-designed chips. There may be constraints dictated by the hardware. For instance, some processors limit a morphological structuring element to be 7×7. Performance measures for coprocessors can be misleading because, although the convolution may be performed very efficiently by a coprocessor, overall performance depends on main-CPU performance for non-coprocessor operations.

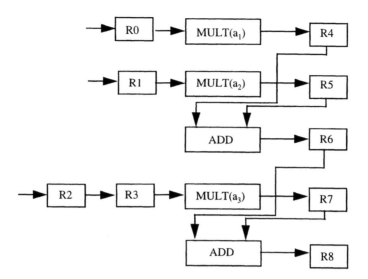

Figure 11.3 Convolution pipeline.

11.2.4 Case study: morphological gradient

One nonlinear approach to edge detection uses the *morphological gradient*, which is defined in terms of moving maximum and moving minimum. Given a window W, the moving maximum (called *flat dilation*) is found at a pixel by translating W to the pixel and then taking the maximum value in the translated window; the moving minimum (called *flat erosion*) is determined analogously. The mophological gradient at a pixel is found by subtracting flat erosion from flat dilation. In slowly varying regions, the difference between the maximum and minimum will be small; in regions of fast changing gray values, the difference will be great. Hence, an edge-detection procedure results from thresholding the morphological gradient.   Figure 11.4 shows a gray-scale chromosome image, its morphological gradient, and a binary edge image resulting from a judiciously chosen threshold applied to the gradient image.

As an example of a special-purpose pipeline, consider the three-point morphological gradient of a digital signal x_1, x_2, x_3, ... A suitable pipeline is shown in Fig. 11.5, where MAX, MIN, and SUB denote processing elements (PEs) capable of maximum, minimum, and subtraction, respectively. From the fourth clock pulse onwards, the flat dilation and flat erosion are given by

$$d_k = x_k \vee x_{k+1} \vee x_{k+2},$$

$$(11.4)$$

$$e_k = x_k \wedge x_{k+1} \wedge x_{k+2} \tag{11.5}$$

and stored in registers R6 and R7, respectively. From the fifth clock pulse onwards, the desired morphological gradient, $g_k = d_k - e_k$, is in register R8 and is sent onto the bus at the next clock pulse. Table 11.4 shows the register contents subsequent to each pulse.

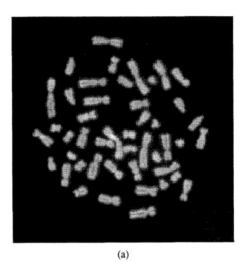

(a)

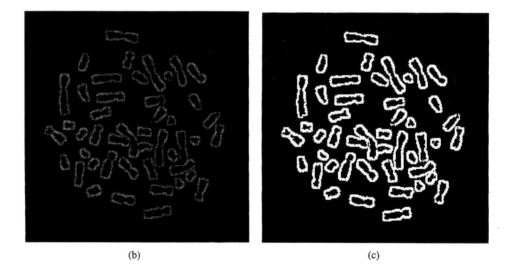

(b)   (c)

Figure 11.4 Morphological gradient: (a) original image; (b) gradient; (c) edge.

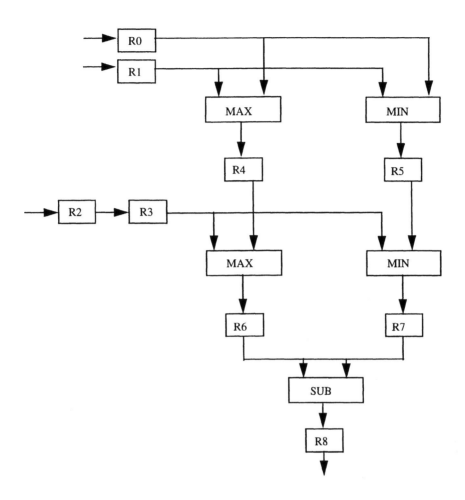

Figure 11.5 Morphological gradient pipeline.

Pulse →	1	2	3	4	5	6
R0	x_1	x_2	x_3	x_4	x_5	x_6
R1	x_2	x_3	x_4	x_5	x_6	x_7
R2	x_3	x_4	x_5	x_6	x_7	x_8
R3		x_3	x_4	x_5	x_6	x_7
R4		$x_1 \vee x_2$	$x_2 \vee x_3$	$x_3 \vee x_4$	$x_4 \vee x_5$	$x_5 \vee x_6$
R5		$x_1 \wedge x_2$	$x_2 \wedge x_3$	$x_3 \wedge x_4$	$x_4 \wedge x_5$	$x_5 \wedge x_6$
R6			d_1	d_2	d_3	d_4
R7			e_1	e_2	e_3	e_4
R8				g_1	g_2	g_3

Table 11.4 Pulse-wise register contents for morphological gradient pipeline.

11.2.5 Superpipelining

Superpipelining, a technique previously used primarily in supercomputers, is becoming a common feature in ordinary processors. Intel's P6 and DEC's Alpha chips have brought this technology to the PC. In superpipelining, each of the n stages is further subdivided into k substages with each substage taking $1/k$ of the pipelining clock cycle. Instructions enter the pipeline every $1/k$ of the pipelining clock cycle. The speedup factor, which is superior to the factor for a normal pipeline, is given by

$$S_n(m) = \frac{n + \dfrac{nk}{m-1}}{1 + \dfrac{nk}{m-1}} \ .$$
(11.6)

For $k = 3$ and $m = 32$, a twelve-stage superpipeline has a speedup factor of 6.09 over a normal twelve-stage pipeline. Thus, it has a speedup factor of 6.09×8.93 = 54.38 over nonpipelined implementation. Superpipelining achieves performance enhancement by employing less stages but splitting the stages into substages. Consequently, superpipelining requires extremely fast circuitry and clock cycles.

11.3 MULTIPLE PROCESSORS

Efficient execution can be achieved by concurrent processing of instructions on multiple processors. The basic structure of a multiple-CPU computer is illustrated in Fig. 11.6. While it might first appear that two CPUs will produce a 50% savings in run-time, such a savings is far from the case. Programming a multiple-CPU computer requires special care so as to make optimum use of the parallel processors and full optimality is impossible except in trivial circumstances. For instance, with two CPUs there are two instruction queues awaiting service and these queues are not operationally independent. Moreover, memory access is still a problem even if each CPU possesses its own set of registers.

To speed memory transfers, one can employ multiple memory modules, thereby allowing concurrent accesses to memory. Such a configuration is illustrated in Fig. 11.7. Because each CPU must be able to access each memory module, a complicated switching apparatus is necessary. Once again, the full benefit of concurrent computation is mitigated by the overhead necessary to operate the more complex system. While Fig. 11.7 illustrates multiple CPUs, the design principle applies to computer networks, where message passing takes place between local memories and CPUs in order to take advantage of the concurrent power of the multiple units.

Combining multiple CPUs, multiple memory modules, and caches yields the design of Fig. 11.8. Each CPU has its own cache, which is itself capable of communicating to all memory modules. While such an architecture appears promising, there are many practical (and difficult) problems associated with it. Communication must be maintained between the processors. Should one processor change the contents of a register in its cache and this change be communicated back to memory, it might occur that another processor operates on stale data prior to the change being communicated to its cache. This *see-through* dilemma is only one of many. Communication issues pertaining to multiple-CPU/multiple-memory architectures are similar to communication issues for networks.

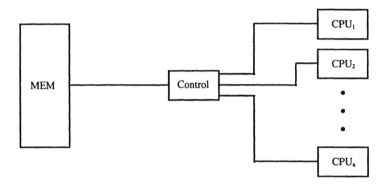

Figure 11.6 Shared memory multiple processor system.

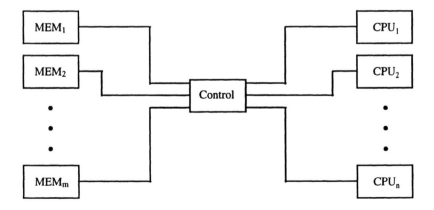

Figure 11.7 Message-passing multiple processor architecture.

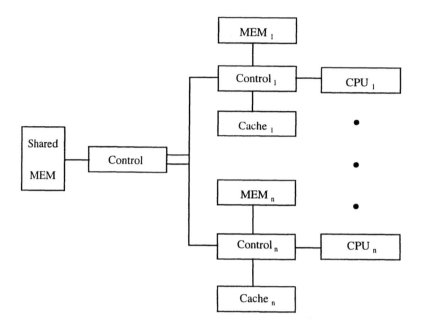

Figure 11.8 A general architecture for multiple processor system.

11.3.1 Program partitioning and scheduling

Sometimes a program can be partitioned into segments that can be executed in parallel and these segments can be appropriately scheduled on a multiprocessor system. The aim of scheduling is to execute the program in the shortest possible time. Optimal scheduling has been shown to be extremely difficult.

A program can be decomposed into one or more segments, with instructions within a segment being executed sequentially. We assume that different segments, if independent, can be concurrently executed on different processors. The number of instructions in a segment is called its *grain size*. A more useful way to measure grain size is the number of CPU cycles needed to execute the segment sequentially. *Fine grain* scheduling involves exploiting instruction-level parallelism, *medium grain* scheduling involves exploiting segment-level parallelism, and *coarse grain* scheduling involves exploiting task-level parallelism. As already stated, a multiprocessor architecture requires communication overhead so that data can be passed from one computer to another and processors can synchronize their operations. *Latency* is a measure of the time taken by a processor to access memory and to synchronize with other processors. The choice between fine, medium, or coarse grain scheduling is dictated by the latency. In general, multiprocessor scheduling starts with fine grain detail of the tasks and a possible schedule. Operations are then collected together to form segments with the aim of reducing latency.

11.3.2 Case study: convolution

The convolution of image A by a 3×3 mask B involves computing sums of the form

$$a_{i-1,j-1} \times b_{1,1} + a_{i-1,j} \times b_{1,2} + a_{i-1,j+1} \times b_{1,3}$$

$$+ a_{i,j-1} \times b_{2,1} + a_{i,j} \times b_{2,2} + a_{i,j+1} \times b_{2,3} \qquad (11.7)$$

$$+ a_{i+1,j-1} \times b_{3,1} + a_{i+1,j} \times b_{3,2} + a_{i+1,j+1} \times b_{3,3} \, .$$

Suppose it takes 100 clock cycles to load two operands, multiply them, and store the result in a register, and suppose addition of two registers takes 10 clock cycles. Then a straightforward sequential schedule will result in $9 \times 100 + 8 \times 10 = 980$ clock cycles.

The computation can be carried out as outlined by the tree structure shown in Fig. 11.9. This tree implements the following sequence of evaluations:

$$P = a_{i-1,j-1} \times b_{1,1} + a_{i-1,j} \times b_{1,2},$$

$$Q = a_{i-1,j+1} \times b_{1,3} + a_{i,j-1} \times b_{2,1},$$

$$R = a_{i,j} \times b_{2,2} + a_{i,j+1} \times b_{2,3},$$

$$S = a_{i+1,j-1} \times b_{3,1} + a_{i+1,j} \times b_{3,2}, \qquad (11.8)$$

$$T = P + Q,$$

$$U = R + S,$$

$$V = U + T,$$

$$\text{Answer} = a_{i+1,j+1} \times b_{3,3} + V.$$

The number of multiplications and additions have not changed. However, fine grain scheduling is straightforward on a nine-processor system (Table 11.5). We have assumed latency of 200 clock cycles. In the last step, note that the result will be transmitted from processor P_9 to P_1 anytime between time 101 and time 721. The final result arrives after 740 clock cycles. We have therefore been able to save 240 clock cycles.

The preceding schedule involves a tremendous waste of the nine-processor system. Processor P_1 has been utilized fully (ignoring latency), P_4 has been utilized only half the time, and the remaining processors are very much underutilized. Even P_1 spends 81% of its time in latency. A better scheme must be found so that (1) processors do almost equal amounts of work and (2) processors spend less time in latency.

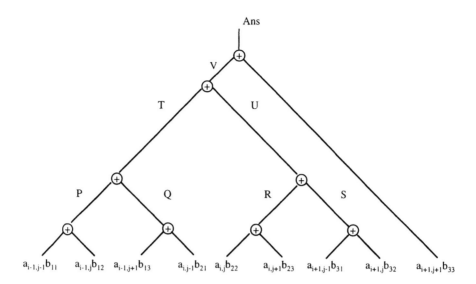

Figure 11.9 Parallel convolution computation at pixel (i, j) by a 3×3 mask.

Time	P_1	P_2	P_3	P_4	P_5	P_6	P_7	P_8	P_9
1-100	$a_{i-1,j-1} \times b_{1,1}$	$a_{i-1,j} \times b_{1,2}$	$a_{i-1,j+1} \times b_{1,3}$	$a_{i,j-1} \times b_{2,1}$	$a_{i,j} \times b_{2,2}$	$a_{i,j+1} \times b_{2,3}$	$a_{i+1,j-1} \times b_{3,1}$	$a_{i+1,j} \times b_{3,2}$	$a_{i+1,j+1} \times b_{3,3}$
101-300	delay		delay		delay		delay		
301-310	P		Q		R		S		
311-510	delay				delay				
511-520	T				U				
521-720	delay								
721-730	V								
731-740	Answer								

Table 11.5 Fine grain schedule of convolution on nine-processor system. Empty cells denote that the corresponding processor is idle. Delay refers to latency.

With *grain packing*, latency and the number of processors are reduced by performing a group of operations on a single processor before exchanging messages. For example, we can carry out all computations for P (two multiplications and one addition) on processor P_1, all computations for Q on processor P_2, all computations for R on processor P_3, and all computations for S on processor P_4. Subsequent computations will be done on these four processors by exchanging data. Table 11.6 shows the timing diagram. The last two steps of the computation are rearranged to further reduce the time:

$$V' = a_{i+1,j+1} \times b_{3,3} + T,$$

$$\text{Answer} = U + V'. \tag{11.9}$$

The total time taken is 630 clock cycles, a reduction of about 33% from the single-processor implementation and about 15% from the nine-processor implementation. Reducing the number of processors has given rise to a better schedule. All processors are utilized at least 33% of the time and P_1 is utilized more than 67% of the time.

Time	P_1	P_2	P_3	P_4
1-100	$a_{i-1,j-1} \times b_{1,1}$	$a_{i-1,j+1} \times b_{1,3}$	$a_{i,j} \times b_{2,2}$	$a_{i+1,j-1} \times b_{3,1}$
101-200	$a_{i-1,j} \times b_{1,2}$	$a_{i,j-1} \times b_{2,1}$	$a_{i,j+1} \times b_{2,3}$	$a_{i+1,j} \times b_{3,2}$
201-210	P	Q	R	S
211-310	delay	$a_{i+1,j+1} \times b_{3,3}$	delay	
311-410	delay		delay	
411-420	T		U	
421-430	V'			
431-620	delay			
621-630	Answer			

Table 11.6 Fine grain schedule of convolution on a four-processor system. Empty cells denote that the corresponding processor is idle. Delay refers to latency.

One can go too far when packing more sequential executions in each processor. Table 11.7 shows the timing for a two-processor implementation. It requires 850 clock cycles. The reduction in latency has been more than offset by the time required for sequential execution. It results in better processor utilization but to the user there is not much gain. The salient point is that the amount of grain packing depends on the grain size of instructions to be packed vis-à-vis latency.

Time	P_1	P_2
1-100	$a_{i-1,j-1} \times b_{1,1}$	$a_{i,j} \times b_{2,2}$
101-200	$a_{i-1,j} \times b_{1,2}$	$a_{i,j+1} \times b_{2,3}$
201-210	P	R
211-310	$a_{i-1,j+1} \times b_{1,3}$	$a_{i+1,j-1} \times b_{3,1}$
311-410	$a_{i,j-1} \times b_{2,1}$	$a_{i+1,j} \times b_{3,2}$
411-420	Q	S
421-430	T	U
431-530	delay	$a_{i+1,j+1} \times b_{3,3}$
531-630	delay	
631-640	V	
641-840	delay	
841-850	Answer	

Table 11.7 Fine grain schedule of convolution on a two-processor system. Empty cells denote that the corresponding processor is idle. Delay refers to latency.

11.4 MASSIVELY PARALLEL ARCHITECTURES

At the high end of real-time imaging systems one often encounters architectures that achieve an extremely high degree of parallelism by employing numerous processors. The number of processors is usually proportional to the width or height (or both) of the image being processed. Different architectures result from different interconnection topologies employed to connect the processors. We discuss three fine-grained SIMD architectures: linear array, mesh, and hypercube.

According to Ref. 3, an operational model of an SIMD computer (Fig. 11.10) is specified by five quantities:

(1) The number of processing elements (PEs).

(2) The set of instructions directly executed by the control unit, including scalar and program flow instructions.

(3) The set of instructions broadcast by the control unit to all PEs for parallel execution.

(4) The set of masking schemes, where each mask enables a subset of PEs to execute the instruction being broadcast.

(5) The set of data routing functions, specifying various inter-PE communication patterns in the interconnection network.

Item 3 determines the complexity of each PE; items 2, 4, and 5 determine the complexity of the control unit. The complexity of item 5 is directly dependent on the topology of the interconnection network. Depending on the number of PEs and the interconnection network topology, one can achieve varying degrees of performance enhancement over the single-processor model.

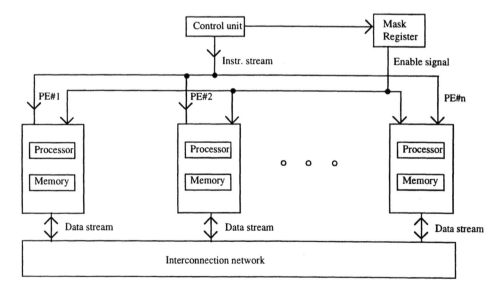

Figure 11.10 An operational model of a SIMD computer.

The simplest interconnection scheme involves the *linear array architecture*. The PEs are arranged in a linear array with each PE connected to its immediate neighbors, east and west (Fig. 11.11). When applied to image processing, there may be one PE per column in the image with each PE being capable of storing an entire column of image data from several images. The image data are sent row-by-row to the PEs and the computed results are stored row-by-row in local memory. Neighborhood information is exchanged between adjacent PEs. This architecture is well-suited for applications requiring processing of $n \times 3$ neighborhoods. The communication overhead is directly related to n.

Broadcast instruction

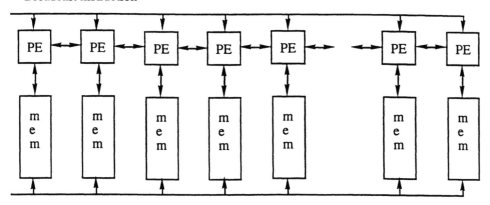

Broadcast address of a row

Figure 11.11 Linear array architecture.

Some applications demand a more complicated exchange of information between PEs. Generalization of a linear array to two dimensions yields the *mesh* interconnection scheme [Fig. 11.12(a)]. Each PE is connected to its neighbors in the east, west, north, and south. A d-dimensional mesh has n^d nodes for some integer n. The degree of each interior node is $2d$ and the degree for those at the boundary and corners is between d and $2d - 1$. To make the degrees of all nodes uniform, additional wiring can be added to obtain an *Illiac mesh* [Fig. 11.12(b)] or a *torus* [Fig. 11.12(c)]. The extra wiring helps reduce communication delays. The maximum communication delay between any two nodes in a torus or an Illiac mesh is half of that in a pure mesh.

The *hypercube* interconnection scheme [Fig. 11.12(d)] allows data communications along a number of other higher dimensional pathways. Some experimental machines (such as the Thinking Machine Corporation's CM-2) have combined the mesh and hypercube interconnection schemes by having each PE in the hypercube be a mesh of simpler PEs. As interconnection schemes get complicated, more complex controllers are needed to update the instruction stream at high speed and to load the image into the local memories.

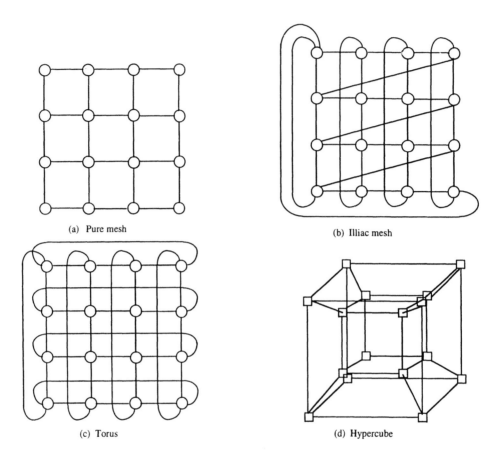

(a) Pure mesh

(b) Illiac mesh

(c) Torus

(d) Hypercube

Figure 11.12 Other commonly employed interconnection schemes in a SIMD computer: (a) mesh, (b) Illiac mesh, (c) torus, and (d) hypercube.

Selection of an interconnection network is dependent on the class of algorithms under consideration, communication delays, and network bandwidth. Some interconnection models (such as hypercube) may be suitable for small problem sizes but impractical for larger problems. As an example, consider adding N numbers on both linear array and mesh architectures. It can be shown that the times taken for the linear array and mesh models are $O(N^{1/2})$ and $O(N^{1/3})$, respectively. Consequently, ignoring communication delays, the mesh architecture is more suitable for addition. It can also be shown that computation times on 3-D mesh and hypercube schemes are $O(N^{1/4})$ and $O(\log_2 N)$, respectively. Even if we consider communication times, the mesh architecture is better than the linear array because there are more interconnections between PEs in the mesh architecture, thereby increasing data locality. The calculations to arrive at the four preceding computation times show the complexity of designing optimal algorithms that take into account the minimal number of processors and communication overhead. This is a primary reason why efficient compilers for SIMD architectures are difficult to design and expensive.

11.5 PERFORMANCE LAWS

When considering parallel architectures one wonders if adding more processors is going to "help." The answer depends on how we define "help." Two common viewpoints in measuring the performance of multiprocessor architectures are: *fixed-load* and *fixed-time*.

The fixed-load model assumes that the problem size is fixed regardless of the number of processors being used. As the number of processors increases, the workload is distributed to more processors for parallel execution. The aim is to perform the computation in the shortest possible time. Letting α be the percentage of code that is sequential in nature, *Amdahl's Law* says that the speedup factor is given by

$$S(n) = \frac{\text{time taken with 1 processor}}{\text{time taken with } n \text{ processors}} = \frac{n}{1 + (n-1)\alpha} \; . \tag{11.10}$$

As n increases, the workload of individual processors decreases even though the total workload is constant. In the limiting case of $n \rightarrow \infty$, the total time taken is the same as the time taken to execute the sequential portions of the code. In terms of speedup, we find that for all n, $S(n) \leq 1/\alpha$. This implies that no matter how many processors we employ, the speedup is limited by the sequential portion of the code. This is known as the *sequential bottleneck* and α is called the *sequential bottleneck factor*. The role of compilers in generating code that has the least amount of sequential portions is crucial to success.

In many applications requiring higher precision the fixed-load model is not appropriate in analyzing performance. In such cases one often assumes that adding more processors can help us in solving bigger problems in the same time. The speedup is thus defined to be

$$S(n) = \frac{\text{size of problem solved with 1 processor in 1 unit time}}{\text{size of problem solved with } n \text{ processors in 1 unit time}} \; . \tag{11.11}$$

Gustafson's Law states that if α is the percentage of code that is sequential in nature, then the speedup factor is given by

$$S(n) = n - \alpha(n-1) \; . \tag{11.12}$$

The decrease in the speedup is not as steep as in the case of fixed-load speedup. Thus, as the number of processors increases, we can achieve scaleable performances. The main idea is to keep all processors busy by increasing the problem size.

11.6 AISI'S AIS PROCESSORS

Applied Intelligent Systems, Inc. (AISI) processors [4] have a fine-grained SIMD architecture that is a one-dimensional linear array of processing elements (Fig. 11.10). The length of the array depends on the model: 64 (AIS-3000), 128 (AIS-3500), or 512 (AIS-4000). In the AIS-4000 there are enough PEs so that the array is as wide as a 512×512 image. There is one PE for each column in the image and enough memory directly connected to each PE to hold entire columns of image data for several distinct images. Image storage allows saving 128 binary images, 16 gray-level images, or various combinations of image word sizes. A single image operation is processed by first broadcasting an instruction to all PEs. Image data are sent row-by-row to the processing elements and results are returned row-by-row back to the memory. Information is exchanged between adjacent processors during processing.

11.6.1 Linear array processing system

There are several components to the processing system shown in Fig. 11.13. A general purpose host microcomputer generally controls the sequence of image processing functions. The host can read or write data directly to the image memory. A controller is connected to a small memory that contains microcode instructions and general data for support of the parallel processing array. The controller has several functions which are: (1) transmit instructions to the parallel processor; (2) supply a high-speed address pattern to the image memory so that the processor array reads and writes the rows of image data according to the address pattern; (3) handle image I/O on an interrupt basis; and (4) directly read or write pixels for individual processing such as for drawing lines in the image or for performing other operations that do not easily lend themselves to parallel processing. As a hardware background task, a camera may be inputting data to a separate buffer on the chip. When an entire line of data has been captured, the processor is interrupted and the data are sent to the image memory.

I/O, parallel processing, and host operations can all occur simultaneously. In practice, they are pipelined according to Fig. 11.14, which shows an overlapped time flow of the three concurrent operations. Generally, the time required for camera input is shorter than the processing time. In the figure, a camera can input image 3 while the parallel processor is operating on the previous image 2 and the host serial processor is performing final operations and outputting results for image 1.

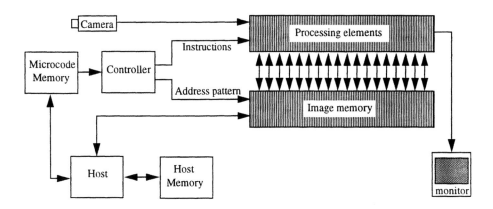

Figure 11.13 Linear fine-grained SIMD array.

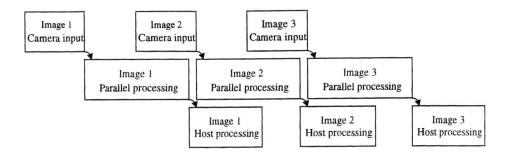

Figure 11.14 Overlapped timing.

Image memory is organized by allocating a set of binary images called *bit planes*. The rows in a bit plane occupy N successive addresses in the image memory, N being the image height. The least significant bits of all rows are grouped together as a bit plane. The next most significant bits are put in a group commencing at an address immediately following the least significant bit plane. Eight such groups are called a *byte plane*.

For real-time image processing, I/O must not occur at the expense of other processing taking place. Hence, buffers dedicated to image I/O are included in the system. In a mesh system, I/O is difficult because data must be loaded into the mesh as a two-dimensional array of numbers. Since there is a limitation on wires to the chips, data must be loaded as a sequence of binary images, where, for example, the first binary image to be loaded corresponds to the least significant bit of a gray-level image. The least significant bits of all words in the array must occupy each point in the mesh before being transferred to local memory associated with each PE; however, image data coming from a camera do not exist in this form and must be transferred to this form by extra hardware in an operation called *corner turning*.

In a linear array, corner turning hardware is straightforward (Fig. 11.15). A row of pixels from a camera is digitized and sent to a byte-wide buffer that is the length of the PE array. When the buffer is full of one line of image data, processing is interrupted and, in 8 cycles, the buffer contents are loaded into the image memory at an address not being used in processing. For an array of 512 PEs, it requires 512 clock pulses to fill the buffer and only 8 cycles to empty the buffer to memory. Since the image input buffers are not used for anything else during processing, processing can continue while the image buffers are being filled. In addition, monitor output is provided by the same buffer. A row of data is loaded from the image memory into the buffer in 8 cycles and the buffer then transmits the row of data to the monitor as a hardware background task.

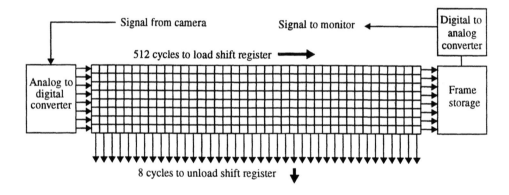

Figure 11.15 Image I/O.

11.6.2 Processing elements

Figure 11.16 shows the details of a single processing element separated into three parts for clarity. PEs have three basic functions: Boolean, neighborhood, and arithmetic. The controller transmits an instruction to registers in the processor chip. The instruction sets up internal data pathways and sets values in a truth table. The controller then sends a sequence of addresses to the image memory along with read or write enable signals.

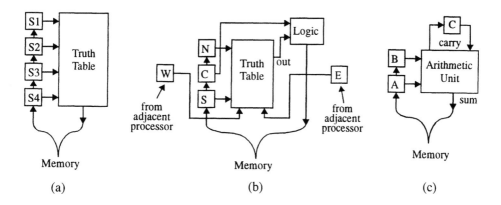

Figure 11.16 Processing element: (a) Boolean; (b) neighborhood; (c) arithmetic.

Internal pathways for the *Boolean function* are shown in Fig. 11.16(a). Four rows of binary data from the image memory corresponding to the controller address are sequentially sent to shift registers S1 through S4. The truth table defines an output for each possible combination of the four sources, S1 through S4. The output of the truth table is read back to the image memory according to a controller-generated address. The controller generates the following address pattern:

> first row of bit plane S1 (read)
> first row of bit plane S2 (read)
> first row of bit plane S3 (read)
> first row of bit plane S4 (read)
> first row of destination bit plane D (write)
>
> second row of bit plane S1 (read)
> second row of bit plane S2 (read)
> second row of bit plane S3 (read)
> second row of bit plane S4 (read)
> second row of destination bit plane D (write)
> $\vdots$
> $\vdots$
> last row of bit plane S1 (read)
> last row of bit plane S2 (read)
> last row of bit plane S3 (read)
> last row of bit plane S4 (read)
> last row of destination bit plane D (write).

Various modifications of the above pattern may be used. In some cases, D might be the same as a source plane. To allow Boolean functions of fewer arguments, one or more of the sources might be missing in the pattern. To accommodate vertical displacements, starting points of the patterns may be other than the first row.

Internal pathways for the *neighborhood function* are shown in Fig. 11.16(b). The PEs are programmed with a truth table that defines the neighborhood transformation. The W (west) and E (east) data come from adjacent PEs; the N (north) and S (south) data come from image memory. The neighborhood is limited to the closest horizontal and vertical neighbors. Larger neighborhoods are formed by using morphological identities for large-neighborhood processing via iterations of small-neighborhood processing or, less efficiently, by separate translations and Boolean operations when sparse large structuring elements are required. The controller generates the following address pattern:

> first row of source bit plane S (read)
> first row of destination bit plane D (read)
> second row of source bit plane S (read)
> second row of destination bit plane D (read)
> ⋮
> last row of source bit plane S (read)
> last row of destination bit plane D (read).

East and west translations are just special cases of the neighborhood function. A horizontal translation of m pixels requires m repetitions of the neighborhood operation; a vertical translation is simply a displacement in the address pattern of the destination and requires no extra overhead.

Internal pathways for an *arithmetic function* are shown in Fig. 11.16(c). Addition is a bit serial operation, where the sum and carry for each successive bit are generated concurrently. Subtraction is also supported but is not illustrated separately in the figure. The carry is stored on the chip since it is immediately used in the next operation. Any word size can be processed but for clarity we illustrate a three-bit add, the sum being four bits. The following address pattern is generated:

> first row of least significant bit plane A1 (read)
> first row of least significant bit plane B1 (read)
> first row of destination bit plane D1 (write)
> first row of next least significant bit plane A2 (read)
> first row of next least significant bit plane B2 (read)
> first row of destination bit plane D2 (write)
> first row of most significant bit plane A3 (read)

first row of most significant bit plane B3 (read)
first row of destination bit plane D3 (write)
first row of destination bit plane D4 (write carry)
second row of least significant bit plane A1 (read)
second row of least significant bit plane B1 (read)
second row of destination bit plane D1 (write)
$\vdots$

11.6.3. Processor operation

Various benchmarks for the AIS-4000 operating at 20 MHz on 512×512 images have been established: the neighborhood function takes 52 µs; an image translation requires $52 \times 512 = 26,624$ µs; a Boolean function with 2, 3, or 4 input arguments requires 78, 104, and 130 µs, respectively; and a byte-wide arithmetic add or subtract requires 780 µs.

The AISI system is programmed in the C language. C alone does not define parallel processing for hardware, so parallel processing is handled by function calls belonging to a *function library*. An image plane in image memory is defined as a special variable type called a *frame*. Image memory can be allocated by a function called frame_new( ). Parallel processor library functions contain frames as arguments. A source frame is always followed by an integer argument that defines a vertical translation, if any. A destination frame is always followed by an integer that defines a horizontal translation.

To illustrate processor operation, consider the subtraction function call subx(). During compile time, subx() generates microcode to form subtract pathways in the PEs shown in Fig. 11.16(c) and to form the appropriate address pattern. These instructions are stored in an instruction buffer in the host computer memory. Image frame definitions together with translations are used to define the start locations of the address patterns. The function looks at the allocation function to find out the number of bits in the words to be subtracted and to form the appropriate address pattern. As a second illustration, consider the dilation function call dilate(). The address pattern specific to dilation is generated and sent to the instruction buffer. At the end of the compilation, the instruction buffer holds all the microcode for the parallel processor for an application. During run time, the host computer transfers the instructions to a first-in-first-out (FIFO) memory that the controller can read. The controller reads instructions and sends some to the parallel processor. It then creates address patterns defined by other instructions. As the controller pulls instructions from the FIFO, and the FIFO is near empty, the host refills the FIFO with new instructions. Meanwhile, the host can perform other tasks. During image I/O, the controller may be interrupted and jump to pre-stored routines in the controller memory that control downloading camera data or uploading monitor data.

11.6.4 Hit-or-miss transform

AISI processors are used for machine vision applications, in particular, for identification. The processors' Boolean and neighborhood functions make them ideal for real-time morphological processing (including sparse large structuring elements) and their arithmetic function facilitates real-time windowed convolution. The present subsection introduces the morphological hit-or-miss transform, which is employed extensively in binary image processing, and the next discusses real-world object recognition using the hit-or-miss transform on the AIS-4000 processor.

Consider a template (defined over a window) consisting of 0s, 1s, and ×s. A binary image operator is defined in the following manner: translate the window to pixel z and define the output image to be 1-valued at z if and only if all 0s and 1s in the template exactly match the input image values at the corresponding pixels in the translated window. All template positions with × are "don't care" pixels. The operator is called the *hit-or-miss transform* and the 0-valued and 1-valued pixels in the template are called the *miss* and *hit* pixels, respectively. The 1-valued and 0-valued pixels are grouped into distinct sets (called *structuring elements*) E and F, respectively, and the template (*structuring pair*) is denoted by (E, F). For instance, the template (E_1, F_1) in Fig. 11.17, where black, white, and gray correspond to 1, 0, and ×, respectively, locates lower-right corners in a binary image. Used together, the four templates shown in Fig. 11.17 find image corners. The manner in which (E_1, F_1) locates lower-right corners is depicted in Fig. 11.18.

The hit-or-miss transform is computed by sending binary image values in the translated window through an AND gate, where pixels in E are sent through uncomplemented and pixels in F are sent through complemented. If $x_1, x_2, ..., x_n$ are the logical image values in the translated window (Fig. 11.19), there exist two sets of variables, $x_{H,1}, x_{H,2}, ..., x_{H,n(H)}$, the *hit variables*, and $x_{M,1}, x_{M,2}, ..., x_{M,n(M)}$, the *miss variables*, and a Boolean function defining the hit-or-miss transform in terms of these by

$$h(x_1, x_2, ..., x_n) = x_{H,1} x_{H,2} ... x_{H,n(H)} x_{M,1}{}^c x_{M,2}{}^c ... x_{M,n(M)}{}^c \quad . \tag{11.13}$$

Figure 11.20(a) shows a digital letter H and Fig. 11.20(b) shows hit-and-miss structuring elements E and F, respectively. Element E is simply the original character and F the character boundary. Assuming both structuring elements are centered at the center pixel of H, the only situation in which E will fit inside the character and F miss the character is when the H is matched exactly, and in this case a 1 will be output at the center of the H. There are two

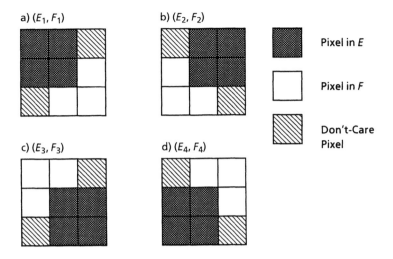

Figure 11.17 Corner-finding structuring pairs.

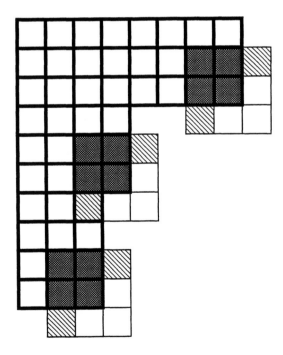

Figure 11.18 Finding lower-right corners.

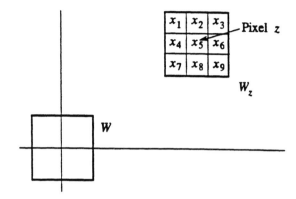

Figure 11.19 Translated window.

Figure 11.20 Hit-or-miss character recognition. (a) letter H; (b) tight-fitting structuring elements.

immediate difficulties with this approach to character identification: (1) even a single-pixel imperfection in the observed character will cause lack of recognition; (2) structuring elements are very large and therefore require much processing time.

Robustness relative to noise is enhanced by using the structuring elements shown in Fig. 11.21(a). If a perfect H occurs, then there will be a 3×3 pixel block located at the center of the observed H; if some pixels are adjoined to the outer edge of the H or if pixels are missing in the inner edge of the H, then a subset of the 3×3 pixel block will appear in the output image, so that recognition is still achieved. The sparse structuring elements shown in Fig. 11.21(b) are subsets of those shown in Fig. 11.21(a) and lessen computation time in addition to achieving robustness. Besides less pixels to process, there is greater robustness relative to missing or adjoined pixels. There is a limit to using structuring elements differing too greatly from the target or deleting pixels in the structuring elements. If carried too far, there will be false identifications. For real-world applications, automated training techniques are used to find satisfactory structuring elements [5].

The hit-or-miss transform is easily programmed on the AISI processor using image translation and the macro AND (performing intersection). For sparse structuring elements consisting of 25 pixels within a 20 × 20 window, there are 25 translation and Boolean operations employed, requiring a total of 8.5 msec for hit-or-miss implementation.

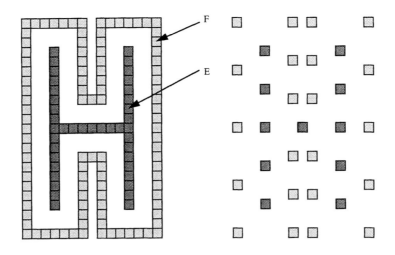

Figure 11.21 Recognition structuring element pairs. (a) loose-fitting structuring elements; (b) sparse structuring elements.

For noisy images, the hit-or-miss transform is often not sufficiently robust, even with sparse structuring elements. Robustness can be improved by taking a rank-order approach. Rather than require that all pixels in the translated structuring elements correctly fit within or without the image, a 1 can be output if and only if at least some predetermined number of structuring-element pixels hit or miss correctly. This *rank-order hit-or-miss transform* can be programmed on the AISI processor and used for classifying highly degraded characters. Total time for a sparse 25-element rank-order hit-or-miss transform on the AISI processor is 18.9 msec. In practice, speeds are usually over four times faster because characters of interest can be located to within one-fourth of the image area.

11.6.5 OCR application

Optical character recognition (OCR) on a fixed, known font is generally not difficult. However, when the image background is extremely cluttered, there can be many false recognitions. Figure 11.22 shows a string of characters partially obscured by a cluttered background. In such a situation it can be beneficial not to apply the hit-or-miss transform directly to a thresholded version of the image, but instead to first transform the image using edge detection.

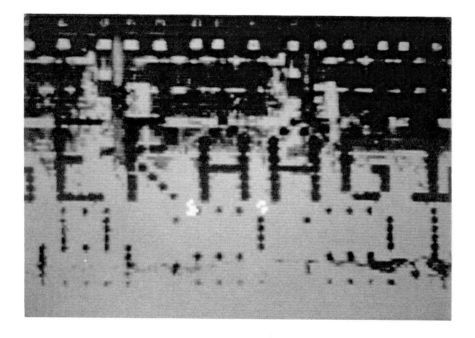

Figure 11.22   Cluttered character image: white blobs mark identification of 'A.'

In standard convolution-based edge detection, a gradient image is constructed by convolving with vertical and horizontal gradient masks, taking absolute values, and then taking the maximum of the vertical and horizontal gradients at a pixel. A different approach is to apply north and south gradient masks in the vertical direction and east and west gradient masks in the horizontal direction, the result being four gradient images rather than two. Each of these gradient images is thresholded at some positive value to create four directional-gradient binary images (directional planes). Figure 11.23 shows north and south gradients resulting from the image of Fig. 11.22, the north and south gradients being shown in white and gray, respectively.

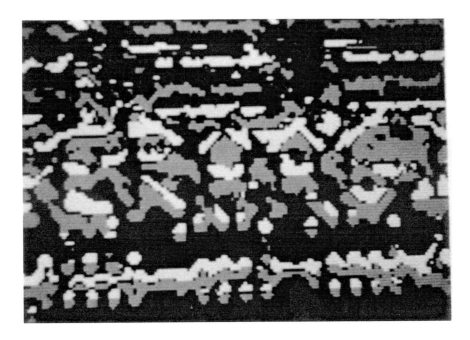

Figure 11.23 North (white) and south (gray) gradients.

An AISI algorithm implemented on the AISI processor applies various sparse order-statistic filters with eight-pixel structuring elements to the four direction planes to find strokes and stroke intersections at various angles. Eight separate stroke features are found and occupy eight binary planes. A sparse rank-order hit-or-miss filter using 25 pixels is applied to the eight separate binary stroke images to identify the character A. The white blobs at the lower left sides of the As in Figure 11.22 show the pixels where 22 of the 25 structuring-element points correctly hit or miss. Figure 11.24 shows an image with a different kind of clutter. The two 0s are identified from the same features, but using a different set of 25 hit-and-miss structuring-element pixels.

11.7 ITI'S MVC 150/40 PROCESSOR

Imaging Technology, Inc. (ITI), Bedford, MA, is the manufacturer of the Modular Vision Computer (MVC) 150/40 product line. This product line spans a wide performance range: from a simple host-based vision system to a full-fledged software configurable, modular, and pipelined vision system [6]. A key advantage of this product line is its fully modular architecture. Special purpose plug-on mezzanine modules can be added at any time to enhance processing speed.

Figure 11.24 Cluttered character image: white blobs mark identification of "O."

11.7.1 Image management

The image manager is a single PCI-bus card that contains reconfigurable image memory; a cross-port switch for image data routing, timing, and area-of-interest (AOI) control; support for color acquisition; and local display. Plug-on mezzanine modules provide the image acquisition and processing capabilities. The block diagram is shown in Fig. 11.25.

The image manager can have up to three quad-ported image memories of 1 Mb or 2 Mb each. These are designated as A0, A1 (Bank A), and B1 (Bank B). Four separate devices can read a quad-ported memory at the same time; however, only one device can write into this memory at any given instant. There is an independent read port, independent write port, and independent DMA read port for each memory. The DMA transfer speed of 67 Mb/sec can be achieved in the burst-mode. The PCI bus is capable of transferring data at $33*4 = 132$ Mb/sec in the burst-mode. The time required to access the source and target memories reduces this rate. If the source or destination is a slower device (like hard disk), then this rate will be considerably slower. Each of the two banks (Bank A and Bank B) can be independently reconfigured in one of the following resolutions: $1K \times 1K \times 8$-bits, $2K \times 512 \times 8$-bits, $4K \times 256 \times 8$-bits, and $8K \times 128 \times 8$-bits. Furthermore, bank A can be reconfigured to represent a single $2K \times 1K \times 8$-bits image. Other features include:

(1) pixel sub-sampling ($\times 1/2$) independently in X and Y at time of image acquisition,

(2) zoom (×1/2, ×2, and ×4) independently in X and Y while using the video bus,

(3) zoom (×2 and ×4) independently in X and Y for display,

(4) bit-plane protection during image acquisition and host access,

(5) single-pixel pan and scroll for each memory port,

(6) unrestricted host access,

(7) 2K × 1K × 2-bits overlay memory.

The quad-ported memory speeds up many iterative operations like background subtraction and frame averaging. We illustrate the frame averaging operation using the memory A0. In the first pass, a frame is grabbed and stored in the memory. While it is being stored, the frame is also being sent to the local display. In the subsequent passes, a frame is grabbed and sent pixel-by-pixel to the ALU. The ALU reads the corresponding pixels from A0 memory, does the summation, and stores the result back to the memory. The updated contents of the memory are also sent to the local display, perhaps via a lookup table (LUT) that performs scaling (Fig. 11.26). This process requires that: (1) the ALU be able to read a pixel value from memory, (2) the ALU be able to write a pixel value into memory, and (3) the local display controller can read a pixel value from memory. These three operations occur concurrently and therefore require multi-ported memory.   In the absence of multi-ported memory, a pipelined sequence of operations will require buffers between each stage of the pipeline to hold the temporary data.

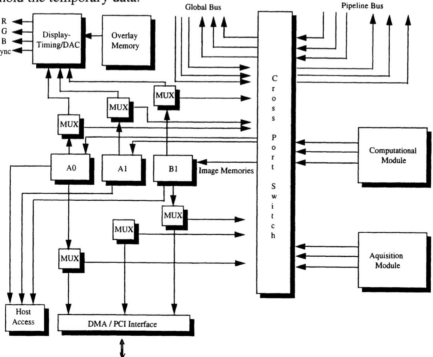

Figure 11.25 Block diagram of image manager.

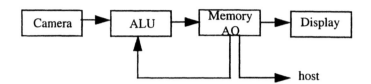

Figure 11.26 Pipelined frame averaging using quad-ported memory.

Depending on applications, there may be a need for more than one image manager. Via software one image manager can be selected as the *master* and the others as *slaves*. The video bus consists of two buses: a 24-bit *global bus* used to transfer data between image managers and a 24-bit *pipeline bus* used to transfer data under software control from one processor to another in a daisy-chained fashion (Fig. 11.27). This allows data paths to be dynamically reconfigured. The video bus is capable of working at 10 MHz, 20 MHz, and 40 MHz, the speed being selected by the software. The software selectable area-of-interest can be in the range 4×1 pixels to 8K × 2K pixels in steps of 4 pixels.

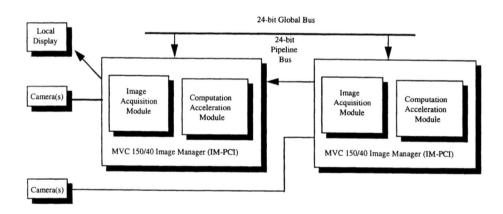

Figure 11.27 Multi-channel configuration.

Plug-on image acquisition modules connect various types of cameras and sensors to the image manager. The specific acquisition module chosen is determined by the camera or sensor being used. An image acquisition module can interface with up to four multiplexed cameras. Plug-and-play interfaces are available for many standard and non-standard RS170/CCIR cameras. For applications requiring more than four cameras or where asynchronous image capture is required, multiple image managers can be connected to each other via the global video bus (Fig. 11.27).

11.7.2 Computational acceleration modules

A computational acceleration module is essentially a special purpose coprocessor. As of this writing, ITI offers six different processing computational modules.

The *Convolver/Arithmetic Logic Unit* is a pipeline processor capable of performing common imaging operations like convolution, edge detection, gray-value scaling, image addition/subtraction/multiplication/division, centroid identification, correlation, and area calculation. The convolution kernel is programmable and can be 4×4 (at 40 MHz pixel rate), 8×4 (at 20 MHz pixel rate), or 8×8 (at 10 MHz pixel rate). It contains eight programmable 8-bit lookup tables (LUTs) which can be bypassed, if needed, by software.

The *Programmable Accelerator* is a digital signal processor based on Texas Instrument's TMS 320C31 chip running at 40 MHz. The processor contains a single cycle ALU with barrel shifter and a parallel multiplier, 64 word (1 word = 32 bits) instruction cache, and two internal 1K-word RAMs. The module contains 4 Mb of image memory, 512 Kb of static memory of programs and variables, and 1 Mb of EEPROM for permanent storage of programs. The module contains debugging facilities.

The *Histogram/Feature Extraction Processor* is essentially two processors running in parallel: the *Histogram/Projection Generator* (*HPG*) generates histograms and computing projections at multiples of $45°$ and the *Feature Extractor* (*FX*) performs real-time statistical analysis of pixel intensities. A feature is identified either by pixel intensity or by a 3×3 binary neighborhood pattern. Up to 16K features can be identified in an image. As a benchmark, this module can simultaneously perform input thresholding, histograming, any 3×3 binary morphological operation, and run-length encoding of a 512×512 gray-scale image in 6.5 msec.

The *Median and Morphological Processor* is capable of performing morphological operations in real time. The following operations can be performed on a 512×512 image in under 6.5 msec: 1×3 or 3×3 median filter; 3×3, 5×5, 7×7, or 9×9 gray-scale erosion; 3×3, 5×5, 7×7, or 9×9 gray-scale dilation; 3×3 or 5×5 gray-scale opening; 3×3 or 5×5 gray-scale closing; and 3×3 or 5×5 top-hat transform. Cascading two of these processors permits up to 17×17 erosion/dilation and 9×9 opening/closing/top-hat.

The *Label and Measurement Processor* is a cascade of two processors capable of performing real-time *blob* (connectivity) analysis. A blob is a connected region whose pixel intensities are above or below a certain threshold value. Many different types of topological and geometric properties can be obtained from blob analysis; some of these are area, perimeter, moments of inertia, rose of direction, histogram, projections, and bounding rectangle.

The *Binary Correlator* is a processor capable of performing template matching and binary morphological operations involving large templates or structuring elements. The structuring element can be configured as 32×32, 64×16, 128×8, 256×4, 512×2, or 1024×1. It is possible to perform independent erosion, dilation, and pattern matching in parallel. On board capabilities also include convergence tests for iterative applications of morphological operations and computation of the morphological *distance function* (yielding a topographic image whose pixel intensities specify the distance of a pixel from the closest edge).

Multiple computational module controllers can be cascaded using the global and pipeline buses. Figure 11.28 shows one such arrangement in which seven computation acceleration modules are in a pipeline.

REFERENCES

1. D. Sinha and E. R. Dougherty, *Understanding Computer-Based Imaging Systems*, SPIE Press, 1998.
2. R. Wilhelm and D. Maurer, *Compiler Design*, Addison Wesley, 1995.
3. K. Hwang, *Advanced Computer Architecture: Parallelism, Scalability, Programmability*, McGraw-Hill, 1993.
4. L. A. Schmitt and S. S. Wilson, "The AIS-5000 parallel processor," *IEEE Transactions on Pattern Analysis and Machine Intelligence*, vol. 10, no. 3, pp. 320-330, May 1988.
5. S. S. Wilson, "Training structuring elements in morphological networks," in *Mathematical Morphology in Image Processing*, E. R. Dougherty, ed., pp. 1-41, Marcel Dekker, 1992.
6. B. Dawson, "Real-time machine vision: A tutorial," *VMEbus Systems*, August, 1995.

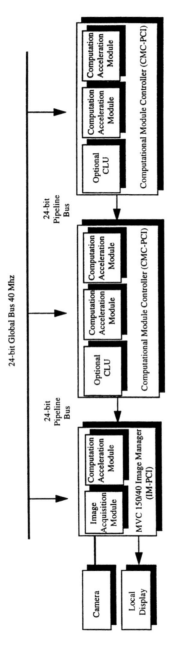

Figure 11.28 A pipeline of seven computation acceleration modules.

INDEX